INTEGRATIVE PHYSIOLOGY IN THE PROTEOMICS AND POST-GENOMICS AGE

INTEGRATIVE PHYSIOLOGY IN THE PROTEOMICS AND POST-GENOMICS AGE

Edited by

WOLFGANG WALZ, PhD

Department of Physiology, University of Saskatchewan,
Saskatoon, Canada

HUMANA PRESS ✳ TOTOWA, NEW JERSEY

© 2005 Humana Press Inc.
999 Riverview Drive, Suite 208
Totowa, New Jersey 07512

humanapress.com

Production Editor: Robin B. Weisberg

Cover design by Patricia F. Cleary

Cover illustration (background) from Fig. 4, Chapter 12, "The Zebrafish As an Integrative Physiology Model," by Alicia E. Novak and Angeles B. Ribera; (foreground) from Fig. 1, Chapter 11, "Repair and Defense Systems at the Epithelial Surface in the Lung" by Pieter S. Hiemstra.

For additional copies, pricing for bulk purchases, and/or information about other Humana titles, contact Humana at the above address or at any of the following numbers: Tel.: 973-256-1699; Fax: 973-256-8341; E-mail: humana@humanapr.com or visit our website at www.humanapress.com

The opinions expressed herein are the views of the authors and may not necessarily reflect the official policy of the National Institute on Drug Abuse or any other parts of the US Department of Health and Human Services. The US Government does not endorse or favor any specific commercial product or company. Trade, proprietary, or company names appearing in this publication are used only because they are considered essential in the context of the studies reported herein.

This publication is printed on acid-free paper. ∞
ANSI Z39.48-1984 (American National Standards Institute) Permanence of Paper for Printed Library Materials.

Printed in the United States of America. 10 9 8 7 6 5 4 3 2 1

eISBN: 1-59259-925-7
Library of Congress Cataloging-in-Publication Data

Integrative physiology in the proteomics and post-genomics age / edited by Wolfgang Walz.
   p. cm.
 Includes bibliographical references and index.
 ISBN 1-58829-315-7 (alk. paper)
 1. Physiology, Experimental. 2. Genomics. 3. Proteomics. 4. Molecular biology. I. Walz, Wolfgang.
 QP43.I585 2005
 571'.072--dc22
                                                         2004020035

PREFACE

There is a perception in the scientific community that the discipline of Physiology is in crisis, or at least, in a phase of profound transition and change. At the root of the problem is confusion between objectives (the biological questions to be solved) and the methods and technologies to be applied. Traditionally, ever since Claude Bernard's concept of the "milieu interieur," Physiology was an integrative science with the prime concern of studying regulatory mechanisms leading to adaptation and homeostasis in the presence of challenges from a dynamic internal and external environment. This study of control mechanisms can be applied on any level of function whether subcellular, cellular, and organ, but reaches its highest level of complexity with the functioning of the body as a whole and its interaction with the external environment. This involves the determination of the interaction of genetic with environmental factors and the resulting integrated body adaptation.

It might seem obvious that in the pursuit of these questions any appropriate combination of techniques on any organizational level could be used. Yet the advent of molecular techniques has resulted in a preoccupation with the problems and challenges inherent in these techniques, sometimes at the expense of the original perspectives and concepts. The many new mechanisms that have been discovered at the molecular level, as well as their economical exploitation, have contributed to a climate of reductionism. However, despite the undeniable and spectacular successes of molecular biology, the lack of, and need for, an integrative perspective has become evident. This is particularly true in many clinical cases of gene and drug therapies that failed despite promising results from defined animal models.

Integrative Physiology in the Proteomics and Post-Genomics Age represents an attempt to highlight the major questions and accomplishments of modern physiological research. It re-addresses the fundamental questions of the classical concept, incorporating information gained from new molecular, genetic, and cellular technologies.

Wolfgang Walz

CONTENTS

CONTRIBUTORS

FRANK W. BOOTH • *Department of Biomedical Sciences, University of Missouri, Columbia, MO*

THEODORE H. BULLOCK • *Department of Neurosciences, University of California at San Diego, La Jolla, CA*

MICHEL CABANAC • *Departement de Physiologie, Universite de Laval, Quebec, Canada*

PENELOPE A. HANSEN • *Faculty of Medicine, Memorial University of Newfoundland, St. John's NL, Canada*

PIETER S. HIEMSTRA • *Department of Pulmonology, Leiden University Medical Center, Leiden, Netherlands*

GARETH LENG • *School of Biomedical and Clinical Laboratory Sciences, University of Edinburgh, Edinburgh, UK*

DAVID L. MATTSON • *Department of Physiology, Medical College of Wisconsin, Milwaukee, WI*

JEAN-PIERRE MONTANI • *Institute of Physiology, University of Fribourg, Switzerland*

P. DARRELL NEUFER • *Department of Cellular and Molecular Physiology, Yale University, New Haven, CT*

ALICIA E. NOVAK • *Department of Physiology and Biophysics, University of Colorado Health Sciences Center, Denver, CO*

ANGELES B. RIBERA • *Department of Physiology and Biophysics, University of Colorado Health Sciences Center, Denver, CO*

JOHN A. RUSSELL • *School of Biomedical and Clinical Laboratory Sciences, University of Edinburgh, Edinburgh, UK*

DEE U. SILVERTHORN • *Integrative Biology, University of Texas, Austin, TX*

MIRCEA STERIADE • *Departement de Physiologie, Universite de Laval, Quebec, Canada*

KENNETH B. STOREY • *Institute of Biochemistry, Carleton University, Ottawa, Canada*

JANET M. STOREY • *Institute of Biochemistry, Carleton University, Ottawa Canada*

MEGHAN M. TAYLOR • *Department of Pharmacological and Physiological Science, Saint Louis University, St. Louis, MO*

BRUCE N. VAN VLIET • *Division of Basic Medical Sciences, Memorial University of Newfoundland, St. John's NL, Canada*

WOLFGANG WALZ • *Department of Physiology, University of Saskatchewan, Saskatoon, Canada*

From Functional Linkage to Integrative Physiology

Wolfgang Walz

1. WHAT IS INTEGRATIVE PHYSIOLOGY?

Integrative science is concerned with the purpose and workings of a larger structure or entity. This, of course, does not exclude an examination of the detailed components of this superstructure—indeed, any other approach would be impractical. In conducting integrative research or inquiry, it is therefore most important to define the question or hypothesis such that the focus is on the role of the superstructure or the larger concept. In the case of animal function, this integrative approach means that any detailed structure or isolated mechanism can be studied by any method deemed opportune. However, the hypothesis or question pursued has to be concerned with the functional role of the superstructure to which the detail is related.

The exact meaning or definition of the superstructure depends on one's perspective. The level of analysis could be a cell organelle, a single cell, a tissue, an organ, or the body. It may be a population of organisms belonging to one species or it may be an entire ecosystem. In dealing with one level in an integrative way, it is not so much an issue which level was selected, but rather it is more important not to lose sight of subsequent levels of organization and to relate the relevance of one's findings to as many levels of organization as possible. Thus, an integrated approach requires a different thought process then if one would describe the mechanism of a single molecule in isolation and then speculates about possible roles. Integrative inquiry demands that the hypothesis or experimental approach consider as many levels of organization as possible. An integrative example is the loss of sodium in excess of chloride during bouts of diarrhea. The resulting compensatory mechanisms range from involvement of sodium excretion to sodium hunger and thirst, initiated by sodium receptors with the involvement of other sensory processes, all of them controlled and integrated by hypothalamic circuits (1). Organizational levels in this example range from ion carriers across membranes to the behavior of the organism.

2. MUST INTEGRATIVE PHYSIOLOGICAL APPROACHES USE SPECIFIC METHODOLOGY?

Superficially, one might assume this to be the case. The use of metabolic cages during whole animal measurements, for example, might seem like a favored method in integrative research. Naturally, methods favored by integrative scientists are usually concerned with

From: *Integrative Physiology in the Proteomics and Post-Genomics Age*
Edited by: W. Walz © Humana Press Inc., Totowa, NJ

whole or intact animal measurements. As mentioned earlier, however, the decisive definition for an integrative vs a more mechanistic study is not the method in use, but rather the study design so long as it relates as many levels of functional organization as possible to the question being pursued. Thus, it is paramount, when defining integrative research, not to exclude all the new methods of molecular biology. In fact, one of the many strengths of contemporary integrative physiology is the inclusion of genomics and proteomics as it applies to larger questions and hypotheses of functional importance among the different levels of organization. This is in fact, the main theme of this monograph and so, several chapters have been dedicated to this approach in order to convince the reader that this will be a most successful approach in the 21st century. This approach is indispensable to future research as one of the central themes in integrative physiology has always been interested in the relationship between influences of the genome and environment in determining the outcome of physiological adaptations and compensatory strategies.

In all likelihood, a successful integrative physiological study or project will involve multiple levels of organization (e.g., isolated cells and whole organs), as well as several different methods. Other methods used in such an integrative project would be of a "classical" nature and might involve measurements of organ metabolism and circulation. Such a project might be too complex to be successfully completed by one major investigator. Therefore, the successful conclusion of a project, which pursues integrative questions involving both the genome and environment, might involve a team of investigators with expertise encompassing the different methods. Such a team effort requires communication of and focus on the pertinent questions of integrative physiology in order to have the team work as cohesive as possible with the least amount of friction. This volume is an effort to help bridge the gap that is inevitable when a group of investigators with different fields of specialization collaborate on an integrative project.

3. HOMEOSTASIS OF THE *MILIEU INTERIEUR*

A central theme of integrative physiology is almost as old as physiology itself: the question of the constancy of the *milieu interieur* or the internal environment of the organism *(2)*. As the organism faces external challenges (e.g., a drop in the ambient temperature) or switches from one function to another (e.g., from rest to physical exercise), its internal parameters are changed. Some of the changes are a direct consequence of the challenge (e.g., lactate production by skeletal muscle activity). Other changes are the result of adaptations to the primary challenge (e.g., shivering of muscles in order to counteract a drop in internal temperatures). As most cell and organ interactions are only functioning in an optimal way within narrow limits of their environmental parameters (e.g., temperature, pH, calcium concentration, oxygen tension), compensatory mechanisms will be activated in a negative feedback loop. The real challenge of integrative physiology is not so much to trace and/or control one parameter in isolation but to monitor as many as can be measured, as there is no one parameter existing in isolation. The challenge is rather to work out the connections between the parameters and the way they influence each other and what these interactions mean for the control circuits and the decision-making processes in the hypothalamus or anywhere else. In other words, there is a hierarchy of control systems, which has to be delineated and put into a functional context. A group of examples are the diverse mechanisms and processes triggered by dehydration, which involve water intake, sodium appetite, as well as cardiovascular responses *(3)*.

4. VERTICAL INTEGRATION

Integrational analysis can involve many different approaches, but not all approaches are equally relevant for integrative physiology. The level of integration with the most significant impact on the understanding of how an organism works is vertical integration. The meaning of vertical integration is illustrated here with an example. In the apical membrane of secretory epithelial tissue, there are several proteins co-localized and functioning cooperatively in conductive chloride transport that are necessary for secretion to occur *(4)*. This cooperation, or functional linkage, involves a level of integration. This linkage can be analyzed directly by using epithelial cell layers or by isolating each participating protein, characterizing their properties and then deducing possible interactions between the different proteins in a meaningful way. This level of integration is horizontal and involves mainly functional linkages and cooperation. The scope of this example changes when one realizes that functional secretion involves control of gene expression, modulation and targeting of proteins. Vertical integration would set these proteins in the context of the organ or whole organism—it would, therefore, involve at least another organizational level above, or in the case of gene regulation, below the epithelial cell type and therefore constitute a process of vertical integration. It would investigate the role of these chloride conductance proteins and their control in the context of fluid secretion. For example, it would investigate how chloride transport is changed in the lungs during exercise and how different homeostatic pathways act directly or indirectly on the chloride conduction and what the consequences might be for lung function *(5)*. Thus, vertical integration is necessary if one wants to relate a specific component (molecule, process) to the overall function of the organism, its control by various circuits and its changes during challenges to the organism. By adding more and more components and increasing the levels, one should be able to get a better and more accurate understanding of the complex homeostatic mechanisms in the organism at rest, during challenges, and in disease states. This is obviously not an easy undertaking and involves a certain level of sophistication in one's thinking. This integrative strategy will build on functional relationships and becomes increasingly more complex as the study progresses. In the very end, a comprehensive understanding of the hierarchy of homeostatic control mechanisms in the whole organism will be the goal.

5. FUNCTION AS A RESULT OF INTERACTION BETWEEN GENES AND ENVIRONMENT

The interaction between genes and the environment is another central theme and is the one that underwent the largest changes in the last decade. A good part of this book is dedicated to this theme. In analyzing the hierarchical organization of homeostatic control, one will inevitably, at one point, be confronted with the relative contributions of gene expression and environmental impact. This is not a straightforward interaction where one can easily assign each factor a percentage of the impact that then adds up to 100% First of all, there is posttranscriptional modification resulting in approx 100,000 different proteins from approx 30,000 to 35,000 genes *(7)*. This leaves much room for modification by cell metabolism or other factors. In addition, the activity of proteins is modulated by their cellular or extracellular environment. Second, another component to the overall complexity involves interaction between genes and environment during development, which can lead to irreversible changes for the duration of the life of the organism *(8)*. DNA methylation and histone modification are examples of mechanisms of environmental influences during development *(9)* that may be passed down through generations, constituting an epigenetic change. The interface between genome and internal or external environment is, at once, a very promising and also very demanding future research area as it involves expertise in molecular biology, as well as in integrative approaches. For a further detailed discussion please refer to Chapter 2 by Mattson.

6. CURRENT OBJECTIVES FOR INTEGRATIVE PHYSIOLOGY

There are two main thrusts for an integrated understanding of the workings of an organism. The first one could be called modeling and prediction. Depending on how many vertical levels it encompasses and how many hierarchical mechanisms are elaborated, it will reach the highest level of understanding of an organism. The goal would be to have a hierarchical control system elaborated in a way, that the functional consequences of any theoretical internal or external change may be predicted by the available information and degree of understanding of the participating interactions. The comparison of a predicted and an actual outcome of an experiment would certainly be the litmus test for the degree of integrative knowledge achieved. The proliferation of genetically engineered animals begs an analysis of the overall functional impact of the presumed change in the gene(s). Compensational changes during development and adult life have to be taken into account in order to get the full functional phenotype for the whole animal. The second one could be called diagnosis and therapy. Clinical science, when dealing with a pathology, will have to rely on a complete view of all the genetic changes, impaired structures, and mechanisms, compensatory and developmental changes involved in a certain pathology. A well-founded therapeutic approach has to take into account all these factors and changed homeostatic circuits and hierarchies. In a way, at least for the human body, the development of a therapeutic approach taking into account the whole integrated function of the diseased body, complete with compensatory mechanisms and changing selectively control parameters in a planned and measured way, would be the highest level of integrated physiology one could expect to achieve. The intervention could combine both, changes in genetics as well as environmental factors to achieve the desired changes.

7. CONCLUSION

It is inevitable that most physiological studies concerned with any mechanism will find evidence regarding functional linkage to other mechanisms on either the same organizational level or levels above and below. Elaboration of such linkages does not in itself represent an integrative physiological study, but it is the first step. A serious investigation into mechanisms of integrative physiology means the delineation of hierarchical control systems and their reaction to internal and external challenges.

REFERENCES

1. Denton, D.A., McKinley, M.J., and Weisinger, R.S. (1996) Hypothalamic integration of body fluid regulation. *Proc. Natl. Acad. Sci. USA* **93,** 7397–7404.
2. Bernard, C. (1865) *L'introduction a l'etude de la medicine experimentale*. Bailliere. Paris.
3. Rowland, N.E. (1998) Brain mechanisms of mammalian fluid homeostasis: insights from use of immediate early gene mapping. *Neurosci. Biobehav. Rev.* **23,** 49–63.

4. Ogura, T., Furukawa, T., Toyozaki, T., et al. (2002) ClC-3B, a novel ClC-3 splicing variant that interacts with EBP50 and facilitates expression of CFTR-regulated ORCC. *FASEB J.* **16,** 863–865.
5. Yager, D., Kamm, R.D., and Drazen, J.M. (1995) Airway wall liquid. Sources and role as an amplifier of bronchoconstriction. *Chest* **107,** 105–110.
6. Rees, J. (2001) Post-genome integrative biology: so that's what they call clinical science. *Clinical Medicine* **1,** 393–400.
7. Cowley, A.W. (2003). Genomics and homeostasis. *Am. J. Physiol.* **284,** R611–R627.
8. Nathanielsz, P.W. and Thornburg, K.L. (2003). Fetal programming: from gene to functional systems—an overview. *J. Physiol.* **547,** 3–4.
9. Li, E. (2002). Chromatin modification and epigenetic reprogramming in mammalian development. *Nature Reviews Genetics* **3,** 662–673.

2

Functional Genomics

David L. Mattson

1. INTRODUCTION

The sequencing of the human genome has provided the scientific community with the "roadmap of life" *(1,2)*, yet the understanding of this map is dependent on the elucidation of the function of the approx 30,000 to 35,000 genes that have been predicted/identified from the sequence of the human genome. As recently reviewed, however, only one-third of these genes have any inferred function ascribed to them, whereas approximately one-sixth or less have a confirmed action *(3,4)*. Moreover, it is estimated that there are as many as 100,000 distinct proteins with presumably distinct biological function. The challenge facing scientists in the post-genome era is to construct a model of the organism that includes gene sequence and expression, protein structure and function, the molecular interactions that occur between proteins and other molecules to create pathways, and the combination of the many complex pathways that results in functioning cells, tissues, and organisms. Only by addressing and meeting this challenge can science utilize the vast potential of the human genome project to link gene to function in health and disease. To begin to address this problem, a combination of automated technologies, novel experimental strategies, and the newly available genomic sequence data are permitting scientists to begin to assemble the individual components of biological systems into a map of the human organism that will provide an understanding of the function of different genes and gene pathways.

The term "genomics" emerged 10 to 15 yr ago and describes the variety of technological and computational approaches employed to examine the structure and function of DNA of large numbers of genes in contrast to the study of a single gene or gene family *(5)*. "Functional" or "physiological genomics" is a term that has been used to describe the multitude of approaches employed to identify and elucidate the functional importance of different genes in physiological and pathophysiological conditions. These approaches broadly encompass scientific disciplines aimed at linking nucleic acid sequence, protein structure, protein function, and complex, integrated biological function.

Function, in biological terms, is understood on many different levels depending on the context *(6)*. A molecular biologist may consider the identification of an unknown gene that encodes a specific cell surface receptor as a functional analysis, but a biochemist might consider function to be the affinity of different ligands for the same receptor. In contrast, a structural biologist may consider the three-dimensional structure of the receptor protein

From: *Integrative Physiology in the Proteomics and Post-Genomics Age*
Edited by: W. Walz © Humana Press Inc., Totowa, NJ

as function, whereas a cell biologist might consider the function of the receptor in terms of the intracellular pathways affected. Similarly, a physiologist might only consider the function of the receptor in terms of the integrated effects of stimulation of the receptor on a particular organ system. Each of these approaches, separately or combined, in the context of assigning function to genomic information, constitute studies in functional genomics. The integration of many or all of these disciplines and approaches, however, along with the computational proficiency to assemble the different pieces into a model of the whole organism most fully defines functional genomics.

Although a large variety of approaches are used to address questions in functional genomics, two general approaches are utilized *(3)*. One approach is a continuation of classical, hypothesis-based research where the importance of a single gene, molecule, or pathway is specifically addressed by the experimental design of the study. A second approach, which takes particular advantage of high-throughput genomic resources, is known as discovery-based research and is performed to determine an unknown gene or genes that are linked to a problem of functional importance. Despite the fact that these two experimental approaches are quite different, they are not necessarily mutually exclusive. For instance, it would be predicted that a discovery-based project that identified a candidate gene as the cause of a disease process would perform a number of very specific, hypothesis-based studies to affirm the functional importance of the candidate gene. The discovery-based and hypothesis-based approaches can therefore complement each other under different circumstances.

Regardless of the experimental approach taken, the combined use of methodologies that manipulate or serve to otherwise describe the genome, along with assays of messenger RNA (mRNA) expression, protein function, protein pathways, cell-signaling pathways, cellular function, tissue function, or organism function have all come to be termed "physiological genomics" or "functional genomics." Moreover, scientists performing functional genomics experiments utilize a wide variety of technologies and resources borrowed from a number of different fields of study in the life sciences. The need to utilize a diverse array of techniques in functional genomics experiments typically leads to large groups of scientists working in collaboration to address a specific question; this also requires biologists to have a working knowledge of the methodology and terminology of each of the different disciplines.

2. GENERAL APPROACHES TO PHYSIOLOGICAL GENOMICS: HYPOTHESIS VS DISCOVERY-BASED RESEARCH

Living organisms are complex biological systems with billions of molecules interacting simultaneously in an organized manner to sustain life. A traditional approach favored by scientists interested in understanding biological systems was to reduce these systems to their simplest common parts. By quantifying the actions and interactions of the individual components, it was hoped that the individual constituents could be reassembled and permit an understanding of the nature of the whole organism. This approach in which the biological role of an individual molecule is systematically addressed is often described as a hypothesis-based or reductionist approach. To address a specific disease process with this scientific approach, biologists first characterize the pathophysiology of a certain condition. Armed with this information, the genetic underpinnings of the disease process can then be identified by first deducing a number of genes/gene products that are likely candidates as the mediators of the disease process. The role of the individual gene is then tested using a variety of methodologies to determine if the pharmacological or genetic deletion or manipulation of the normal gene leads to phenotypes consistent with the disease process. Finally, if there is a positive association between the disease phenotype and the blockade of the gene product, further study is performed to understand the particular genetic mutations responsible for the pathological phenotype.

This popular "reductionist" approach to biological science has revealed an enormous amount of information regarding the mechanisms whereby biological systems function. One weakness of this approach, however, has been the inability to fully understand the complicated nature of the integrated organism because of insufficient knowledge regarding the innumerable complex interactions that occur in an organism and the inability of currently available technology to reconstruct an organism from its individual components.

The public and privately funded projects to sequence the human genome and the evolution in technology that has driven and arisen from these projects have led to an alteration in this traditional experimental paradigm. Although many researchers continue a hypothesis-based approach to research in which experiments are specifically designed to determine the functional role of an individual gene or gene product, the genomic revolution has permitted scientists to begin discovery-based studies in which large parts of the entire genome are scanned to determine the genetic basis of disease. In discovery-based research, scientists utilize approaches and technologies that permit the entire genome to be scanned or randomly altered, and functional experiments are then performed to determine the association of the affected genes with disease. Two examples of discovery-based research are the genetic linkage analysis of quantitative trait loci (QTL) and mutagenesis screens.

In the QTL segregation analysis, two inbred animal strains, one normal and one affected with the disease of interest (the phenotype), are initially bred *(7–9)*. An example of this breeding strategy in inbred rats is presented in Fig. 1, with each parent homozygous for a different allele at a representative genetic locus (AA or BB). The offspring in the first generation of this genetic cross, known as the F_1 generation, are identical because both parental strains are inbred; this is represented by heterozygosity, AB, at the representative locus. Further intercrossing within the F_1 generation produces an F_2 generation, in which the genetic information from the two parental strains has been scrambled due to recombination during meiosis *(9)*. These rats in the F_2 generation are then carefully phenotyped for traits associated with the disease. In addition, a systematic, genome-wide scan is performed on each rat of the F_2 generation utilizing anonymous genetic markers that are polymorphic between the two parental strains. A mathematical analysis, known as a linkage analysis, is then performed on the data to determine if the inheritance of the genetic markers correlates with a given quantitative trait in the F_2 generation. The linkage analysis utilizes many genetic markers distributed throughout the genome and provides a powerful tool for the discovery of novel genes or gene pathways in animal models of disease.

A second example of discovery-based research in physiological genomics are the mutagenesis projects that have been carried out in mice, zebrafish, and other species. An example of this approach, as performed in zebrafish to screen for recessive mutations, is schematized in Fig. 2 *(10)*. In this particular mutagenesis screen, a parental male zebrafish was exposed to ethylnitrosurea (ENU), a mutagenic agent. The dose of ENU, an alkylating agent that acts as a powerful mutagen, is titered to induce random point mutations in the spermatogonia of the treated animal. Subsequent breeding of the mutagenized male to

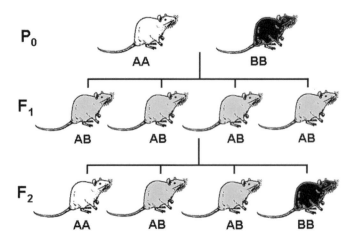

Fig. 1. Schematic illustrating an F_2 intercross. Two inbred parental strains (P_0) with homozygous genotypes for different alleles are initially crossed to yield an F_1 generation. The F_1 animals will be identical, but intercrossing of the F_1 generation yields an F_2 generation with the segregation of different alleles. Phenotyping and genotyping of the F_2 animals allows the mapping of quantitative trait loci. See text for details.

a normal female yields an F_1 generation. The F_1 generation of this cross was then interbred to yield heterozygote carriers of the mutated genes in the F_2 generation. The F_2 generation was then interbred to yield homozygotic and heterozygotic carriers of the mutations in the F_3 generation. In this mutagenesis study, careful screening of the F_3 embryos was then performed. Any signs of mutations were judged to be owing to a single-gene, recessive mutation if it was evident in 25% of the growing embryos and larvae. Once a mutation has been identified and confirmed, further phenotyping and genotyping protocols are performed to identify the gene or genes affected that lead to this altered trait *(10)*.

Both the QTL analysis and the mutagenesis screen are examples of discovery-based science in which there are no initial assumptions in regard to the gene, molecule, cell, or tissue type affected. The entire experiment is performed to discover novel genes or gene pathways important in normal function and in cases of pathology. This approach contrasts with hypothesis-based research where the importance of a specific gene or molecule is specifically addressed by the experimental design of the study. Regardless of the experimental approach, the combined use of methodologies that manipulate or serve to otherwise describe the genome, along with functional assays of protein, pathway, cellular, tissue, or organism function have all come to be termed "physiological" or "functional genomics."

3. MODELS AND APPROACHES COMMONLY USED IN PHYSIOLOGICAL GENOMICS EXPERIMENTS

As one might expect, there is a diverse array of methodology and technology utilized as well as a large assortment of different species that are studied. Examples of different methodologies employed include expression profiling, proteomics, chromosome transfer, gene deletion, gene delivery, gene silencing, mutagenesis, and a variety of gene-mapping studies. These types of experiments are performed in a number of species including rats, mice, primates, dogs, yeast, zebrafish, and *C. elegans*. These approaches and models have been

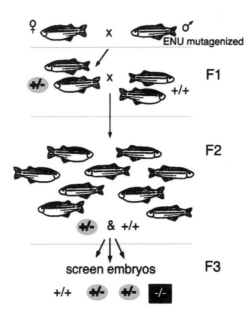

Fig. 2. Outline of zebrafish mutagenesis screens. *N*-Ethyl-*N*-nitrosourea (ENU) was used to mutagenize spermatogonia of parental males. Outcrosses were performed with wild-type females to produce an F_1 generation, each F_1 fish possessing a unique set of mutations. Sibling F_1 matings created the F_2 generation, and the mutations were driven to homozygosity in the F_3 embryos. (Reproduced from ref. *10* with permission.)

utilized to explore the basis of both monogenetic and polygenetic disease processes. A comprehensive review of all of the techniques and methodologies employed in physiological genomics studies is beyond the scope of this chapter. Outlined here are a number of examples of approaches successfully utilized in functional genomics applications. Also included is a concluding section that describes a general approach to integrate these approaches in order to begin to link genes with function. This summary is by no means exhaustive, but is designed to provide the reader with a sense of the approaches currently in use. Those interested in more specific details and a further discussion of the different methodologies are referred to the source material referenced throughout the chapter.

4. GENE-MAPPING STUDIES

Gene mapping refers to the localization of a disease gene or genes on the gene map of the organism of interest. A number of approaches are used to map disease genes, but the basis for any mapping study is the phenomenon of recombination which produces new arrangements of genes. During meiosis, pieces of DNA are exchanged between homologous chromosomes by crossing-over. Recombination that occurs between two distinct positions or loci on a gene map results in a new combination of alleles (different forms of the same gene).

The position of genetic loci on the gene map is determined through the use of genetic markers *(11)*. Genetic markers are typically short-tandem repeats of DNA sequences or single nucleotide polymorphisms that can be readily detected by the polymerase chain reaction. Other genetic markers are genes with known phenotypic expression and restriction fragment length polymorphisms. The combination of alleles on the genetic map, known as the genotype, is then correlated to the measured function in the animals, known as the phenotype. A mathematical analysis of the relationship between the genotype and phenotype allows one to determine the degree of association of a certain marker with the genotype of interest.

A large proportion of the genetic variation that underlies disease phenotypes is caused by complex, polygenic interactions and is controlled by loci that have quantitative effects on the phenotype of interest. A phenotype that can be quantified, or a quantitative trait, is one that has a measurable variation owing to genetic and/or environmental influences. This variation can consist of discrete values or can have a continuous distribution. Importantly, quantitative traits contrast with qualitative traits, which are either present or absent (e.g., cystic fibrosis or sickle cell anemia). A QTL is a genetic locus that contains alleles that affect this genetic variation. These genetic loci are identified through a statistical analysis of the complex traits. Typically, the traits are affected by environmental as well as genetic factors.

Genetic-mapping strategies in experimental animals generally involve specific animal breeding approaches in which a genetic cross of two inbred strains (usually a normal or unaffected strain and a strain that demonstrates a disease phenotype) is made to examine the correlation between genetic loci and the disease phenotype. An example of such an approach is provided in Fig. 1. Mapping studies in human families will typically involve the close genotypic and phenotypic examination of a family whose members are affected with a particular disease. Coarse mapping of genetic loci to a chromosome segment using standard breeding strategies in animals or in human family studies will normally narrow a region of interest to 10 to 30 centimorgans (cM). A centimorgan is a unit of measure of recombination frequency; 1 cM equals a 1% chance that a marker at one genetic locus will be separated from a marker at a second locus owing to crossing over in a single generation. In humans, 1 cM is equivalent to approx 1 million base pairs, and would be predicted to contain 11 to 12 separate genes. Coarse mapping would, therefore, narrow a region to between 100 and 300 genes. Although this is an enormous reduction from the approx 35,000 potential genes, the number of possible genes affecting the phenotype of interest is too large to begin a gene-by-gene functional evaluation. A more detailed analysis, often termed "fine mapping" is then employed *(11,12)*.

5. CHROMOSOME TRANSFER: CONSOMIC AND CONGENIC RATS

As described previously, a co-segregation analysis and mapping of QTL is largely a coarse-mapping function that can narrow a region of interest (ROI) in the genome to a section of DNA of approx 30 cM *(11,12)*. Since a section of genomic DNA of this size would be predicted to have up to 300 genes, additional work is needed to reduce the number of candidate genes in the ROI. To further narrow ROI, a number of different breeding approaches can be used in mice and/or rats. These breeding approaches include a number of different backcrossing and intercrossing strategies that produce advanced intercross lines, recombinant inbred strains, congenic strains, or consomic strains, to name a few *(11)*. The reader is referred to the literature for a more comprehensive review of these different methodologies *(7,9,11,12)*. This section focuses on the use of consomic strains, although each of the previously described approaches has proven valuable.

Inbred consomic strains are developed by marker-assisted breeding to insert a complete chromosome from one inbred strain into the genetic complement of a recipient strain *(4,5,7)*.

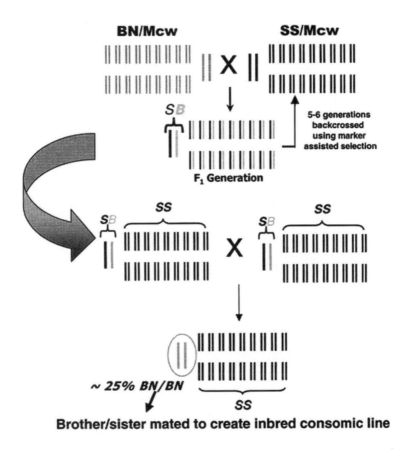

BN/Mcw   **SS/Mcw**

5-6 generations
backcrossed
using marker
assisted selection

S_B

F_1 Generation

SS   SS

S_B S_B

X

~ 25% BN/BN

SS

Brother/sister mated to create inbred consomic line

Fig. 3. The breeding and genotyping scheme required to develop an inbred consomic line. See text for details. The genome for the normotensive, salt-resistant BN/Mcw (Brown Norway) rat is shown in gray and the genome for the hypertensive SS/Mcw (Dahl Salt-Sensitive) rat is shown in black. (Reproduced from ref. 5 with permission.)

The breeding strategy to generate an inbred consomic rat line is illustrated in Fig. 3, in which an individual chromosome from the BN/Mcw (Brown Norway) rat is transferred to the SS/Mcw (Dahl Salt-Sensitive) background with the aid of marker-assisted breeding. In this strategy, the F_1 generation of an intercross between these two inbred parental lines is backcrossed to one of the parents, the SS/Mcw. Through the use of a total genome scan, which indicates which alleles are from which parental strain, subsequent progeny can be selected that possess the greatest numbers of homozygous alleles for the parental strain to which the backcross is occurring, the SS/Mcw, while maintaining the heterozygosity for the targeted chromosome. This strategy is employed with five to six subsequent backcrosses to the SS/Mcw parental strain until the progeny are isogenic for all SS/Mcw chromosomes except for the targeted chromosome, which is heterozygous. These rats can then be intercrossed to derive an F_2 generation, which has one chromosome homozygous for the BN/Mcw, whereas the rest of the genetic complement is homozygous for the SS/Mcw genetic background.

Once these consomic lines have been derived, they can be bred to provide stable inbred lines for physiological study. These consomic lines can be used to validate the significance of a QTL discovered in a mapping study, they permit the localization of physiological traits to a chromosome without the necessity for further genotyping, they provide a genetically well-defined strain for physiological study, and they provide a basis for further fine mapping by the creation of congenic strains *(4,5,7,13)*. Congenic strains are developed by integrating an individual piece of a chromosome containing a genomic region with genes of potential interest into the genomic background of the recipient strain *(8,9)*. The creation of a set of congenic strains with overlapping regions can then be utilized to further refine the genetic region containing the QTL of interest. Although this and other breeding strategies can serve to narrow the QTL of interest to a genetic locus of 1–5 cM, there are still a relatively large number of possible candidate genes in the ROI. Further study is then necessary to identify the causative gene or genes.

The process of identifying the genes responsible for diseases in a QTL can utilize a number of different resources depending on the model *(11)*. Positional cloning involves the use of linkage mapping to determine the location of a disease gene by employing additional genetic markers in the genetic ROI. This ROI is narrowed by relating genotype to phenotype with high resolution by utilizing congenic strains, recombinant inbred strains, or other breeding strategies. The DNA from the ROIcan then be isolated with the use of various genomic resources and the disease gene can be sequenced with the mutation potentially identified.

6. GENE-EXPRESSION PROFILING

Changes in gene expression underlie many biological phenomena. The quantification of changes in gene expression by Northern blotting, *in situ* hybridization, reverse transcription coupled to the polymerase chain reaction, ribonuclease protection assays, and other methodologies have provided invaluable insight into the importance of mRNA changes during a variety of physiological maneuvers and under different pathophysiological conditions. These methods are limited for functional genomics applications, however, in that they are targeted to a single mRNA species which has been designated by the investigator as worthy of study. The advent of DNA microarray technology now permits the evaluation of gene expression changes in large numbers of genes in a single experiment *(14–16)*.

The basic scheme behind a microarray experiment is quite simple and is schematized in Fig. 4 *(14)*. Total RNA is isolated from two individual biological samples (organs, tissues, cells), which are to be compared. The total mRNA from each sample is reverse transcribed into individual complementary DNAs (cDNAs), and the cDNA from the different samples are labeled with different fluorescent dyes. The samples are then hybridized to a microarray slide which typically contains thousands of individual spots. Each spot on the array contains a single nucleic acid probe that is complementary to a known mRNA. The nucleic acid probes are cDNA or oligonucleotide probes that have been robotically arrayed on the glass slide. Following the hybridization and appropriate washing steps, a confocal laser scanner that can differentiate between the two labeling dyes is used to quantify the binding of the cDNAs from the different samples to the individual spots. Because the identity of the probe in each spot is known, and the samples have been labeled with different dyes, the ratio of intensity of binding to the different spots can then be used to assess the relative expression of individual transcripts in each sample.

This revolutionary technology permits the investigator to determine which genes are active in a specific cell or tissue at a specific time. Variations in mRNA expression can serve as

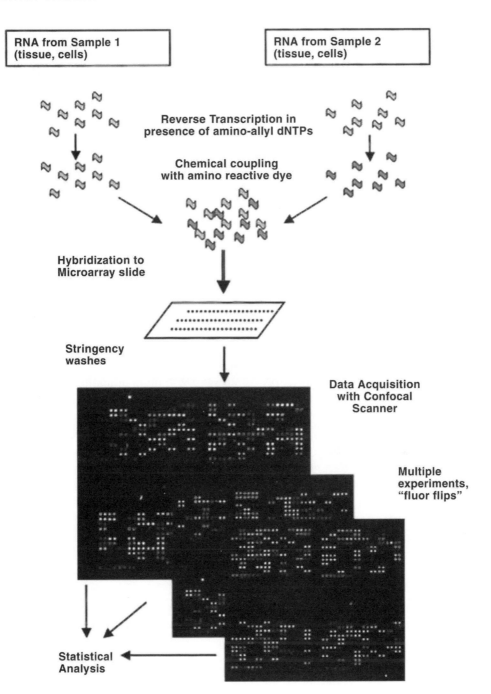

Fig. 4. Schematic demonstrating a DNA microarray experiment. Two RNA samples are reverse transcribed and labeled with fluorescent tags. The samples are then hybridized to a microarray slide and scanned to acquire the expression data. See text for details. (Reproduced from ref. *14* with permission.)

an important indicator of disease or predisposition to disease. By comparing gene expression patterns between cells or tissues from different environments, such as normal tissue compared to diseased tissue, specific genes, or gene pathways that play a role in pathophysiological processes can be identified. In the same manner, evaluation of expression profiles between parental and consomic or congenic animals can help identify the genes or gene pathways responsible for different disease processes *(17)*.

Although this technology permits the rapid analysis of mRNA expression in different biological samples, it is important to note that the etiology of all disease processes cannot be explained on the basis of changes in RNA expression alone. Moreover, mRNA expression can be increased or decreased as a result of the disease process; expression profiling studies should therefore be carefully designed to discriminate between the changes that are causative in the disease process and those changes that result from the disease. Finally, it is important to verify the results obtained by expression profiling at the level of protein and biological function to confirm that the gene in question is indeed responsible for the observed change in function.

7. PROTEOMICS

Proteins are involved at some level in all aspects of biological function; the examination of the entire complement of proteins in a biological system has therefore become a rapidly expanding and evolving field of research. The field of study that addresses gene function at the level of proteins is known as proteomics. Proteomic studies include experiments aimed at understanding protein–protein interactions, studies designed to perform the large-scale identification of proteins and their numerous posttranslational modifications, and studies planned to quantify the individual levels of different proteins in health and disease. Each of these endeavors, but especially the identification and quantification of different proteins under different pathological conditions, can complement the other methodologies described in this chapter to address questions of scientific importance in functional genomic studies.

The identification and quantification of unknown proteins in a complex mixture is a multistep procedure that is approached using a number of different strategies depending upon the application *(18–20)*. All of these procedures require initial steps to isolate the proteins from the tissue; a separate step involves the separation of the complex protein mixture into individual components; and the final step is then taken to identify the unknown protein or proteins. Although there are a variety of different approaches utilized to address this question, one commonly utilized procedure for protein identification is the coupling of two-dimensional (2D) gel electrophoresis with mass spectrometry to identify proteins expressed under different physiological conditions. The 2D gel electrophoresis process separates proteins based on two parameters, isoelectric point (charge) and molecular weight, in a polyacrylamide gel. The gel is then stained to reveal individual spots that contain a single protein with a unique isoelectric point and molecular weight. The intensity of staining of the individual spots may be quantified to provide an index of the amount of that particular protein present in the sample. The individual spots on the gel are then excised, digested with a protease such as trypsin, and subjected to mass spectrometry analysis. The mass spectrometry methodologies can then be used to identify the unknown protein using a process known as peptide mapping. Through this technique, the unknown protein, which has been digested with the protease, can be identified based on the composition of its individual peptides. This identification is made by comparing the peptide composition of the unknown protein to the composition of proteins previously identified and stored in a database.

The use of 2D gel electrophoresis for protein separation, however, is problematic because there is often comigration of more than one protein to a single spot on a gel. In addition, only the most abundantly expressed proteins are resolved and detected by 2D gel electrophoresis so only a relatively small subset of the total cellular complement of proteins is detected by this technique. As an alternative to methods that employ 2D gel electrophoresis in the separation step, mass spectrometry methodologies have been coupled with liquid chromatography techniques to identify differentially expressed proteins. The application of different liquid chromatography separation strategies has greatly increased the number of proteins that can be identified.

Recent technological advancements have led to the possibility that unknown proteins can not only be identified but that the protein content of cells and tissues can be determined. Proteins have been incorporated with stable isotopes or labeled with different marker agents that enable the quantification of protein content by mass spectrometry. This approach takes advantage of the ability of mass spectrometry to differentiate chemicals of the same composition that are composed of stable isotopes of different masses; the ratio of the signal for such pairs can be used to calculate the abundance of each. This exciting approach, along with chip-based methods to determine differential expression of proteins, and strategies employed to examine protein–protein interactions and posttranslational modifications of proteins are some of the challenges facing the proteomics field *(18,20)*.

Proteomics has an important place in the functional genomics approach to understand systems biology. The connection between the genes and gene expression and the function of biological systems and whole organisms are the proteins. A description of the proteome, the quantification of protein levels throughout development and during the cell cycle, and an understanding of protein pathways are all important components if the chart linking genes to biological function is to be assembled.

8. GENE TARGETING

Once a candidate gene has been identified, it is still imperative that the investigator confirm the role of this gene in the physiological or pathophysiological process under investigation. One experimental approach utilized in functional genomic studies to confirm the action of a candidate gene following a gene-mapping, mutagenesis, or other discovery-based study is gene targeting. In one of the most commonly used gene-targeting approaches, homologous recombination is performed in germ line-competent stem cells to disrupt a targeted gene in mice *(21,22)*. Gene-disruption studies, also known as gene "knockout" experiments, are utilized to demonstrate the influence of complete deletion of a gene product. This approach permits scientists to examine function in animals which completely lack the functional gene product. Studies of this type can, therefore, be used to confirm the importance of a specific gene in biological processes. One other approach taken with homologous recombination strategies is to create animals in which a specific gene is duplicated or other genes are inserted into the genome of mice. Using this strategy, the functional influence of elevated levels of a native gene product, or the influence of elevated levels of a mutant gene product can be examined.

This innovative approach has led a large number of physiology labs to begin using mice as a model system. One concern with the use of gene-targeted mice, particularly the null mutant models, is that the targeted gene has been deleted from conception. There may, therefore, be developmental abnormalities in the affected animals or other components of the same or other biological pathways may be up- or downregulated to compensate for the genetic manipulation. In addition, the gene of interest will be manipulated in every cell of the animal's body, so the importance of a single molecular species in an individual cell type can be difficult to discern. To address this problem, strategies have been conceived that permit the knockout of a gene in a particular tissue. One approach to create a tissue-specific knockout utilizes the Cre/loxP system *(23)*. Cre is a 38 kDa recombinase protein from a bacteriophage that mediates site-specific recombination between loxP sites (two 13 bp inverted repeats separated by an 8 bp spacer region). By combining this approach with a tissue-specific promoter, it is possible to create a mouse with a tissue-specific gene deletion. Moreover, this system can be combined with a tetracycline-sensitive element that will permit a tissue-specific deletion that can be controlled by the presence (administration) of tetracycline to the mouse *(24–26)*. Using these strategies, the timing of the tissue-specific knockout condition can be controlled permitting the examination of genetic knockout animals that have had the gene product present throughout development and early life.

The advent of this revolutionary technology has permitted the examination of gene function in a large number of different fields of physiological study. The limiting factor to date in the use of gene-targeting strategies is that the mice are the only mammal from which germ line-competent stem cells have been isolated and successfully cultured. Although different strategies have been successfully used to create transgenic *(9)* and even knockout rat (with enu mutagenesis) models *(27),* the mouse is the only species that has been successfully and widely used for gene-targeting studies. Because mice have a small body size and most physiologists have limited experience in the study of function in these animals, the initial impact of this technology has been restricted to in vitro studies and simple, noninvasive, in vivo studies. Moreover, the availability of these models was limited to investigators with access to a transgenic facility and laboratory budgets that permitted the development of gene-targeted models. As such, the widespread application of mouse knockout and gene-targeted models to physiological questions has been somewhat limited. In the past several years, however, numerous technological advancements have occurred permitting the more thorough evaluation of physiological function in vivo in mice. Moreover, the Jackson Laboratory (www.jax.org/) and a number of other commercial vendors have served as repositories for gene targeted mice that can be purchased for study by the general scientific community.

9. SIRNA/ANTISENSE/PHARMACOLOGICAL APPROACHES

As discussed previously, gene targeting by homologous recombination has been successfully used to manipulate gene function in mice. Unfortunately, gene-targeting experiments are expensive and time-consuming and cannot be readily performed in all species. An alternative strategy to examine or confirm the functional importance of a candidate gene is an approach that uses standard pharmacological agents (i.e., receptor agonists and antagonists or enzyme inhibitors) to manipulate a particular gene product or pathway. This fairly straightforward approach can serve as a final confirmation of the importance of a particular molecule in normal function or in a disease process. The greatest weakness of this traditional pharmacological approach, however, is that it is largely dependent on the availability, specificity, and selectivity of reagents specific for the candidate gene. It is, therefore, only possible to employ this approach when interested in examining a gene product or a biological system that has already been described in some detail and for which the appropriate reagents to manipulate the system have been developed. Because discovery research, by definition, should lead to the detection of novel genes and pathways, experimental approaches that can be applied to a greater number of possible candidate genes is required.

To circumvent the problems associated with the use of pharmacological agents, both small interference RNA (siRNA *[28–30]*) and antisense oligonucleotide technology *(31)* have been developed and utilized to selectively target genes or gene products for which standard pharmacological tools are not available. These reagents are small strands of RNA or DNA that are complementary to the mRNA of the targeted gene. These agents bind to the target RNA and create a double stranded complex. The double-stranded complex leads to either the arrest of translation *(31)* or cleavage of the double-stranded complex *(28–30)*; regardless of the mechanism of action, the end result is gene silencing. This approach theoretically permits an investigator to create his or her own designer drugs to target any mRNA for which the sequence if known. The use of siRNA or antisense, along with the pharmacological approach described earlier, permits a phenotypic analysis both before and after loss of function in an otherwise normal animal without the often confounding developmental abnormalities that can occur in models in which a target gene has been deleted from conception.

10. BIOINFORMATICS

An additional and very important component of functional genomics studies is bioinformatics. Bioinformatics is the information science that employs computer technology, databases, and web-based resources to decipher and organize the vast amounts of information generated from genomics and proteomics projects *(32)*. For example, bioinformatics scientists develop algorithms and other data management tools which facilitate functional genomics research by permitting the examination of large numbers of genes, proteins, and interacting networks of proteins. For any of the above described techniques and methodologies, if a high-throughput operation is desired, suitable data-processing, data-storage and -retrieval, and data-visualization systems must be in place. One of the most important tasks of the bioinformatics field is this critical task. As the volume and the complexity of the data sets increase, the role of bioinformatics will only increase. Without this critical resource, physiological genomics applications will not be possible.

11. COMPARATIVE GENOMICS

Animal models of disease are generally chosen based on pathophysiological characteristics shared with human disease. It is important, however, to ensure that the genetic basis for disease in the animal models correlates with the genetic basis of disease in humans. As such, comparative gene-mapping strategies, which are largely successful owing to the efforts of bioinformatics scientists, have been utilized to demonstrate that there are regions of the human genome that are evolutionarily conserved with other mammalian species. In fact, it has been demonstrated that QTLs for obesity, autoimmune disease, and hypertension lie in evolutionarily conserved regions in the rat, mouse, and human *(7)*. These comparative mapping data justify the use of rodent and other species as models for human disease. Moreover, these data indicate that scientists can utilize different species as they prepare to attach function to the genome. For instance, a mapping study performed in rats may indicate that a certain gene is present in a QTL judged important in a disease of interest. In the absence of reagents to manipulate the function of this gene in the rat, gene-targeting strategies could be justified in the mouse to demonstrate the importance of this gene to the disease phenotype. Finally, comparative mapping strategies could be utilized to demonstrate that a given QTL is conserved between the mouse, rat, and human to justify the experimental approach.

12. HIGH-THROUGHPUT PHENOTYPING

The term "phenotyping" has been used throughout this chapter to refer to the quantification of the observable or visible properties of an organism. As such, an organism's phenotype is either its physical appearance and makeup or a specific trait. The ability to accurately and quickly measure a given phenotype is extremely important in functional genomics. All of the previously described techniques and methodologies are useful tools in functional genomic studies, but an experimental design which contains the most highly automated genotyping or expression array process will still be limited by the ability to accurately quantify the phenotype of the organisms to be studied. Some simple phenotypes such as height, weight, eye color, and coat color are easily measured and therefore may not impose a significant impediment to research progress. The quantification of more complex phenotypes, however, can pose a particular difficulty to those interested in quantifying complex traits. Examples of disease processes that require careful measurement include cardiovascular disease, diabetes, cancer, respiratory disease, sleep disorders, and other neurological disorders. The ability to accurately phenotype large numbers of animals for complex traits is therefore a tremendous challenge facing investigators participating in functional genomics studies.

As it has been recently noted, one of the greatest problems with obtaining accurate phenotypic information and comparing this information is the large number of different approaches to quantify similar phenotypes *(33)*. Because phenotyping is often times based on methods that vary between different fields of study and different laboratories, there is a great need to systematically standardize phenotyping protocols. The scientific community will only be able to take advantage of the enormous wealth of genotypic data when comprehensive and truly high-throughput systems for phenotyping have been established. One such effort, known as the Mouse Phenome Project is headquartered at the Jackson Laboratory *(34)*. In this project, quantifiable phenotypic characterization of a defined set of mouse strains by industrial and academic laboratories has begun with the data deposited in an accessible database (www.jax.org/phenome).

A number of groups have attempted to circumvent phenotyping limitations by developing new strategies and methods to measure functional parameters in different experimental models. The advent of gene manipulation in the mouse has led to a large effort to adapt functional measurements to this small mammal. Similarly, the relative advantages of performing ENU mutagenesis in zebrafish have by necessity led to advances in assays of function in these tiny animals. Several examples of groups developing phenotyping strategies that increase throughput in different models are the screening protocols developed in the National Institutes of Health Program for Genomic Applications to quantify heart, lung, and blood phenotypes in mice (http://pga.jax.org/) and rats (http://pga.mcw.edu/). Technological advances have, therefore, made the measurement of different indices of physiological function feasible even in minute animals and embryos. Despite the significant efforts and advances that have increased the ability to characterize complex phenotypes in small animals (e.g., rats, mice, and zebrafish), however, the accurate quantification and measurement of different parameters in these small animals is fraught with difficulty. An extensive effort to increase the throughput of this important portion of the functional genomics strategy to a point that is equivalent to that which can be reached by many of the genomics and proteomics technologies described above will be required before functional genomics studies can reach their full potential.

13. INTEGRATIVE PHYSIOLOGY: PUTTING THE PIECES TOGETHER

It is clear that no single scientific field or methodology by itself will be sufficient to connect genes with proteins, cell function, organ and system function, and the intact organism. Instead, a combination of methodologies borrowed from different scientific disciplines will likely best serve the needs of an investigator. To fully integrate genomic data and functional data, we will initially require a structured list of genes and gene products. Following that enormous task, quantitative data regarding the complex interactions in cells will be desired; for instance, whole-scale analysis of mRNA and protein levels at different times throughout a cell's life and in different cell types will be necessary. Furthermore, knowledge in regard to the interactions between different gene products (i.e., rate constants, affinities, etc.) with subsequent mathematical modeling of the thousands of different parameters that vary at different times throughout the life of different cells is needed to begin to understand function in a single cell or cell type. Once that level of complexity has been mastered, the interactions among the trillions of cells and the myriad signaling mechanisms and pathways that combine to make up an organism must be similarly analyzed and quantified to be understood *(5,19,35–37)*.

Although there are significant efforts underway to understand and quantify gene expression and proteins on a large scale, the assembly of these individual parts into functional cells and into individual organisms will require a much greater understanding of cellular and systems physiology than is currently recognized. As a logical next step in the quest to link genes with function, the quantitative description of functional parameters in humans and in experimental models is being considered. In fact, as discussed previously, a database of physiological parameters from different mouse strains has recently been established *(34)*. Despite this effort, however, the complex interactions among cells, organs, and organ systems in the intact organism must still be described and quantified in order to be placed into proper context in terms of organism function.

An early attempt to describe and quantify the numerous positive and negative feedback systems involved in the regulation of the circulation was made more than 30 yr ago by Guyton and colleagues *(38,39)*. As depicted in Fig. 5, the complex interrelationship among blood vessels, the heart, the lungs, the autonomic nervous system, the kidneys, numerous endocrine controllers of blood pressure, the metabolic needs of tissues, and a host of other factors were included in the model. Using a systems analysis approach, the model is based on the quantitative analysis of positive and negative feedback systems, the temporal response of the different feedback pathways, and the relationship between the different pathways. This quantitative analysis of circulatory function, which involved approx 400 physiological phenomena and their interrelationships, was gradually constructed over a 12-yr period using the knowledge available at that time *(38,39)*. Each of the blocks in the diagram is represented by one or more mathematical equations that were derived from experiments

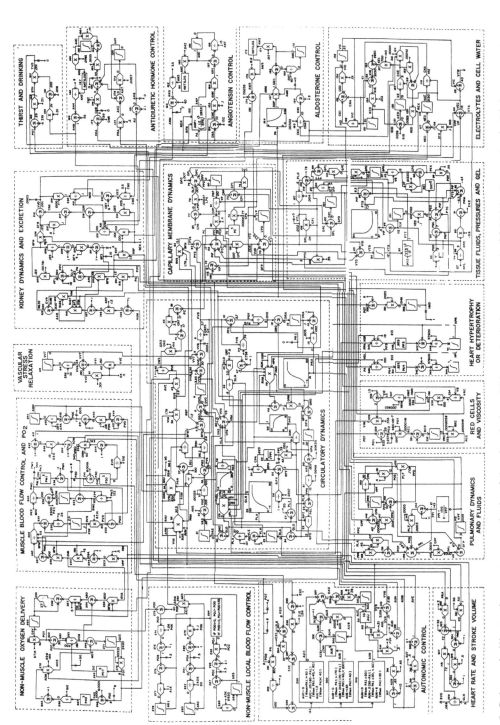

Fig. 5. Schematic illustrating systems analysis of the circulation. This analysis is composed of 354 blocks, each of which represents one or more mathematical equations describing the circulation. In general, each block represents the quantitative results of research performed by one or more investigators. (Reproduced from ref. 39 with permission.)

performed by different investigators. This type of complex modeling of quantitative physiological responses is one approach that could be utilized as part of the effort to understand complex biology from gene to function. When one considers the complexity of this diagram in the context of the roughly 35,000 genes, 100,000 proteins, and the trillions of cells that make up the body, the construction of a model that relates all of the genes to total organism function is an intimidating undertaking.

To perform this task, the basic quantitative information regarding the numerous interactions that occur between different organ systems in the normal control of the body's systems must be determined and systematically described. A far greater emphasis will need to be placed on the careful acquisition of these basic, quantitative types of physiological data. This information, once obtained, will then need to be analyzed and integrated for the construction of complex mathematical models by computational biologists, which should provide the final link between genes and function.

14. SUMMARY AND CONCLUSIONS

The sequencing of the human genome has provided 21st century physiologists with an unprecedented opportunity to explore and understand integrated organism physiology. The rapid evolution of genomic, proteomic, bioinformatic, and computational biology resources along with a renewed interest and recognition of the need and importance of quantitative measurements of cell and organism function places physiologists in an enviable position. Science is poised to begin the process of understanding the complex relationship between the approx 35,000 genes and organism structure and function. To accomplish this monumental charge, physiology as a field must step up to the task. This will likely require physiologists to utilize resources from multiple fields of investigation (genomics, proteomics, computational sciences, etc.), to adapt established techniques and develop new methodologies that enable true high-throughput phenotyping, to develop models and model systems that can provide functional information on a system-wide or even an organism-wide basis, and to begin thinking of new ways to describe and quantify physiological and pathophysiological processes. To accomplish this task, the future physiologist will likely work in a large collaborative group, embrace cutting-edge technology, and perform experiments in which enormous amounts of data are generated. At the same time, this physiologist will still be required to carefully design well-controlled studies, to produce high-quality, quantitative data, and to perform a rigorous statistical analysis to substantiate his or her conclusions. The future is bright for integrative physiologists; it is now up to the individual scientists to take advantage of this tremendous research opportunity.

REFERENCES

1. Lander, E.S., Linton, L.M., Birren, B., et al. (2001) International Human Genome Sequencing Consortium. Initial sequencing and analysis of the human genome. *Nature* **409**, 860–921.
2. Venter, J.C., Adams, M.D., Myers, E.W., et al. (2001) The sequence of the human genome. *Science* **291**, 1304–1351
3. Glueck, S.B. and Dzau, V.J. (2002) Physiological genomics: implications in hypertension research. *Hypertension* **39**, 310–315.
4. Roman, R.J., Cowley, A.W. Jr., Greene, A., Kwitek, A.E., Tonellato, P.J., and Jacob, H.J. (2002) Consomic rats for the identification of genes and pathways underlying cardiovascular disease. *Cold Spring Harbor Symposia on Quantitative Biology* **67**, 309–315.

5. Cowley, A.W. Jr. (2003) Genomics and homeostasis. *Am. J. Physiol.* **284**, R611–R627.
6. Vukmirovic, O.G. and Tilghman, S.M. (2000) Exploring genome space. *Nature* **405**, 820–822.
7. Jacob, H.J. and Kwitek, A.E. (2002) Rat genetics:attaching physiology and pharmacology to the genome. *Nature Reviews Genetics* **3**, 33–42.
8. Kreutz, R. and Hubner, N. (2002) Congenic rat strains are important tools for the genetic dissection of essential hypertension. *Seminars in Nephrology* **22**, 135–147.
9. Rapp, J.P. (2000) Genetic analysis of inherited hypertension in the rat. *Physiol. Rev.* **80**, 135–72.
10. Warren, K.S. and Fishman, M.C. (1998) Physiological genomics: mutant screens in zebrafish. *Am. J. Physiol.* **275**, H1–H7.
11. Members of the Complex Trait Consortium. (2003) The nature and identification of quantitative trait loci: a community's view. *Nature Reviews Genetics.* **4**, 911–916.
12. Glazier, A.M., Nadeau, J.H., and Aitman, T.J. (2002) Finding genes that underlie complex traits. *Science* **298**, 2345–2349.
13. Greene, A.S. (2002) Application of physiological genomics to the microcirculation. *Microcirculation* **9**, 3–12.
14. Arcellano-Panlilio, M. and Robbins, S.M. (2002) Cutting-edge technology I. Global gene expression profiling using DNA microarrays. *Am. J. Physiol.* **282**, G397–G402.
15. Ruijter, J.M., Van Kampen, A.H.C., and Baas, F. (2002) Statistical evaluation of SAGE libraries: consequences for experimental design. *Physiol. Genomics* **11**, 37–44.
16. Schena, M., Shalon, D., Davis, R.W., and Brown, P.O. (1995) Quantitative monitoring of gene expression patterns with a complementary DNA microarray. *Science* **270**, 467–470.
17. Liang, M., Yuan, B., Rute, E., et al. (2002) Renal medullary genes in salt-sensitive hypertension: a chromosomal substitution and cDNA microarray study. *Physiol. Genomics* **8**, 139–149.
18. Aebersold, R. and Mann, M. (2003) Mass spectrometry-based proteomics. *Nature* **422**, 198–207.
19. Bader, G.D., Heilbut, A., Andrews, B., Tyers, M., Hughes, T., and Boone, C. (2003) Functional genomics and proteomics: charting a multidimensional map of the yeast cell. *Trends in Cell Biology* **13**, 344–356.
20. Patterson, S.D. and Aebersold, R.H. (2003) Proteomics: the first decade and beyond. *Nature Genetics* **33**, 311–323.
21. Coffman, T.M. (1998) Gene targeting in physiological investigations: studies of the renin–angiotensin system. *Am. J. Physiol.* **274**, F999–F1005.
22. Smithies, O. (1997) A mouse view of hypertension. *Hypertension* **30**, 1318–1324.
23. Sauer, B. (1993) Manipulation of transgenes by site-specific recombination: use of Cre recombinase. *Methods of Enzymology* **225**, 890–900.
24. Furth, P.A., St. Onge, L., Boger, H., et al. (1994) Temporal control of gene expression in transgenic mice by a tetracycline-responsive promoter. *Proc. Natl. Acad. Sci.* **91**, 9302–9306.
25. Gossen, M., Freundlieb, S., Bender, G., Muller, G., Hillen, W., and Bujard, H. (1995) Transcriptional activation by tetracyclines in mammalian cells. *Science* **268**, 1766–1769.
26. Gossen, M. and Bujard, H. (1992) Tight control of gene expression in mammalian cells by tetracycline-responsive promoters. *Proc. Natl. Acad. Sci.* **89**, 5547–5551.
27. Zan, Y., Haag, J.D., Chen, K.S., et al. (2003) Production of knockout rats using ENU mutagenesis and a yeast-based screening assay. *Nature Biotechnology* **21**, 645–651.
28. Dykxhoorn, D.M., Novina, C.D., and Sharp, P.A. (2003) Killing the messenger: short RNAs that silence gene expression. *Nature Reviews* **4**, 457–467.
29. Skipper, M. (2003) Have our dreams been shattered? *Nature Reviews Genetics* **4**, 671.
30. Tijsterman, M., Ketting, R.F., and Plasterk, R.H.A. (2002) The genetics of RNA silencing. *Annu. Rev. Genet.* **36**, 489–519.
31. Scalzitti, J.M., and Hensler, J.G. (1999) Design and efficacy of serotonin-2A receptor antisense oligodeoxynucleotide. *Methods Enzymology* **314**, 76–89.
32. Kanehisa, M. and Bork, P. (2003) Bioinformatics in the post-sequence era. *Nature Genetics* **33**, 305–310.

33. Freimer, N. and Sabatti, C. (2003) The human phenome project. *Nature Genetics* **34**, 15–21.
34. Bogue, M. (2003) Mouse Phenome Project: understanding human biology through mouse genetics and genomics. *J. Appl. Physiol.* **95**, 1335–1337.
35. Bassingthwaighte, J.B. (2000) Strategies for the physiome project. *Annals of Biomedical Engineering* **28**, 1043–1058.
36. Kitano, H. (2003) Computational systems biology. *Nature* **420**, 206–210.
37. Hood, L. and Galas, D. (2003) The digital code of DNA. *Nature* **421**, 444–448.
38. Guyton, A.C., Coleman, T.G., Cowley, A.W. Jr., Manning, R.D. Jr., Norman, R.A. Jr., and Ferguson, J.D. (1974) A systems analysis approach to understanding long-range arterial blood pressure control and hypertension. *Circ. Res.* **35**, 159–176.
39. Guyton, A.C., Coleman, T.G., and Granger, H.J. (1972) Circulation: overall regulation. *Ann. Rev. Physiol.* **34**, 13–46.

3

Electrolytes and Acid–Base Physiology

Meghan M. Taylor

1. CONSISTENCY OF THE EXTRACELLULAR FLUID

The extracellular space is composed of two compartments, blood vessels and interstitial space, which are separated merely by highly permeable capillary walls. Therefore, the chemical makeups of the fluids in these spaces are nearly identical. The major differences in components between these compartments arises from the fact that blood vessels are almost completely impermeable to proteins, especially albumin, and consequently protein concentrations in plasma are higher than those in interstitial fluid.

The extracellular fluid (ECF) is a solution containing small ions, organic molecules derived from metbolism and proteins. Sodium and chloride ions are by far the most abundant ions in the ECF (Table 1). Other ions include calcium, potassium, hydrogen, and bicarbonate. The osmolality, the ratio of solute to solvent, of the ECF is maintained at approx 295 mOsmol/L in a normal individual. The body judiciously regulates overall ECF volume and composition. Although no ion is regulated independently of other ECF components, the focus here is on the regulation of hydrogen ion levels, or pH.

The body has developed a series of intricate and coordinated mechanisms to maintain, with few minor deviations, the pH of extracellular fluids at approx 7.4. When these mechanisms are altered under pathological conditions the pH of extracellular fluid can fall to 7.0 or rise as high as 7.8, both leading to dire consequences. Homeostasis requires that the rate of acid (or hydrogen ion) intake and production equals the amount of acid excreted. The delicate acid–base balance is maintained by coordinated respiratory and renal mechanisms as well as intracellular and extracellular buffers.

2. BUFFERING SYSTEMS CONTROLLING ACID–BASE BALANCE

The body uses a variety of buffers to prevent rapid and extreme changes in extracellular pH. After an acid load, the blood pH of an animal will return to normal within a day even though only one-fourth of the acid load has been excreted by the kidneys, indicating the importance of buffers in the maintenance of normal pH. Plasma proteins are an important buffering system because they contain multiple negative charges owing to dissociated carboxyl groups on amino acids. Hemoglobin in red bloods cells is also a major buffer molecule, again because of its multiple negative charges. In addition, hemoglobin is extremely abundant and therefore has a higher buffering capacity than plasma proteins. Small organic

From: *Integrative Physiology in the Proteomics and Post-Genomics Age*
Edited by: W. Walz © Humana Press Inc., Totowa, NJ

Table 1
Components of Extracellular Fluid

Component	Concentration
Na+	140 mmol/L
Cl$^-$	110 mmol/L
HCO3$^-$	20 mmol/L
K+	4 mmol/L
Ca2+	2.5 mmol/L
H+	0.04 mmol/L
Protein	6.5 g/dL

anions such as phosphate and sulfate also act as buffers. However, the most important buffering system is the bicarbonate/carbon dioxide (HCO_3^-/CO_2) system.

The end product of the bicarbonate system is CO_2, which is rapidly regulated by respiration. The equilibrium reaction of the bicarbonate system is shown in the equation below. The reaction between H_2CO_3 and $CO_2 + H_2O$ is catalyzed by carbonic anhydrase (CA). Although CO_2 levels are regulated by the rate of respiration, the buffering capacity of this system resides in the body's store of HCO_3^-. When acids ingested or formed by metabolism are buffered by HCO_3^-, H_2CO_3 is formed and the levels of HCO_3^- fall. The increasing amounts of H_2CO_3 are then converted to CO_2 and water. As CO_2 levels rise, the rate of respiration is increased to maintain a normal CO_2 tension (P_{CO2}) in the plasma.

$$H^+ + HCO_3^- \Leftrightarrow H_2CO_3 \overset{CA}{\Leftrightarrow} CO_2 + H_2O$$

The buffering capacity of the system would be rapidly exhausted if the consumed HCO_3^- was not restored. This is accomplished by the kidneys, which excrete H^+ into the urine and return HCO_3^- to plasma. In addition, excess base can also be buffered by the HCO_3^-/CO_2 system.

$$OH^- + CO_2 \overset{CA}{\rightarrow} HCO_3^-$$

This reaction is also catalyzed by carbonic anhydrase. The excess HCO_3^- formed must then be eliminated by the kidneys.

The relationship between components of a buffer system can be described by the Henderson-Hasselbalch equation. The general reaction for buffer systems is:

$$HA \leftrightarrow H^+ + A^-$$

By the law of mass action, at equilibrium the relationship between the products and associated acid remain constant. Each buffer is described by a unique equilibrium constant, K.

$$K = [H^+][A^-]/[HA]$$

This equation can be rearranged to solve for the H^+ concentration (expressed as pH, which equals the negative log of the H^+ concentration). The resultant equation is termed the Henderson-Hasselbalch equation.

$$pH = pK + \log([A^-]/[HA])$$

The buffering capacity of a weak acid is maximal when the A^- concentration is equal to the HA concentration (then pH = pK).

For the HCO_3^-/CO_2 system the equation can be rewritten as:

$$pH = 6.1 + \log([HCO_3^-]/[CO_2])$$

At physiologic pH (7.4), this buffer system is far from its optimal buffering capacity (pH 6.1). In fact, at pH 7.4, the ratio of $[HCO_3^-]/[CO_2]$ is 20:1. The effectiveness of this system to buffer changes in hydrogen ion concentrations, however, is derived from the rapid regulation of plasma CO_2 levels by the pulmonary system.

The concentration of CO_2 is difficult to directly measure in a clinical setting. Instead, the partial pressure of CO_2 (in mmHg) within the blood is assessed. The Henderson-Hasselbalch equation for the HCO_3^-/CO_2 system can then be further rewritten by substituting the partial pressure (P_{CO2}) and the solubility coefficient of CO_2 (0.031 mmol/L/mmHg) for the concentration of the gas.

$$pH = 6.1 + \log([HCO_3^-]/0.031\, P_{CO2})$$

The Henderson-Hasselbalch equation shows that plasma pH can be altered by changing the normal ratios of HCO_3^-/CO_2. Changes in HCO_3^- are brought about by alterations in renal retention or secretion of HCO_3^-, whereas changes in CO_2 are brought about through changes in respiration.

3. PULMONARY REGULATION OF CO_2 LEVELS

The pulmonary regulation of CO_2 levels allows the body to rapidly regulate changes in pH. During metabolic acidosis, when there is an excess H^+ concentration, plasma levels of HCO_3^- fall as the molecule is consumed to buffer the extra acid. This leads to the formation of H_2CO_3, which is rapidly converted to CO_2 and water by carbonic anhydrase. It is thought that changes in P_{CO2} are sensed by chemoreceptors within the central nervous system that can lead to changes in ventilation. The respiratory response to the elevated P_{CO2} induced by metabolic acidosis is hyperventilation. By increasing their tidal volume, the acidotic patient dilutes the CO_2 levels in alveolar air leading to decreased CO_2 in plasma. In this manner, the HCO_3^-/CO_2 system effectively buffers an increased acid load.

In contrast, the respiratory response to metabolic alkalosis, excess base or loss of acid, leads to hypoventilation. Hypoventilation elevates P_{CO2} which is converted to H^+ and HCO_3^-, thus elevating acid levels in plasma. The respiratory compensation to alkalosis, however, is much less effective than the response to acidosis. The decreased minute volume of respiration induced by alkalosis also decreases the partial pressure of oxygen which is opposed by feedback mechanisms that counteract hypoxia.

4. RENAL REGULATION OF HCO_3^- LEVELS

The basic mechanisms of acid and base movement across the renal tubular cells are illustrated in Fig. 1. Under normal acid–base conditions in both the proximal and distal tubule, filtered H^+ is actively secreted into the tubular lumen and HCO_3^- is returned to the plasma. The mechanisms used to accomplish these tasks, however, are slightly different in these two sections of the nephron.

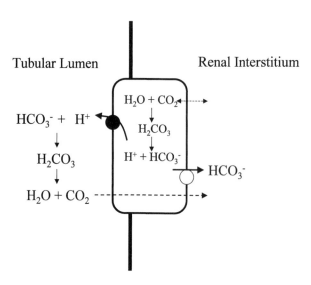

Fig. 1. The basic mechanisms of acid and base movement across the renal tubular cell. As H^+ is actively secreted into the tubular lumen, the H^+ concentration within the tubular cell is decreased. This leads to increased production in H^+ and HCO_3^- from water and CO_2, which can freely pass through cell membranes. The increasing cellular levels of HCO_3^- lead to movement of HCO_3^- across the basolateral membrane of the cell into the renal interstitium. In this basic manner, the kidneys excrete excess acid and reabsorb most of the filtered bicarbonate.

4.1. Bicarbonate Reabsorption

Approximately 90 to 95% of all filtered HCO_3^- is reabsorbed in the proximal tubule. Here, H^+ ions are secreted into the tubular lumen in exchange for Na^+ ions via the Na^+/H^+ exchanger. Because of the active secretion of H^+ ions, lumenal HCO_3^- is converted by carbonic anhydrase to water and CO_2, which is freely permeable to the renal epithelial cell. CO_2 is then free to move either across the tubular cell into plasma or it can remain in the renal epithelial cell where it can be converted to more H^+ to be secreted and HCO_3^-, which is passively transported into the renal interstitum by HCO_3^- transporters in the basolateral membrane. Because this process requires carbonic anhydrase, inhibitors of the enzyme such as cyanide, azide, sulfonamides, and other diuretics and natriuretics lead to decreased HCO_3^- reabsorption and increased HCO_3^- and Na^+ excretion in the urine. Despite the proximal tubule's high reabsorptive capacity, it is unable generate enough acidity to remove all the filtered HCO_3^- from the lumen.

The distal tubule has the capability to reclaim the remaining 5 to 10% of the filtered HCO_3^-. In order to do this, the distal tubule must be able to secrete H^+ against a steep concentration gradient (approx 1000:1 or plasma pH 7.4 vs urine pH 4.5). This is accomplished by a H^+-ATPase in the lumenal membrane of distal tubule epithelial cells.

Under normal conditions (plasma HCO_3^- concentration: 24 mmol/L; glomerular filtration rate: 130 mL/min), almost all of the filtered HCO_3^- is reabsorbed and, therefore, very little appears in the urine. However, when plasma HCO_3^- concentrations rise above 25 mmol/L, the plasma threshold for HCO_3^-, the molecule begins to appear in urine. The basic mechanism for excreting excess HCO_3^- is by "spillover" into the urine.

4.2. Regulation of Bicarbonate Reabsorption in the Proximal and Distal Tubule

There are several factors that alter H^+ secretion and thus HCO_3^- reabsorption.

One of the most important factors is the systemic acid–base balance. Low plasma pH or acidemia is a direct stimulus for H^+ secretion. Low pH stimulates the Na^+/H^+ exchanger in the proximal tubule and increases the activity of the H^+-ATPase in the distal tubule. Alkalosis produces the opposite effect to decrease H^+ secretion, and thus also lowers HCO_3^- reabsorption.

The plasma CO_2 content also directly effects H^+ secretion. Increased plasma P_{CO2} (respiratory acidosis) increases HCO_3^- reabsorption effectively by shifting the plasma HCO_3^- threshold. Conversely, low P_{CO2} (respiratory alkalosis) decreases the plasma HCO_3^- threshold leading to increased HCO_3^- excretion.

Potassium, and thus aldosterone, is another factor that regulates bicarbonate reabsorbion. It is thought that the intracellular, as opposed to plasma, K^+ concentration is the important regulator of HCO_3^-, however, the intracellular levels rise and fall in proportion to plasma K^+ levels. Potassium depletion leads to intracellular acidosis resulting in H^+ excretion and HCO_3^- reabsorption. Potassium excess leads to decreased HCO_3^- reabsorption. This is because K^+ and H^+ compete for Na^+ exchange. Therefore, increased K^+ excretion leads to decreased H^+ excretion and decreased HCO_3^- reabsorption.

The final factor affecting HCO_3^- reabsorption is the volume status of the body. Hypovolemia, particularly that caused by a drop in Cl^- concentration, leads to increased HCO_3^- reabsorption. This is because H^+ is lost to conserve Na^+, and thus volume, when no Cl^- is available for reabsorption with Na^+. However, in this circumstance, the Cl^- concentration seems to be more important than volume. If the volume status is returned to normal, the kidney will continue to conserve HCO_3^- until Cl^- levels return to normal. During hypervolemia the opposite is seen; HCO_3^- reabsorption is decreased.

4.3. Bicarbonate Production in the Collecting Duct

The kidneys not only regulate HCO_3^- levels by reabsorbing almost all that is filtered, but also are responsible for replenishing stores of the buffer molecule that are depleted during the buffering process. The bicarbonate system alone cannot excrete enough acid to maintain plasma pH at 7.4. The average daily acid intake from food and production through metabolism is approx 100 mmol/L/d and the HCO_3^- system only eliminates about 1 mmol H^+/L/d. Therefore, H+ must be excreted in a form other than the dissociated acid to maintain a homeostatic pH. This is accomplished by the formation of titratable acidity (H^+ bound to buffers in the urine such as HPO_4^{2-} and SO_4^{2-}; Fig. 2).

All buffer anions that are filtered at the glomerulus can augment the excretion of H^+ into the urine. The maximal titration of urine buffers to their acid forms depends on the presence of an acidic urine. For example, HPO_4^{2-} is the primary urinary buffer. In the plasma at pH 7.4 the ratio of $HPO_4^{2-}/H_2PO_4^-$ is 4:1. However, in the urine at pH 4.5 almost all of the phosphate is in the associated acid ($H_2PO_4^-$) form.

Some of the same factors that alter HCO_3^- reabsorption also alter the formation of titratable acidity by changing urine pH through differential H^+ secretion. Again, the acid–base status of the body affects the formation of titratable acidity. Acidosis, increased P_{CO2} and low plasma K^+ increase H^+ secretion and therefore titratable acidity formation. Likewise, alkalosis, low P_{CO2}, and high K^+ all decrease H^+ secretion and the formation of titratable

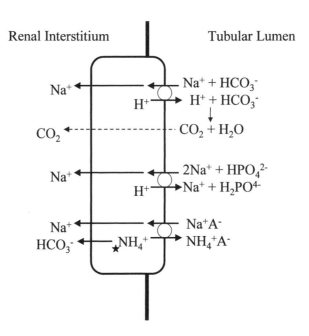

Fig. 2. The formation of titratable acidity. The bicarbonate system alone cannot eliminate the entire acid load ingested and produced each day. Therefore, buffer anions in the urine such as HPO_4^{2-} and SO_4^{2-} can bind free H^+ ions resulting in an associated acid. The formation of these associated acids, however, is limited by the urinary pH. Tubular cells are able to further increase acid secretion by generating and excreting NH_4^+ (the metabolism of glutamine within renal epithelial cells results in NH_4^+ and HCO_3^-).

acidity. Mineralocorticoids and 17-hydroxy glucocorticoids operate exclusively in the collecting duct to alter H^+, Na^+, and K^+ excretion. Increased mineralocorticoid levels increase the renal H^+ and K^+ secretion capacity while reabsorbing Na^+. Likewise, low mineralocorticoid levels inhibit H^+ and K^+ secretion while increasing Na^+ excretion. Thus, diseases such as Cushing's syndrome (excess production of glucocorticoids) or Addison's disease (deficient glucocorticoid production) are often accompanied by alterations in acid–base balance.

Collecting duct cells excrete H^+ into the tubular lumen at normal pH via activation of the H^+-ATPase and HCO_3^- is reabsorbed in exchange for Cl^-. During alkalosis, however, the collecting duct can essentially work in reverse, secreting HCO_3^- into the urine in exchange for Cl^- and actively pumping H^+ into the plasma to elevate urine pH. Even so, the proximal tubule is much more important in alkalosis, during which it simply fails to reabsorb all filtered HCO_3^-.

The collecting duct has developed an efficient mechanism to counteract acidosis. The concentration of buffer anions and the minimum attainable urine pH set limits on the amount of H^+ ions that can be excreted as titratable acidity. To overcome these obstacles, the collecting duct cells generate ammonium ions (NH_4^+) that are excreted in the urine. The formation and excretion of NH_4^+ helps replace the HCO_3^- that is consumed by urea production in the liver (Fig. 3). The glutamine molecules generated in the liver by urea formation can be converted by glutaminase to glutamate and NH_4^+. Glutamate can further be metabolized by glutamic dehydrogenase to α-ketogluterate and NH_4^+. HCO_3^- can then be produced from α-

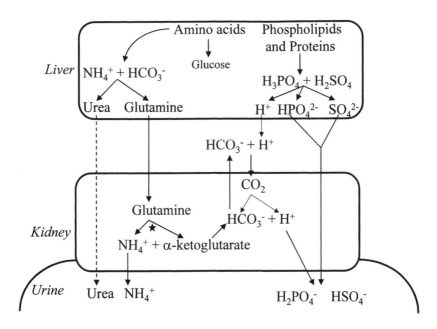

Fig. 3. Bicarbonate generation in the collecting duct. Metabolism of amino acids within the liver results in NH_4^+ and HCO_3^-. These can then be used to generate glutamine or urea, which is excreted by the kidney. Glutamine can be metabolized within the kidney resulting in the formation of NH_4^+ and α-ketoglutarate. The NH_4^+ is excreted while the α-ketoglutarate is further broken down to HCO_3^- and H^+. The HCO_3^- is returned to plasma and the H+ is excreted into the urine where it is buffered by titratable acids. Thus, the metabolism of glutamine by the kidney regenerates HCO_3^- that would otherwise be consumed during the production of urea. (Glutamine is metabolized by glutaminase and glutamic dehydrogenase.)

ketogluterate and transported into plasma to replenish stores. Thus, when the kidney metabolizes glutamine it restores the HCO_3^- that was lost in the formation and excretion of urea.

Under normal conditions, the kidney excretes approximately equal amounts of titratable acidity and NH_4^+. During chronic acidosis, gluconeogenic enzymes, such as glutaminase, are increased within the kidney. Thus, although the amounts of excreted titratable acidity and NH_4^+ are both increased, NH_4^+ excretion is elevated to a greater extent leading to increased plasma HCO_3^- concentrations. Conversely, chronic alkalosis decreases glutamine metabolism and, therefore, decreases NH_4^+ and HCO_3^- production. Hence, the kidney is able to control acid–base balance by autoregulating the production of new HCO_3^-.

5. DISRUPTIONS OF ACID–BASE HOMEOSTASIS

Maintenance of acid–base balance is more than just maintenance of pH. According to the Henderson-Hasselbalch equation, normal pH could be maintained at any number of P_{CO2} values as long as the ratio of HCO_3^- to CO_2 remained unchanged. Therefore, we must distinguish between elevated H^+ concentration (acidemia) and acidosis, which may not involve acidemia. And similarly, we must distinguish between alkemia and alkalosis. Acid–base balance is most easily detected by examining the HCO_3^-/CO_2 relationship, which shifts following changes in acid–base conditions.

5.1. Respiratory Acidosis

Changes in pulmonary ventilation can rapidly alter plasma pH. Respiratory acidosis is the result of elevated CO_2 levels caused by lung disease, drugs, or any illness that depresses the respiratory system. Elevated P_{CO2} decreases the ratio of HCO_3^-/CO_2 by shifting the equilibrium reaction and leads to acidosis. The changes in equilibrium are shown, with the primary incident indicated by a ⇑.

$$CO_2 + H_2O \rightarrow H_2CO_3 \rightarrow H^+ + HCO_3^-$$
$$\Uparrow \qquad\qquad \uparrow \qquad \uparrow \qquad \uparrow$$

The initial defense against elevated CO_2 levels falls to plasma and tissue buffers that can rapidly bind H^+, and thus prevent drastic alterations in pH. The renal compensatory response to respiratory acidosis, which takes 24 to 48 h to commence, is to increase H^+ secretion in the proximal and distal tubules. The increased H^+ secretion results in elevated HCO_3^- reabsorption and generation despite the already elevated HCO_3^- levels. The kidney, therefore, produces compensating metabolic alkalosis. The increased HCO_3^- levels buffer the elevated H^+ concentration. However, the correction is never complete and the acidosis persists until respiration returns to normal.

5.2. Respiratory Alkalosis

Respiratory alkalosis develops secondarily to hyperventilation. Hyperventilation can be induced by high altitudes (low atmospheric O_2 content); stimulation of the central nervous system by infections, drugs, or tumors; or a respirator set inappropriately high. Hyperventilation lowers P_{CO2} and increases the ratio of HCO_3^-/CO_2.

$$CO_2 + H_2O \leftarrow H_2CO_3 \leftarrow H^+ + HCO_3^-$$
$$\Downarrow \qquad\qquad \downarrow \qquad \downarrow \qquad \downarrow$$

The renal response to respiratory alkalosis is rapid. The tubules merely fail to reabsorb all the HCO_3^- filtered. Consequently, H^+ secretion is decreased and plasma pH begins to fall toward normal. Note that although HCO_3^- levels are already low in respiratory alkalosis, the renal compensation further decreases the HCO_3^- concentration.

5.3. Metabolic Acidosis

Metabolic acidosis is caused by an apparent increase in plasma H^+ concentration. There are four potential mechanisms by which the H^+ levels can be elevated: increased acid intake, increased acid production, decreased acid secretion by the kidneys, or decreased plasma HCO_3^- levels with no change in H^+ concentration. In the first three cases, the excess H^+ is buffered by HCO_3^- leading to a drop in HCO_3^- levels and a concomitant increase in H_2CO_3 levels and consequently CO_2 and H_2O. The elevated CO_2 levels are rapidly disposed through respiration.

$$H^+ + HCO_3^- \rightarrow H_2CO_3 \rightarrow CO_2 + H_2O$$
$$\Downarrow \qquad \uparrow \qquad\qquad \downarrow \qquad \downarrow$$

If the primary cause of metabolic acidosis is decreased plasma HCO_3^- levels, then the HCO_3^- losses are partially replaced by conversion of CO_2 resulting in elevated H^+ concentration.

$$H^+ + HCO_3^- \rightarrow H_2CO_3 \rightarrow CO_2 + H_2O$$
$$\uparrow \qquad \Downarrow \qquad\quad \downarrow \qquad\quad \downarrow$$

No matter the cause of metabolic acidosis, the hallmark is elevated H^+ levels with decreased HCO_3^- concentrations. This characteristic makes metabolic acidosis distinguishable from respiratory acidosis, which includes elevated levels of both H^+ and HCO_3^-. In metabolic acidosis, the fall in plasma HCO_3^- induced by the buffering of H^+ must be compensated for to maintain electroneutrality. Consequently, plasma levels of chloride or the anionic weak base form of an acid (e.g., HPO_4^{2-}) will rise. In fact, examination of the relative Cl^- and HCO_3^- levels (the anion gap) can help establish the etiology of the metabolic acidosis.

The anion gap is a means of determining the levels of anions in plasma other than Cl^- and HCO_3^-. This is done by subtracting the levels of the two major anions from the concentration of the major cation, Na^+.

$$\text{Anion Gap} = \left[Na^+ \right] - \left(\left[Cl^- \right] + \left[HCO_3^- \right] \right)$$

Under normal acid–base and renal circumstances the gap is less than or equal to 12 mmol/L. Anions that are involved in this gap include phosphate, sulfate, lactate, oxalate, pyruvate, and other products of metabolism. Patients with metabolic acidosis and no renal impairment will also have a normal anion gap of less than 12 mmol/L. Some common causes of metabolic acidosis that do not alter the anion gap include diarrhea, carbonic anhydrase inhibitors (diuretics), low aldosterone levels, or high parathyroid levels. Patients with metabolic acidosis and renal impairment have an anion gap of greater than 12 mmol/L. Some common causes of metabolic acidosis with increased anion gap include diabetic acid ketosis, starvation, methanol poisoning, lactic acidosis, and renal failure.

The initial response to metabolic acidosis in patients with and without changes in unmeasured anions is hyperventilation (respiratory alkalosis) to remove the excess CO_2 formed from the buffering of H^+ by HCO_3^-. The respiratory response decreases H^+ levels, although not back to normal. However, hyperventilation also further decreases HCO_3^- levels.

The correction of metabolic acidosis is dependent on a return of H^+ and HCO_3^- levels to normal. The kidneys increase plasma HCO_3^- concentration by conserving all filtered HCO_3^- and producing new HCO_3^- through the excretion of NH_4^+ and the metabolism of glutamine. The kidney also increases plasma pH by excreting H^+ in exchange for Na^+ and by increasing the excretion of titratable acidity. Under conditions where renal function is not impaired, these mechanisms are enough to correct the metabolic acidosis. However, when renal function is impaired (anion gap > 12 mmol/L), the chronic acidosis will persist until the original problem is corrected.

5.4. Metabolic Alkalosis

Metabolic alkalosis results from an apparent increase in plasma HCO_3^- levels. Metabolic alkalosis can be caused by excessive HCO_3^- intake, retention or production, or by excessive loss of acid (usually through vomiting).

$$H^+ + HCO_3^- \rightarrow H_2CO_3 \rightarrow CO_2 + H_2O$$
$$\downarrow \qquad \Uparrow \qquad\quad \uparrow \qquad\quad \uparrow$$

In the case of excess HCO_3^- concentrations as drawn above, the primary compensatory action is to increase the excretion of HCO_3^- by the kidneys.

$$H^+ + HCO_3^- \rightarrow H_2CO_3 \rightarrow CO_2 + H_2O$$

When metabolic alkalosis is caused by loss of acid, the initial response is hypoventilation to attempt to elevate plasma H^+ levels to normal. This mechanism, however, also increases HCO_3^- levels and is limited by the physiological response to hypoxia and is, therefore, not sufficient to correct the alkalosis. The renal excretion of the excess formed HCO_3^- is again necessary to correct the alkalosis. Because metabolic alkalosis is commonly accompanied by changes in ECF volume and Na^+ and Cl^- levels (as is the case for vomiting, diuretics and Cushing's syndrome), clinically, the illness is corrected by restoring normal ECF volume and Cl^- concentration rather than treating the acid–base disturbance itself. Again, note that metabolic alkalosis is characterized by an increase in HCO_3^- concentration with a fall in H^+ levels and can be distinguished from respiratory alkalosis, which includes decreased HCO_3^- and H^+ levels.

5.5. Summary of Acid–Base Disturbances

The characteristics of the individual acid–base disturbances as well as some causes of acid–base alterations are shown in Table 2. Although the disorders have been presented in isolation here, it is important to realize that mixed acid–base disturbances can occur (e.g., respiratory acidosis with metabolic acidosis). A normal body can efficiently handle changes in acid–base balance and restore homeostasis fairly rapidly. Chronic impairments in acid–base homeostasis, however, are usually accompanied by or caused by disease and often require treatment of the underlying disease before the acid–base disorder can be corrected. For further details on basic mechanisms of acid–base balance and acid–base disorders see refs. *1* and *2*.

6. ADVANCES IN OUR UNDERSTANDING OF ACID–BASE BALANCE THROUGH GENOMICS AND PROTEOMICS

Molecular tools have permitted the study of the minutest component of the acid–base homeostatic machinery. It is imperative, however, that we use the knowledge gained from the molecular insights and integrate them into the current understanding of the physiological mechanism. This section presents new insights gained from molecular techniques used to study three different mechanisms involved in the maintenance of acid–base balance.

6.1. Genomic and Non-Genomic Actions of Aldosterone in Acid–Base Balance

The overall function of aldosterone is to spare Na^+ and thus ECF volume by exchanging Na^+ ions in the distal tubule, as well as other target organs, for different cations, namely K^+ and H^+. The classical or genomic actions of the steroid (reviewed in ref. *3*) are elicited by binding of the steroid to its cytosolic receptor, which is then translocated to the nucleus. The activated mineralocorticoid receptor binds to DNA response elements along with nuclear complexes that then regulates transcription of major effectors of transepithelial Na^+ transport, including ion channels and mitochondrial enzymes necessary for the production of ATP. This process takes 30 to 60 min to initiate but can have effects on ion flux that last for days.

Table 2
Ion Changes in Simple Acid–Base Disorders

Disorder	Plasma H^+	Plasma HCO_3^-	P_{CO2}	Causes
Respiratory acidosis				
Uncompensated	↑↑	↑	↑	Hypoventilation
Compensated	↑	↑↑	↑	Lung disease
Respiratory alkalosis				
Uncompensated	↓↓	↓	↓	Hyperventilation
Compensated	↓	↓↓	↓	CNS tumors/infection
				High altitude
Metabolic acidosis				
Uncompensated	↑↑	↓↓	—	Diarrhea
Compensated	↑	↓↓↓	↓	CA inhibitors
				Renal failure
				Diabetic acid ketosis
Metabolic alkalosis				
Uncompensated	↓↓	↑↑	—	Vomiting
Compensated	↓	↑↑↑	↑	Diuretics

CNS, central nervous system; CA, carbonic anhydrase.

Because aldosterone increases active Na^+ reabsorption, an electrochemical gradient is established across the tubular epithelium that facilitates transfer of positive ions from the renal tubular cells into the urine. As explained previously, K^+ and H^+ compete for the second cation position in the exchanger that moves Na^+ back into plasma. Therefore, under normal acid–base conditions when the concentration of K^+ is 100 times that of H+ (*see* Table 1), potassium is the major ion that is exchanged for Na^+. However, as the plasma concentration of H^+ rises during acidemia, it binds in increasing amounts to the Na^+ exchanger. This mechanism is, therefore, able to lower the H^+ levels and is important in the maintenance of acid–base balance.

Glucocorticoids (cortisol and corticosterone) have equal affinity for the mineralocorticoid receptor. Because the concentration of glucocorticoids in plasma is much higher than that of aldosterone, it may seem surprising that aldosterone can compete for and bind its own receptor efficiently. The specificity of the receptor, however, lies in the enzyme 11β-hydroxysteroid dehydrogenase, which converts glucocorticoids to metabolites with little affinity for the mineralocorticoid receptor. Indeed, 11β-hydroxysteroid dehydrogenase is active in tubular cells that express the mineralocorticoid receptor. However, during situations of glucocorticoid excess, such as Cushing's syndrome, 11 β-hydroxysteroid dehydrogenase cannot metabolize glucocorticoids fast enough to prevent "spillover" activation of mineralocorticoid receptors.

The genomic actions of aldosterone are to increase the number and function of ion channels and exchangers in tubular epithelial cells, particularly in the distal tubule. These transcriptional effects are long lasting but require at least 30 min to produce results. However, evidence has been accumulating that aldosterone, along with many other steroids, can exert rapid, non-genomic effects. These non-genomic effects of aldosterone occur within a few minutes and do not require transcription or translation. In addition, the rapid actions of aldosterone have been shown to directly effect cellular H^+ and HCO_3^- transport (*3–5*).

Aldosterone activates the Na^+/H^+ exchanger in a variety of cell types including renal epithelial cells. This activation occurs within minutes and is not affected by inhibitors of transcription and translation, suggesting that the action is non-genomic (*4,5*). The rapid action of aldosterone on Na^+/H^+ exchange has been hypothesized to play a role in the regulation of epithelial Na^+ reabsorption and the intracellular pH-induced regulation of K^+ excretion. In addition, aldosterone rapidly increases H^+ conductance, permitting H^+ to flow back into the tubular cell (*4,5*). Although this action may seem to counteract the actions of the Na^+/H^+ exchanger, it in fact enhances the function of exchanger whose limiting factor is the intracellular H^+ concentration. Because the ratio of Na^+ in the tubular filtrate to H^+ in the epithelial cell is quite large, the increased H^+ conductance in renal epithelial cells induced by aldosterone actually facilitates the process of Na^+ reabsorption.

Despite the action of aldosterone to increase H^+ conductance, the stimulation of the Na^+/H^+ antiporter results in intracellular alkalinization both in cultured cells and in whole kidney preparations. One would expect, then, that there would be concomitant changes in HCO_3^- reabsorption. Indeed, aldosterone induces a rapid inhibition of HCO_3^- absorption in the distal tubule. Normally, the activation of the Na^+/H^+ exchanger leads to reabsorption of tubular HCO_3^-, but the effect of aldosterone to increase the intracellular pH may reverse this activity to maintain intracellular electrochemical homeostasis.

The rapid actions of aldosterone are not thought to be mediated by the classical mineralocorticoid type I receptor (reviewed in ref. *3*). Binding studies demonstrate the presence of membrane bound aldosterone receptors. These sites have different binding characteristics than the cytosolic mineralocorticoid receptor. Additionally, membrane impermeable versions of aldosterone still initiate rapid actions without inducing the genomic actions of the steroid. The finding that glucocorticoids, which bind and activate the cytosolic mineralocorticoid receptor as described above, cannot bind to this membrane receptor and cannot activate rapid changes in ion flux further supports the uniqueness of a membrane-bound aldosterone receptor. A final line of evidence that supports a unique membrane aldosterone receptor is that mineralocorticoid receptor knockout mice, which have no genomic aldosterone actions, do display rapid changes in cellular activity when aldosterone is applied.

Physiologically and clinically, the genomic and non-genomic actions of aldosterone have a dramatic impact on acid–base balance. Under normal physiologic conditions, the body efficiently regulates ECF pH and ion balance. However, during clinical conditions when aldosterone is low, metabolic acidosis can occur because of the loss of short-term and longer term H^+ excretion and Na^+ reabsorption. And, excess aldosterone and/or glucocorticoids can lead to metabolic alkalosis. A major clinical issue to address is the acid–base status during the use of aldosterone antagonists to treat hypertension. The most common aldosterone antagonist in use clinically is spironolactone. Although spironolactone is an effective antagonist for the classical mineralocorticoid receptor, it does not appear to inhibit the rapid, non-genomic actions of alsosterone. Chronic metabolic acidosis has been reported in a small number of patients receiving spironolactone therapy. It is conceivable, however, that more major imbalances in acid–base homeostasis are not seen in the presence of the antagonist because of the remaining non-genomic actions of aldosterone. New aldosterone antagonists are being developed currently that have been reported to alter both genomic and non-genomic actions of aldosterone. In patients treated with these new drugs, it will be imperative to examine their acid–base balance.

6.2. Nitric Oxide and Acid–Base Balance

The discovery that nitric oxide (NO) was produced by cells and had a dramatic impact on cellular function represented a major advance in cellular biology. Indeed, the significance of the discovery of NO as a signaling molecule was recognized with the awarding of the Nobel Prize in medicine. There is now a growing body of evidence that NO plays an essential role in renal physiology including that of acid–base balance.

Initial studies of the function of renal NO in acid–base balance have been confusing. Some studies claim that NO enhances the excretion of Na^+ and HCO_3^- and increases urine volume while decreasing H^+ secretion. It is thought that these actions of NO are mediated by decreased activity of the Na^+/H^+ exchanger in the proximal and distal tubules and decreased activity of the H^+-ATPase in the collecting duct *(6)*. Other studies, however, have shown the complete opposite effects of NO on ion channels and movement *(7,8)*. The reasons for these discrepancies may lie in the methods of assessment (cultured cells, tubule segments or in vivo studies) and the doses of NO donors or inhibitors of NO synthesis used. One study showed a bell-shaped dose response curve to NO, indicating that at low NO concentrations (μM) a stimulation of the Na^+/H^+ exchanger was seen, whereas at high NO concentrations (mM) an inhibition was reported *(7)*. It has now been determined that the physiological NO concentration within the proximal tubule is approx 110 nm, suggesting that endogenous renal NO may stimulate H^+ efflux and urine acidification. To clear the confusion surrounding the physiologic role of endogenous NO in renal acid–base handling, mice lacking the genes necessary to generate NO were examined *(7,8)*.

NO is produced by the metabolism of L-arginine, a process that is catalyzed by the enzyme nitric oxide synthase (NOS). There are now three known isoforms of this enzyme, all of which are found within the kidney (reviewed in ref. *2*). Neuronal NOS (nNOS) was the first isoform to be purified and cloned. This enzyme, named for the location from which it was cloned, is expressed constitutively in many tissue types and is regulated by Ca^{2+}/calmodulin. Within the kidney, nNOS is primarily expressed in the epithelial cells of the macula densa and in cells of the collecting duct. Endothelial NOS (eNOS) has a more restricted localization than nNOS, but is found in renal vascular endothelial cells. The third isoform, inducible NOS (iNOS) is not constitutively expressed like nNOS and eNOS, but the transcriptional expression of this enzyme must be "induced" by activating factors, of which cytokines are the best characterized. Most systems within the body contain iNOS and the enzyme is widely expressed throughout the kidney.

Alterations in acid–base balance were examined in knockout mice lacking each one of these NOS enzymes. Not surprisingly, because eNOS is not produced within tubular cells, the eNOS knockout mice did not display alterations in renal Na^+, H^+ or HCO_3^- handling (7). Importantly however, endothelial NO also influences blood pressure and, therefore, changes in renal blood flow and interstitial pressure could mask any possible actions of eNOS to regulate acid–base balance within the kidney. On the other hand, both nNOS and iNOS knockout mice displayed changes in H^+ and HCO_3^- balance *(7,8)*. The most dramatic changes were seen in nNOS knockouts that had lower blood pH, P_{CO_2}, and HCO_3^- concentrations compared to control animals. The nNOS knockout animals, therefore, displayed chronic metabolic acidosis. These changes in plasma were a consequence of decreased renal H^+

excretion in the urine presumably due to decreased activity of the Na^+/H^+ exchanger and, consequently, decreased HCO_3^- reabsorption. Similar changes in H^+ and HCO_3^- reabsorption were also seen in iNOS knockout mice, but these changes were much smaller in magnitude and did not result in acidosis. These findings suggest that NO upregulates proximal tubule transport of H^+ and HCO_3^- under physiological conditions and is an important regulator of acid–base balance.

6.3. Molecular Identity and Regulation of Bicarbonate Transporters

Rapid HCO_3^- transport across plasma membranes is crucial for the equilibrium of CO_2 and HCO_3^- and, therefore, acid–base balance. There are two families of bicarbonate transporters whose activities are coordinated with those of cation exchangers and other regulators of pH to maintain acid–base homeostasis. One family of bicarbonate transporters mediates the cotransport of Na^+ and HCO_3^- and is aptly termed the sodium-bicarbonate cotransporters or NBCs (reviewed in ref. 9). There are currently four NBC isoforms (NBC1, 2, 3, and 4) and two NBC-related proteins (anion exchanger 4 [AE4] and sodium-dependent chloride-bicarbonate exchanger [NCBE]). Owing to differential splicing of the same gene product, there are two variants of NBC1. The kidney variant of NBC1 is localized to the basolateral side of proximal tubule cells. NBC2 was initially cloned from retinal cells, but is also produced in cortical collecting duct cells where it co-localizes with H^+-ATPase. NBC3 is highly expressed in neural tissue as well as in the kidney, where it is localized to the medullary collecting duct. NBC4 is moderately expressed only in the distal tubule. AE4 was cloned based on similar DNA sequences with the anion exchanger family of bicarbonate transporters. Despite this, however, AE4 is highly homologous to NBC1 and so is placed in the NBC family of transporters. The NCBE was cloned based on sequence similarity to NBC1, however, it is now known that this transporter is more closely related to NBC2 and 3.

The NBC-mediated transport of Na^+ and HCO_3^- occurs against a concentration gradient for both ions. The driving force for Na^+ and HCO_3^- efflux is the net movement of negative ions from the negatively charge cell interior to the extracellular space. Stoichiometric studies predict that three HCO_3^- molecules are moved by NBC1 for every one Na^+ molecule extruded from the cell.

The regulation of NBC activity is just beginning to be understood. NBC activity is stimulated by activators of protein kinase C and inhibited by activators of protein kinase A and cyclic adenosine monophosphate (cAMP). Elevated cAMP can act on NBCs to reverse the direction of HCO_3^- transport from efflux to influx. In addition, cAMP effects HCO_3^- transporters indirectly through the inhibition of the Na^+/H^+ exchanger. Interestingly, full function of NBC requires the activation of carbonic anhydrase, which physically interacts with the transporter's cytoplasmic tail. Inhibitors of carbonic anhydrase activity such as acetazolamide are known to decrease HCO_3^- reabsorption in the proximal tubule. Originally, this was attributed to the inhibition of apical membrane-bound carbonic anhydrases, which catalyze the conversion of HCO_3^- to CO_2. It is now, however, believed that the inhibition of HCO_3^- transport by carbonic anhydrase inhibitors is the result of the decreased activity of cytosolic carobonic anhydrase, which catalyzes the formation of HCO_3^-. By inhibiting this cytosolic carbonic anhydrase, the availability of HCO_3^- for transport is reduced, thereby accounting for the decreased biocarbonate reabsorption in the presence of carbonic anhydrase inhibitors.

The regulation of NBCs during various acid–base disorders is beginning to be studied. Metabolic acidosis increases the activity of NBCs in the proximal tubule most likely through protein modification (phosphorylation, subcellular localization) and not a change in transporter numbers. Acidosis does, however, increase the number of NBCs in the distal tubule and collecting duct. The increased number and function of NBCs in acidosis are important for increased acid excretion and elevated HCO_3^- reabsorption to buffer the acidemia.

Proximal renal tubular acidosis is a disease caused by decreased reabsorption of bicarbonate in the proximal tubule resulting in renal bicarbonate wasting and metabolic acidosis. Single-base mutations within the genomic sequence of NBC1 that decrease transporter activity have been identified in patients with proximal renal tubular acidosis. These findings along with alterations in NBC activity in Cushing's syndrome and hypokalemia (low K^+) suggest that proper NBC function is necessary to maintain acid–base homeostasis.

The second family of bicarbonate transporters, termed anion exchangers (AE), promotes the exchange of HCO_3^- and Cl^- (reviewed in ref. *10*). This diverse family includes members that are Na^+-independent and Na^+-dependent, many of which are found in the kidney. A prototypical member of this family, AE1 is located on the basolateral side of the collecting duct, where is functions in cooperation with the H^+-ATPase to mediate acid excretion. AE1 moves monovalent anions in a sequential manner. A chloride or bicarbonate ion binds to one surface of the AE1, which induces a conformational change in the protein that allows movement of the anion across the cell membrane. Following release of the anion, another anion is free to bind AE1. Upon anion binding, AE1 then returns to the initial conformation and moves the second ion in the opposite direction of the first across the plasma membrane. Because the chloride concentration is higher on the outside of the cell and is usually much higher than the HCO_3^- gradient, the anion exchange is frequently Cl^- influx and HCO_3^- efflux. However, depending on other cellular activities, this process can also work in reverse or transport two of the same anions.

Like NBCs, AEs require carbonic anhydrase binding and function for activity. Inhibition of carbonic anhydrase or loss of carbonic anhydrase binding can reduce Cl^-/HCO_3^- exchange dramatically. In addition to carbonic anhydrase, AE exchangers are also regulated by pH via direct binding of H^+ ions to the transporter. Cells have low Cl^-/HCO_3^- exchange rates under normal conditions, but small increases in intracellular pH (decreased H^+ concentration) rapidly increase anion exchange activity. AE function may also be regulated by phosphorylation, as this has been noted to occur on several serine, threonine, and tyrosine residues. However, phosphorylation has not yet been demonstrated to regulate AE activity.

Genetic mutations in AE1 have been linked to distal renal tubular acidosis (RTA), a disease with similar consequences as proximal RTA. One set of point mutations in the AE1 gene or a truncation of the 3' terminal of the gene are linked to autosomal dominant forms of distal RTA. The mechanism by which AE1 mutations lead to dominant distal RTA is largely unknown as the mutant AE1 proteins are localized properly in the plasma membrane and display only a minor reduction in activity. A different set of point mutations in the AE1 gene is linked to a recessive form of distal RTA that is common in southeast Asia and New Guinea. The recessive disorder leads to both inhibition of AE1 protein transport to the plasma membrane as well as decreased activity of the exchanger itself. It is, therefore, apparent that Cl^-/HCO_3^- exchangers are necessary for maintenance of acid–base homeostasis.

7. SUMMARY

The body meticulously regulates the composition of ECF. Here, we have examined the coordinated mechanisms used to regulate acid–base balance. The primary buffering system in the body is the HCO_3^-/CO_2 system. CO_2 levels can be rapidly regulated through changes in respiration, while the kidneys can alter the excretion/reabsorption of both HCO_3^- and H^+. Certain drugs and diseases can lead to disruptions in acid–base homeostasis by altering the equilibrium of the HCO_3^-/CO_2 buffer system or the renal and respiratory mechanisms controlling it. In the modern age of genomics and proteomics it is imperative that we study the molecular components of the acid–base regulatory system and integrate this knowledge into our current understanding of acid–base physiology. Three individual areas of molecular research on components of the acid–base system were described. Aldosterone has been shown to have both genomic and rapid, non-genomic effects on H^+ transport. The signaling molecule NO has been shown to be important in acid–base physiology through the aberrant acid–base transport in transgenic animals lacking NO synthases. Molecular dissection of bicarbonate transporters gave new insights into the regulation of HCO_3^- levels and several disorders of acid–base balance. In conclusion, the molecular examination of mechanisms involved in acid–base balance will not only further our understanding of the overall physiology of acid–base regulation but also provide insights into the causes of acid–base disorders and potential methods to treat them.

REFERENCES

1. Schafer, J.A. (1998) The role of the kidney is acid-base balance.I In *Essential Medical Physiology* (Johnson, L.R., ed.), Lippincott-Raven, Philadelphia, PA, pp. 403–414.
2. Brenner, B.M. (ed.) (2000) *The Kidney*. W.B. Saunders Company, Philadelphia, PA.
3. Falkenstein, E., Tillmann, H.C., Christ, M., Feuring, M. and Wehling, M. (2000) Multiple actions of steroid hormones-a focus on rapid, non-genomic effects. *Pharamacol. Rev.* **52**:513–555.
4. Good, D.W., George, T., and Watts, B.A. III. (2002) Aldosterone inhibits HCO_3^- absorption via a nongenomic pathway in medullary thick ascending limb. *Am. J. Physiol.* **283**:F699–F706.
5. Gekle, M., Silbernagel, S., and Wunsch, S. (1998) Non-genomic action of the mineralocorticoid aldosterone on cytosolic sodium in cultured kidney cells. *J. Physiol.* **511**:255–263.
6. Garvin, J.L. and Hong N.J. (1999) Nitric oxide inhibits sodium/hydrogen exchange activity in the thick ascending limb. *Am. J. Physiol.* **277**:F377–F382.
7. Wang, T. (2002) Role of iNOS and eNOS in modulating proximal tubule transport and acid–base balance. *Am. J. Physiol.* **283**: F658–F662.
8. Wang, T., Inglis, F.M., and Kalb R.G. (2000) Defective fluid and HCO3- absorption in proximal tubule of neuronal nitric oxide synthase-knockout mice. *Am. J. Physiol.* **279**:F518–F524.
9. Soleimani, M. (2002) Na+:HCO_3^- cotransporters (NBC): Expression and regulation in the kidney. *J. Nephrol.* **15**:S32–S40.
10. Alper, S.L., Darman, R.B., Chernova, M.N. and Dahl, N.K. (2002) The AE gene family of Cl-/HCO3- exchangers. *J. Nephrol.* **15**:S41–S53.

4

Circulation and Fluid Volume Control

Bruce N. Van Vliet and Jean-Pierre Montani

1. INTRODUCTION

1.1. The Kidney, Circulation, and Integrative Physiology

The kidney and circulation provide an especially rich source of examples of integrative physiology. The regulation of blood pressure (BP), for example, involves many disparate tissues, ranging from the heart and vasculature to the brain, adrenal gland, and kidney. BP regulation also involves multiple layers of physiological organization—from the biophysics of renal transport to the regulation of flows and resistances to the overall architecture of the fluid volume control system. The value of a truly "integrative" approach is evident from the many properties and features of such systems that could not have been discovered even by the most detailed examination of their individual components.

As we begin this century and millennium, technological advancements are not only leading to new avenues of investigation and additional layers of knowledge, but also an explosive growth in physiological information and literature. The task of integrating the information—forming a comprehensive understanding of each system and the interactions between systems—will be a major challenge. Meeting this challenge will require new efforts and new approaches, but also an awareness of what has been done and what is possible.

1.2. Arthur C. Guyton's Systems Analysis of Circulatory Dynamics and Their Control

To instill a sense of what has and can be done, we can do no better than revisit some of our greatest successes. It is therefore both appropriate and timely to dedicate this chapter to the memory of Arthur C. Guyton (1919–2003) *(1–3)*, exceptional in each of his many roles as scientist, inventor, teacher, and mentor, and one of the most accomplished and truly integrative physiologists of the past century. Although widely respected for his *Textbook of Medical Physiology*, he also made a number of seminal contributions to cardiovascular physiology (Table 1) and effectively fathered the modern-day field of quantitative physiological systems analysis *(4–7)*.

Guyton's "quantitative systems analysis" approach not only led him to consider the most important aspects of various components of circulatory control, but especially how they fit together and interacted. Thus, to him, cardiac output represented the outcome of an interac-

From: *Integrative Physiology in the Proteomics and Post-Genomics Age*
Edited by: W. Walz © Humana Press Inc., Totowa, NJ

Table 1
Selected Accomplishments of Arthur C. Guyton

Influential books published

- *Textbook of Medical Physiology* (10 editions) and related texts.
- *Circulatory Physiology: Cardiac Output and Its Regulation* (1973).
- *Circulatory Physiology II: Dynamics and Control of the Body Fluids* (1975).
- *Circulatory Physiology III: Arterial Pressure and Hypertension* (1980).

Important concepts and ideas championed by Dr. Guyton

- Dominant role of the kidney, pressure natriuresis, and the renal body fluid feedback mechanism in long-term blood pressure control.
- Interstitial fluid dynamics, including concept of a negative interstitial fluid pressure.
- Role of venous return and mean circulatory filling pressure in regulating cardiac output.
- Safety factor for pulmonary edema.
- Whole body autoregulation.
- Graphical analysis of physiological regulation (especially cardiac output and blood pressure).
- Quantitative computer modeling of physiological systems.

tion of venous return and cardiac function, total peripheral resistance represented, in part, a consequence of the effects of the effects of BP on tissue blood flow (i.e., autoregulation), and the long-term BP level itself represented an interplay between renal perfusion pressure and renal excretory function. Once revolutionary ideas, these and other concepts have proven their value over time and now penetrate the very fabric of modern cardiovascular physiology.

The quantitative nature of Guyton's systems analysis was conspicuously manifest in a "large circulatory model" (Fig. 1) in which several hundred equations were used to quantify different components of the circulation and their control. Although model parameters were largely based on empirical values, the overall architecture of the model was based on insight and intuition, trial and error, and repeated comparison and testing against experimental data. In addition to its ability to simulate the behaviour of the cardiovascular system (e.g., refs. 8 and 9), each version of the model essentially represented an explicitly stated theory or hypothesis: "How would a circulatory system of this design behave?" As Dr. Guyton pointed out, the most helpful contribution of the model was when it failed to correctly predict an empirical outcome, since that clearly indicated a limitation in our understanding of the system.

Dr. Guyton's influence lives on, his theories permeating our text books, his approach embedded in the careers and research programs of the many scientists who trained in his department and laboratory. Nevertheless, in an era of rapid advances in highly focused areas of cellular physiology, molecular biology, and genetics, there has perhaps never been greater need for his desire and ability to integrate the pieces into a quantitative and comprehensive model of the whole. In the remainder of this chapter, we focus on two concepts of cardiovascular regulation that arose from Guyton's systems analysis and remain fundamentally important today. These are the dominant role of the renal body fluid feedback mechanism in setting the long-term BP level, and the concept of "whole body autoregulation," by which the total peripheral resistance is considered to be, in part, a response to the BP level (and not the other way around). If nothing else, this review may serve to illustrate the role of an integrative approach in understanding physiological systems, and the contribution of "quantitative systems analysis" to this end.

2. THE RENAL BODY-FLUID FEEDBACK MECHANISM

Of the various concepts championed by Guyton, perhaps the most influential was the dominance of the kidney in setting the long-term BP level. The basis for this concept was the renal body-fluid feedback mechanism (RBFFM) illustrated in Fig. 2. At the heart of the RBFFM lies the "pressure natriuresis relationship," which describes the link between BP, salt intake, and salt excretion.

The *acute* pressure natriuresis relationship is determined in isolated kidneys (e.g., refs. *10* and *11*), anesthetized animals (e.g., ref. *12*), or conscious animals (e.g., ref. *13*) by varying the renal perfusion pressure and observing the resultant change in the rate of renal salt excretion. As shown in Fig. 3, acute manipulations of renal perfusion pressure elicit a corresponding change in the renal excretion of salt and water, with increases in BP causing increased sodium excretion (i.e., natriuresis) and reductions in BP causing reduced sodium excretion or even complete cessation of urine flow at low BP levels. The mechanisms underlying this phenomena have been reviewed elsewhere *(14,15)*.

It is important to realize that the acute pressure–natriuresis relationship also predicts the long-term level of BP that would arise for a given level of salt intake if the acute relationship was fixed and did not change with time. However, the acute pressure–natriuresis relationship is not fixed. It changes in response to many factors including changes in dietary salt intake. Thus, another approach is used to assess the chronic relationship between BP, salt intake, and renal salt excretion.

The chronic pressure–natriuresis relationship is measured in a different manner by imposing a level of salt intake on a subject for several days until salt balance is established (i.e., until the rate of salt intake is matched by the rate of renal salt excretion) and then measuring the resultant BP level. This process is repeated for a variety of salt intakes, each level of salt intake contributing one point to the chronic pressure–natriuresis relationship, also known as the "chronic renal function curve" (Fig. 3). In contrast to the acute pressure–natriuresis mechanism, which represents a property of the renal tissues that can be demonstrated even in isolated perfused kidneys, the chronic pressure–natriuresis relationship represents the performance of the entire RBFFM at equilibrium: that is, after the many control mechanisms affecting renal function and BP have exerted their influence, after salt balance has been established, and after BP has stabilized. Because the chronic renal function curve describes the pressure–natriuresis relationship when salt intake and renal salt excretion are equal, the *y*-axis on a chronic renal function curve simultaneously represents both salt intake and salt excretion.

Several characteristics of the renal function curve are of fundamental importance. First, on each curve there is an equilibrium point corresponding to the one level of BP that can maintain salt excretion at the level of salt intake. In Fig. 3, this is represented by point A in the case of either the acute or chronic renal function curve at normal salt intake, point B in the case of the acute renal function curve under the condition of high salt intake, and point C in the case of the chronic renal function curve at high salt intake. BPs above the equilibrium

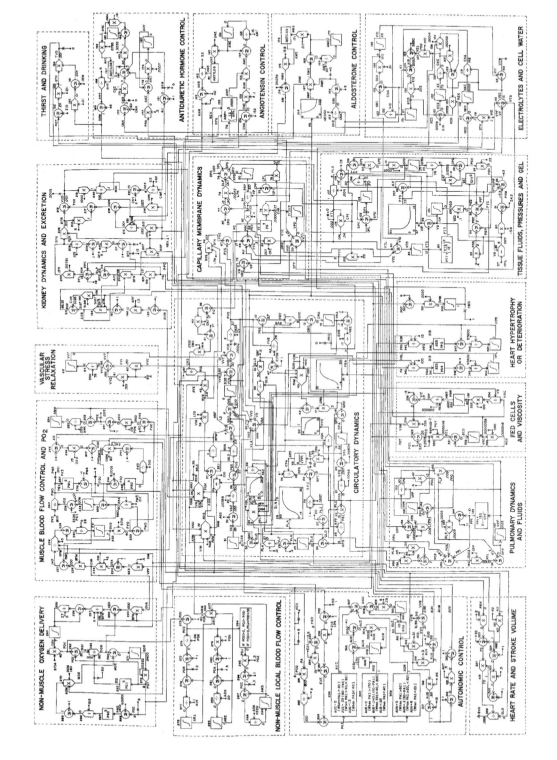

level will raise renal salt excretion above the rate of salt intake leading to a slow and progressive fall in of extracellular fluid (ECF) volume and cardiac output (Fig. 2). Given sufficient time, these changes would return the BP to the equilibrium level at which salt balance would again be achieved. Conversely, BP levels below the equilibrium value will lead to salt retention and a slow rise in cardiac output which, in time, would slowly raise the BP back to the equilibrium level. In this manner, one can appreciate the principle of how the RBFFM operates as a negative feedback controller of the equilibrium BP level (Fig. 2). It is important to note that this equilibrium BP level is set only by the shape and position of the chronic renal function curve and the level of salt intake *(4)*.

A second important characteristic of the renal function curve is its slope. Although the acute pressure–natriuresis relationship is relatively shallow, the slope of the chronic renal function curve is remarkably steep in most individuals (Fig. 3). The steepness of the chronic renal function curve reflects the property of salt insensitivity—a lack of change of the long-term BP level despite changes in salt intake. The steepness of the chronic renal function curve, relative to the acute pressure–natriuresis relationship, is thought to be largely the result of the actions of the renin-angiotensin system (RAS *[16–18]*). This system acts to facilitate salt excretion at high levels of salt intake, and facilitate salt retention at low levels of salt intake, thereby allowing salt balance to be re-established with little or no change in the equilibrium BP level. These effects of the RAS are time-dependent and require an intact circulation, and are therefore not apparent in the acute pressure–natriuresis curve of isolated kidneys. In chronic situations in which the RAS is unresponsive to a change in salt intake, BP becomes salt sensitive, corresponding to a renal function curve with a shallow slope, reminiscent of the acute pressure–natriuresis relationship.

A third and final characteristic of the chronic renal function curve to mention is its position along the BP axis. Repositioning the curve along the BP axis shifts the equilibrium BP level that the system will defend. A shift of the curve to the left lowers the equilibrium BP level, causing increased salt excretion until BP reaches the new equilibrium level, whereas shifting the curve to the right will produce a state of hypertension (Fig. 4). In the case of a curve with a shallow slope (salt sensitivity), hypertension can also be produced by increasing the level of salt intake (moving the equilibrium BP point along the curve to the right). In some cases, hypertension may be associated with a combination of a rightward shift and reduced slope of the renal function curve (Fig. 4).

Fig. 1. *(continued from facing page)* Guyton's large circulatory model *(5)*. The core of the large circulatory model consisted of equations calculating the pressures, volumes, and flows within different segments of the heart and circulation. Each additional section added to the complexity and realism of the control system. For example, an autonomic control section allowed cardiovascular dynamics to be affected by reflex adjustments of sympathetic tone (e.g., baroreflex), sections on renal dynamics and excretion and thirst and drinking allowed circulatory volumes to be governed by sodium and water balance. Other sections added specific fluid compartments, regional circulations, and additional control mechanisms (e.g., antidiuretic hormone, angiotensin, and aldosterone systems). Atrial natriuretic peptide was included in the 1992 version of the model. (From ref. 5 with permission.)

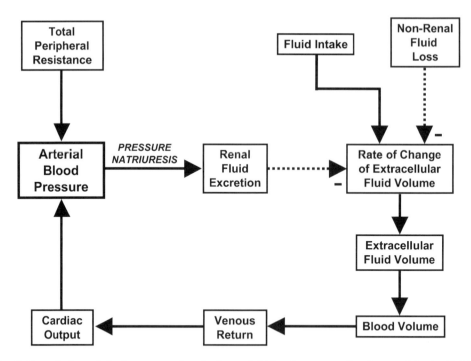

Fig. 2. Regulation of the long-term level of arterial blood pressure by the renal body-fluid feed-back mechanism. (Modified from ref. *4*.)

3. DOMINANCE OF THE RBFFM IN CONTROLLING THE LONG-TERM BP LEVEL: ITS IMPLICATIONS AND SIGNIFICANCE

3.1. Effectiveness of the RBFFM as a BP Controller

Of the many feedback mechanisms affecting BP, the RBFFM is believed to be the most powerful, capable of dominating all other mechanisms *(4,5)*. The dominance of this mechanism in controlling the long-term BP level arises from the cumulative, progressive nature of its actions. As long as the BP remains above the equilibrium level predicted by the chronic renal function curve, renal sodium excretion will remain high. Because the ensuing loss of ECF volume is progressive, the longer the arterial pressure remains above the renal set point, the greater the loss of ECF volume will be. Thus, the RBFFM is unusual among BP regulatory mechanisms in that its effectiveness grows with time. Given sufficient time, the RBFFM should theoretically be capable of completely correcting any perturbation in BP. This ability to completely correct a disturbance in arterial pressure is what Guyton often referred to as the "infinite gain" of the RBFFM, "gain" being a control systems term for the power or effectiveness of a regulatory mechanism.

The power of the RBFFM in BP control has many important implications for long-term BP control and hypertension. Several are discussed here.

3.1.1. Predictive Value of the Chronic Pressure–Natriuresis Relationship

A conclusion reached by Guyton's systems analysis is that at equilibrium (i.e., salt and water balance), the long-term BP level is solely determined by the value predicted by the intersection of the chronic pressure–natriuresis relationship and the level of salt intake (Fig. 3 *[4]*). In many respects, this principle is self-evident because this is the only BP level at

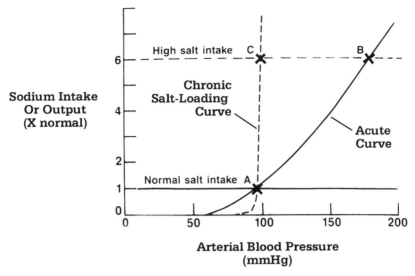

Fig. 3. The acute and chronic pressure–natriuresis relationships. The solid curve (acute curve) illustrates typical changes in salt excretion that quickly arise following acute changes in renal perfusion pressure in an isolated kidney. The dashed curve (chronic salt loading curve) indicates the long-term blood pressure level associated with salt balance (i.e., salt excretion = salt intake) at various levels of salt intake in intact animals. The solid and dashed horizontal lines indicate normal and six times the normal levels of salt intake, respectively. Point A indicates the equilibrium point (the blood pressure level at which salt balance can be achieved) for both the acute and chronic pressure–natriuresis relationships under conditions of normal salt intake. Points B and C indicate the equilibrium point for the acute and chronic pressure–natriuresis relationships, respectively, when salt intake is six times the normal. (Modified from ref. *4*.)

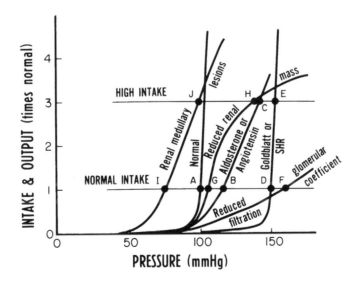

Fig. 4. Effect of hypertension on the chronic renal function curve. Four different kinds of renal function curve associated with hypertension (reduced renal mass, aldosterone or angiotensin infusion, Goldblatt hypertension [i.e,. renal stenosis] or spontaneously hypertensive rats, and reduced glomerular filtration coefficient) and one form associated with hypotension (owing to renal medullary lesions) are compared with the normal curve. Equilibrium points for normal and elevated (three times normal) levels of salt intake are indicated on each curve. (From Guyton ref. *4* with permission.)

which salt balance can be achieved. Any other BP level will lead to a change in salt and water balance that will slowly drive the BP back to the predicted level. Thus, a shift in the chronic pressure–natriuresis relationship will lead to a change in the long-term BP level. In turn, any change in the long-term BP level can be considered to reflect an underlying shift of the chronic renal function curve (assuming salt intake is approximately constant).

3.1.2. Dominant Role of the Chronic Pressure–Natriuresis Relationship in Hypertension

A closely related implication is the prediction that hypertension must reflect a change in the chronic renal function curve. Thus, Guyton was well known for championing the concept of the "dominant role played by the kidney in hypertension." Support for his view of the dominant role of the kidney and RBFFM in causing hypertension comes from several sources of which we mention just a few.

3.1.2.1. SUPPORT FOR THE DOMINANT ROLE OF THE KIDNEY AND PRESSURE–NATRIURESIS MECHANISM

Shortly after Guyton's prediction of a dominant role of the kidney in BP control was published, his prediction was tested by transplantation studies in various genetically hypertensive rat strains (SHR, SHRSP, Dahl, Milan, Prague) *(19–21)*. Recent studies have refined the approach further, circumventing potential pitfalls such as rejection phenomena, the need for immunosuppressants, and the use of indirect methods of BP assessment *(22)*. In such experiments, the BP level of animals receiving a transplanted kidney was shown to be strongly influenced by the BP level of the donor animal: implantation of a hypertensive kidney elevated the BP level of normotensive animals, whereas implantation of a normotensive kidney reduced the BP of hypertensive animals. This general finding has also been supported by transplantation studies in humans *(23,24)*. Recent studies have utilized the transplantation approach to define the contribution of the kidney to the hypertension associated with a considerably smaller number of hypertensive alleles located on a segment of rat chromosome-1 *(25,26)*. The role of the kidney in human genetic hypertension has been supported by observations of altered renal handling of sodium by hypertensive kidneys *(27)* and by our recent understanding of genetic disorders that directly affect renal transport mechanisms and lead to hypertension *(28–30)*.

The kidney is also directly implicated in many well-known forms of acquired hypertension, including that associated with renal stenosis (Goldblatt hypertension), ureteral obstruction, renal wrap hypertension, reduced renal mass, infusion of various substances into the renal artery (e.g., ref. *31*) or medulla (e.g., ref. *32*), and a various renal pathologies. The involvement of the kidney in acquired forms of BP change has also been demonstrated using the transplantation approach. For example, the persistent correction of hypertension induced by high doses of an angiotensin-converting enzyme inhibitor in genetically hypertensive rats has recently been shown to be transferable to untreated rats by renal transplantation *(33)*.

Above all else, support for a dominant role of the kidney and renal function curve in setting the BP level in hypertension has come from a series of studies in which a resetting of the chronic renal function curve to higher pressures has been found to be a consistent feature of all models of hypertension examined (Fig. 4) *(34)*. Again, this once controversial concept appears almost self-evident, since without a resetting of the chronic renal function curve, hypertension would result in an increased rate of renal fluid and water excretion until BP returned to normal.

In summary, the concept of the dominant role of the kidney (or RBFFM, or pressure natriuresis relationship) in hypertension is supported by a wide variety of data. In its most basic form, this principle simply means that all forms of hypertension must include a mechanism that allows the renal function curve to be reset so that normal sodium and fluid balance can be maintained at the elevated level of BP.

3.1.3. Non-Causes of Hypertension

A third important implication has to do with the "non-causes" of hypertension. Based on the principles discussed previously, non-causes would include any factor that failed to affect the chronic renal function curve (assuming salt intake remains approximately normal). A particularly important example is that this would exclude a contribution from changes in vascular resistance that fail to influence the renal circulation (or chronic renal function curve). Thus, although an increase in the resistance of the renal microcirculation, renal artery, or aorta above the kidney (Guyton's "resistance axis of hypertension" [35]) will shift the renal function curve rightward, resulting in a higher long-term BP level, an increase in the resistance to other parts of the arterial circulation do not directly affect the renal function curve, and are therefore presumed to be unable to influence the long-term BP level.

A comparison of the predicted effects of doubling vascular resistance in renal vs nonrenal circulations is provided in computer simulations of Figs. 5 and 6. Doubling the nonrenal vascular resistance transiently raises the BP level but rapidly leads to compensatory mechanisms, including suppression of the RAS and sympathetic system and marked increases in renal salt and water excretion (Fig. 5). Consequently, the long-term BP is unchanged despite the initial doubling of resistance in the nonrenal vasculature. This result fits well with the example frequently given by Guyton of war veterans who had undergone amputation of all four limbs (thereby elevating their peripheral resistance by approx 20%) yet lacked hypertension. In marked contrast, doubling of the renal vascular resistance results in a prompt marked elevation of BP mediated initially by a rise in angiotensin-II and subsequently by increases in fluid volumes and cardiac output (Fig. 6). As discussed in a later section, these responses are facilitated by the phenomena of whole body autoregulation.

A dramatic example also occurs in the presence of a large arteriovenous (AV) fistula (e.g., a communication between the abdominal aorta and vena cava). The creation of a large AV fistula is associated with a profound fall in peripheral resistance and BP. With time, however, renal salt and water retention leads to elevation of cardiac output and a restoration of BP to approximately normal despite the maintenance of a profoundly reduced peripheral resistance (36). This demonstrates the ability of the RBFFM to control BP without regard for the level of peripheral resistance. In rats with abdominal AV fistulas, treatment with deoxycorticosterone acetate and salt (DOCA salt) can produce considerable hypertension despite the fact that peripheral resistance remained well below the level seen in control rats (37). Thus, although hypertension is typically accompanied by an elevation of vascular resistance, it is not actually required for hypertension to occur. Indeed, as discussed in the subsequent section on whole body autoregulation, Guyton's system analysis led to a second provocative conclusion that vascular resistance was largely a consequence, and not a cause, of the long-term BP level.

3.1.4. The Important Role of Nonrenal Tissues in Hypertension

An important misconception is that by acknowledging a dominant role for the kidney in long-term BP control we are excluding a role for other tissues. On the contrary, the important role played by the kidney actually empowers other tissues with the ability to influence the long-term BP level (4,5,38). The only caveat to this is that to do so, nonrenal tissues must in some way reset renal function to agree with the new pressure level. That is, nonrenal

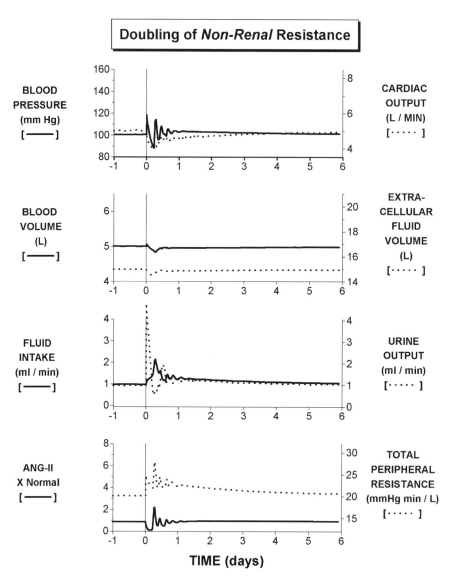

Fig. 5. Computer simulation of the response to doubling of the vascular resistance in all vascular beds other than the renal circulation. The increase in nonrenal resistance evokes a transient rise in blood pressure, which is rapidly compensated for, resulting in no change in the long-term blood pressure level. The simulation was run using the 1992 version of Guyton's large circulatory model.

tissues must, in some manner, affect the chronic renal function curve. The adrenal gland, for example, is a well-appreciated example in which a nonrenal tissue (in this case, via aldosterone) can influence the long-term BP level through their influence on the kidney. The case for an influence of the central nervous system (CNS) on the RBFFM is discussed next.

3.1.4.1. ROLE OF NEURAL MECHANISMS

Guyton spent much of his early research career investigating neural mechanisms of BP control *(38)*. In a 1972 review of his systems analysis of the cardiovascular system *(5)*, he

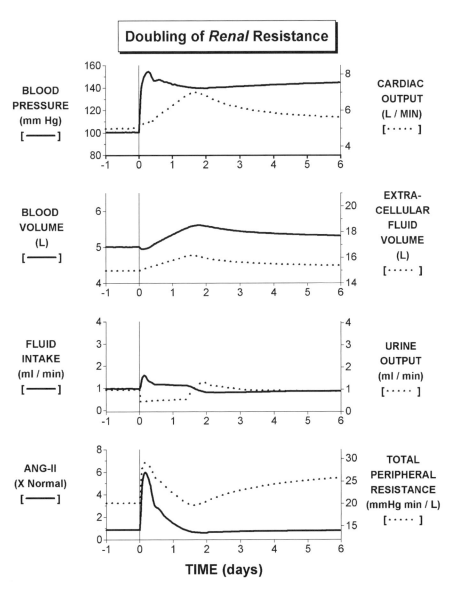

Fig. 6. Computer simulation of the response to doubling the renal vascular resistance. The increase in renal (afferent arteriolar) vascular resistance evokes a prompt and sustained rise in blood pressure, which is accompanied by volume loading and whole body autoregulation of blood flow. The simulation was run using the 1992 version of Guyton's large circulatory model.

pointed out the potential for the CNS to influence the renal body fluid feedback control of the long-term BP level through several pathways including the renal sympathetic nerves. The physiology of the renal nerves and their role in hypertension has subsequently been widely investigated *(39–41)*.

3.1.4.1.1. Effects of the Renal Nerves on Renal Function. As illustrated in Fig. 7, renal nerve activity has three direct actions on the kidney that may influence the long-term BP level. First, renal nerve activity facilitates and/or directly stimulates the release of renin from the

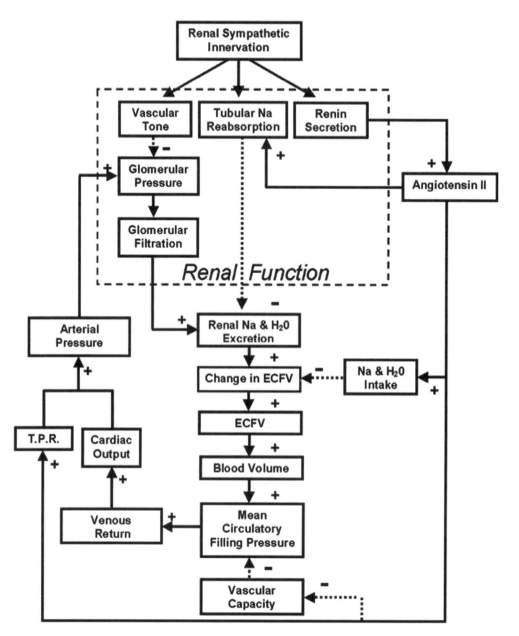

Fig. 7. Effects of the renal sympathetic nerves on renal function and blood pressure control. (Modified from ref. *96*).

granular cells of the juxtaglomerular apparatus. This can influence the long-term BP level through the formation of angiotensin peptides such as angiotensin-II, whose actions include vasoconstriction, reduced renal sodium excretion, and a stimulus for the secretion of aldosterone, itself a potent antinatriuretic hormone. Second, renal nerve activity can directly stimulate the tubular reabsorption of sodium, thereby promoting an increase in ECF volume and BP. Third, renal nerve activity can evoke renal vasoconstriction (e.g., ref. 41), thereby diminishing the rate of glomerular filtration and increasing the rate of tubular reabsorption of salt and water (via reduced peritubular capillary pressures).

3.1.4.1.2. Effect of the Renal Nerves on the Pressure–Natriuresis Mechanism. The effect of the renal nerves on renal salt excretion represents a shift of the acute pressure–natriuresis relationship along the BP axis. For example, increases in renal sympathetic nerve activity (RSNA) shift the pressure–natriuresis relationship rightward along the BP axis *(42,43)*. Overtime, a sustained shift would be expected to promote renal salt retention and fluid volume expansion until BP rises to meet the new set point level. In this manner, the sympathetic system could direct the kidneys to shift the BP level to a higher level. Once the new equilibrium BP level were achieved, renal salt excretion would return to normal, and salt balance would again be achieved despite the sustained shift of the pressure–natriuresis relationship.

It is interesting to consider that little or no change in the rate of renal sodium excretion would occur if the pressure–natriuresis relationship and BP were to both shift at the same time and by a similar amount. This may occur in fact occur during the course of the transient fluctuations in BP and RSNA that occur throughout the day. Although experimentally imposed changes in renal perfusion pressure leads to corresponding changes in renal salt and water excretion (i.e., pressure natriuresis) in anesthetized animals (e.g., ref. *12*) and conscious animals (e.g., ref. *13*), spontaneous fluctuations in BP in conscious animals are generally not associated with corresponding changes in renal salt and water excretion *(13,44)*. However, spontaneous fluctuations in BP do lead to pressure–natriuresis in denervated kidneys *(13,44)*, and pressure–natriuresis associated with arousal is amplified following autonomic blockade *(45)*. These results suggest that RSNA may act to reset the pressure–natriuresis relationship, thereby minimizing changes in renal salt excretion associated with transient changes in BP. Such an arrangement is presumed to be highly advantageous in wild populations in which dietary salt intake is limited, and its conservation directly impacts on survival and reproduction.

3.1.4.1.3. The Renal Nerves and the Long-Term BP Level. Two important parameters define the ability of the renal nerves to influence the long-term BP level: the magnitude of their effect on the pressure–natriuresis mechanism and how well it is sustained with time. The renal nerves normally provide a considerable restraint on renal sodium excretion as evidenced by the up to threefold increase in renal sodium excretion that can be produced by reflex unloading of RSNA in conscious animals *(46–49)*. If such an effect were well sustained, renal denervation would be expected to have a profound effect on the long-term BP level. However, in conscious dogs in which one of the two kidneys is chronically denervated, the 24-h sodium excretion of the denervated kidney is typically no more than about 12% greater than that of the innervated kidney (50–54). This is far less than the threefold difference seen following acute reductions in RSNA, and suggests that the renal nerves are most effective in mediating short-term resetting of the pressure–natriuresis relationship. Nevertheless, even a modest effect on sodium excretion, if sustained, is expected to have significant consequences for the long-term BP level.

The impact of the renal nerves on long-term BP control has been investigated by studying the consequences of renal denervation. Whereas comparison of chronic renal-denervated and sham-operated animals has generally not revealed differences in the BP level under control conditions, there have been several exceptions *(55,56)*. In a recent comparison of renal-denervated and sham-operated rats, for example, renal denervation was associated with significantly (approx 8–10 mmHg) lower 24-h telemetered BP level *(55)*. In hypertension, renal denervation has been shown to reduce the final level of BP achieved in several models including spontaneously hypertensive rats *(57,58)*, obesity-induced hypertension in dogs *(59)*, and phenol-induced hypertension *(60,61)*. Renal denervation has been reported to prevent salt-loading hypertension in rabbits *(62)*, and increases in BP caused by central infusion of angiotensin-II in rats maintained on a high salt diet *(63)*, although it does not appear to greatly impact on salt-induced hypertension in the Dahl rat model *(64,65)*. Renal denervation was reported to reduce the final BP level in the two-kidney one-clip and one-kidney one-clip rat Goldblatt model *(66,67)*, although negative results for the one-kidney one-clip model have also been reported *(68,69)*.

In addition to the role played by renal sympathetic fibers, renal afferent (sensory) fibers are thought to provide critical contributions in several models including the one-kidney one-clip Goldblatt hypertension *(70)* and phenol-induced hypertension models *(60,61)*. It is notable that in man, lumbar sympathectomy was employed for many years as a method with which to relieve hypertension. Although the surgery resulted in a pronounced initial decrease in BP, follow-up studies demonstrated the effects to be poorly sustained (possibly because of reinnervation) and the approach was abandoned *(71)*.

4. WHOLE BODY AUTOREGULATION

Whole body autoregulation is an important concept that explains the role of tissue blood flow in setting the total peripheral resistance (TPR). The concept of blood flow autoregulation is most widely understood in the context of the local or regional circulations whereby mechanisms intrinsic to the tissue or organ make adjustments to the local vascular resistance such that: (a) blood flow occurs at a level that is appropriate for the metabolic needs of the organ or tissue, and (b) this level of blood flow is held relatively constant despite moderate increases or decreases in the perfusion pressure. Alternately, autoregulation can also be identified as a positive correlation between an imposed change in BP or flow and the resultant change in vascular resistance, the increase in resistance serving to restrain the increase in blood flow that would otherwise occur. Autoregulatory adjustments of local vascular resistance can be clearly demonstrated in isolated tissues *(72)* and, to some extent, even in isolated vessels (e.g., ref. *73*). Although autoregulation is presumed to involve myogenic control of the vascular smooth muscle in combination with feedback provided by local metabolism, metabolites, or oxygen delivery *(74–76)*, the mechanisms contributing to blood flow autoregulation vary from tissue to tissue and remain incompletely understood.

Because autoregulation influences the control of vascular resistance in virtually all tissues within the systemic circulation (it is absent in the pulmonary circulation), it is not surprising that autoregulation should be evident in the pressure-resistance and flow-resistance relationships of the entire circulation. In dogs and rats in which reflexes and other regulatory systems have been disabled, progressive expansion of the blood volume leads to increases in cardiac output (CO) and BP, which are accompanied by an increase in the TPR, whereas withdrawal of blood leads to reductions in CO and BP that are accompanied by a fall in TPR *(76–81)*. Such changes in TPR reflect the underlying whole body autoregulation of blood flow: essentially the summation of the local autoregulatory efforts in different tissues and organs to sustain local blood flow at a normal and appropriate level despite changes in the BP.

Whole body autoregulation is not apparent during acute manipulations of BP or CO in the normal circulation in which circulatory reflexes (e.g., baroreflexes) and other regulatory mechanisms remain intact (e.g., ref. *80*). Indeed, in such circumstances, manipulations of BP or CO lead to changes in TPR in the opposite direction, reflecting the dominance of other mechanisms, including the passive effects of BP on resistance vessels (e.g., increased distending pressure lowers resistance) and control mechanisms such as the baroreflex that adjust vascular resistance in an effort to hold BP near the normal level (e.g., elevated BP evoking reflex vasodilation). In time, however, local autoregulatory mechanisms become increasingly evident in circulatory control. In conscious, chronically instrumented dogs in which blood withdrawal was used to lower BP by 25% for at least 8 h, whole body autoregulatory increases in TPR became apparent within 1 to 7 h following the reduction in BP *(80)*. In dogs made salt sensitive by reducing renal mass to 30% of normal, whole body autoregulatory adjustments of TPR slowly became apparent over several days of volume loading with saline (Fig. 8 *[82,83]*).

In considering the time course of whole body autoregulation, it is important to keep in mind that it is only one of a number of mechanisms that may simultaneously influence vascular resistance. At any point in time, the relationship between CO and TPR will reflect the summation of the effects of mechanisms promoting a positive (e.g., autoregulation) vs negative relationship (e.g., baroreflex, pressure-induced distension) in all the regions of the system circulation. Thus, the transition from negative to positive CO–TPR relationships reflects the shifting effectiveness of the underlying mechanisms, with autoregulation becoming more effective (and circulatory reflexes becoming less effective *[4]*), with time.

The precise contribution of whole body autoregulation in setting the long-term level of TPR is difficult to assess for several reasons. First, as discussed previously, relationships between flow and resistance represent the net effect of a number of control mechanisms, not only autoregulation. Second, the effectiveness of autoregulation and other phenomena affecting blood flow are often time-dependent. And third, because there is no selective way of blocking autoregulation, the phenomena is difficult to manipulate experimentally. Despite these limitations, several forms of indirect evidence suggest autoregulation is important in setting the long-term level of tissue blood flow and vascular resistance. In patients and rats in which there is a narrowing (coarctation) of a segment of the aorta, the BP is considerably higher above the coarctation than it is below. Nevertheless, blood flow above and below the coarctation are similar *(84–86)* suggesting that local autoregulatory mechanisms have the capability of adjusting tissue blood flow to appropriate levels even in the face of large and chronic changes in the BP level. Such long-term forms of blood flow autoregulation appear to include a component of altered structural vascular resistance in which the number and caliber of blood vessels is affected. Indeed, this may be the explanation for the increases in structural vascular resistance and rarefaction *(87)* of blood vessels in hypertension subjects, and for the normalizing of vascular resistance in regional vasculatures in which BP has been lowered by restriction of the arterial supply *(88)*.

5. WHOLE BODY AUTOREGULATION: IMPLICATIONS AND SIGNIFICANCE

The concept of whole body autoregulation of blood flow has several important implications for circulatory control, which are briefly discussed next.

5.1. Autoregulation May Contribute to the Elevated Level of Vascular Resistance in Salt-Loading Hypertension

In studies of conscious, chronically instrumented dogs in which salt sensitivity was induced by reducing renal mass to 30% of normal *(82,83)*, whole body autoregulation of TPR slowly became apparent over several days. As shown in Fig. 8, a sustained infusion of saline resulted in a prompt rise in BP that was initially mediated by an increase in CO *(82)*. Subse-

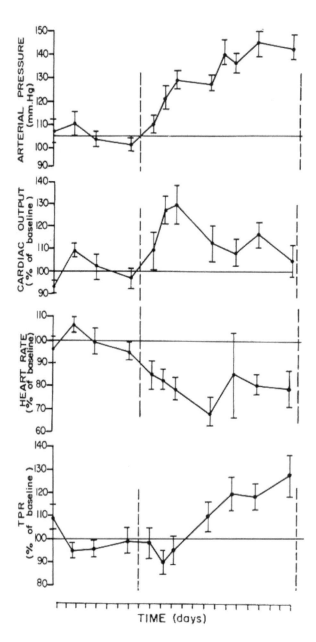

Fig. 8. Time course of hemodynamic effects induced by volume loading in salt-sensitive dogs. Volume loading consisted of an infusion of isotonic saline (191 mL/kg/d). Salt sensitivity was induced by reducing renal mass to approximately one-third of normal. TPR, total peripheral resistance. (From ref. *82* with permission.)

quently, peripheral resistance slowly rose (whole body autoregulation) and CO progressively fell (the predicted effect of increased TPR to decrease venous return and CO *[6]*), such that the increase in BP was eventually sustained by the increase in TPR. Similarly, in the Dahl strain of genetically salt-sensitive rats, salt loading led to hypertension that was initially caused by an increase in CO but eventually sustained by slow increases in TPR *(89–92)*. Slow increases in TPR were also observed to occur during salt loading in dogs made salt sensitive by a chronic infusion of angiotensin-II *(93,94)*. In regular (i.e., salt-insensitive) dogs and in the Dahl strain of salt-resistant rats, although salt loading does not lead to an increase in BP, it does produce an initial rise in CO counter-balanced by a fall in TPR *(89,90,94)*. In time, CO and TPR returned to normal levels *(89,91,92)* as one might expect to occur in the presence of whole body autoregulation but a normal (salt-insensitive) renal function curve.

Although a general discussion of the role of whole body autoregulation in essentially hypertension is well beyond the scope of this chapter, it is worth noting that a very slow transition from elevated CO to elevated peripheral resistance has been described in a number of longitudinal hemodynamic studies of patients with essential hypertension (see, e.g., Table 1 in ref. *95*). Thus, one view of the increased TPR that accompanies hypertension is that, as is the case with volume loading, it is an essential response to the elevated BP in order to prevent the over perfusion of the tissues that would otherwise occur. The precise mechanisms by which whole body autoregulation may occur remain to be worked out and strongly merit further study.

5.2. Autoregulation Does Not Cause Long-Term Changes in BP or Hypertension

Because whole body autoregulation of blood flow is an important determinant of TPR, whole body autoregulation may be presumed to have an important effect on the long-term BP level. However, this is generally thought not to be the case. As discussed previously, the RBFFM is the dominant controller of the long-term BP level. As shown in Fig. 2, TPR (and therefore autoregulation) lies outside of the main feedback loop of the RBFFM. Consequently, the RBFFM is expected to act to regulate the long-term BP level irrespective of the level of TPR. In other words, the RBFFM sets the long-term level of BP, the product of CO and TPR; whole body autoregulation is simply responsible for adjusting the balance of CO and TPR in a manner that provides an appropriate level of tissue perfusion.

5.3. Autoregulation Prevents the Large Changes in ECF Volume, Blood Volume, and Cardiac Output Otherwise Required to Achieve a Change in BP

Although whole body autoregulation of TPR does not directly influence the long-term BP level, it strongly influences which combination of TPR and CO will be used to achieve a given BP level, and the extent of fluid accumulation required to achieve the increase in CO. To illustrate this role of whole body autoregulation, we have used Guyton's large circulatory model to simulate the response to volume loading caused by a sixfold increase in salt intake in a salt-sensitive individual (caused by reduced renal mass) in the presence and absence of blood-flow autoregulation (Figs. 9 and 10). The simulation demonstrates a valuable benefit of mathematical modeling: the ability to conduct experiments on the model that we (so far) have no way of conducting in real life.

In the presence of normal autoregulation (Fig. 9), a high level of salt intake leads to an increase in fluid volumes (blood volume and ECF volume) associated with a rise in CO and BP and a fall in TPR (pressure-induced distension of the vasculature). In time, however, TPR progressively rises (whole body autoregulation), leading to reductions in the volume loading and CO required to sustain the elevated BP level. Thus, in the long-term, the increase in salt intake leads to salt-sensitive hypertension with rather minor changes in fluid volumes or cardiac output.

In contrast, Fig. 10 shows the results of the same simulation repeated with blood-flow autoregulation removed. In this situation, increases in fluid volumes, CO, and BP occur

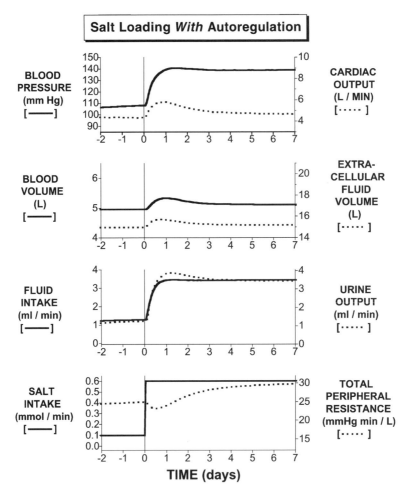

Fig. 9. Computer simulation of the hemodynamic response to volume loading in salt-sensitive subjects with blood-flow autoregulation intact. The simulation was run using the 1992 version of Guyton's large circulatory model. Salt sensitivity was created by reducing renal mass to 30% of normal. Volume loading was commenced at time 0 by increasing salt intake to six times the normal levels. The progressive increase in total peripheral resistance (whole body autoregulation) slowly reduces the volume loading and cardiac output required to sustain blood pressure at the new equilibrium level.

without any increase in TPR (which in fact falls, largely because of unopposed pressure-induced distension of the vasculature). In the absence of an increase in TPR, marked increases in fluid volumes and CO are required to elevate the BP toward the equilibrium level. Indeed, equilibrium is not achieved within the time course of this simulation: by the end of 1 wk on salt, fluid volumes, CO, and BP continue to rise, and BP has not yet reached the equilibrium level that it achieved in the presence of autoregulation (Fig. 9).

This simulation illustrates that whole body autoregulation does more than ensure that regional blood flow occurs at a rate that meets the metabolic needs of the tissues. When fluid volumes are changed, it also serves to minimize the changes in fluid volume and cardiac output that are required to affect the long-term BP level. Because the ability of changes in

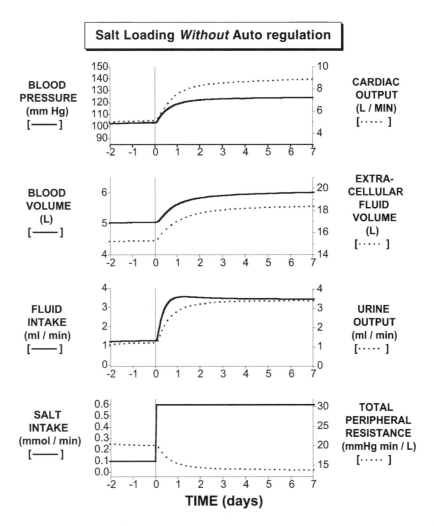

Fig. 10. Repeat of the computer simulation of the response to volume loading shown in Fig. 9 after disabling mechanisms of blood-flow autoregulation. In the absence of autoregulation, the rise in blood pressure induces a sustained fall in total peripheral resistance. Blood pressure does not reach the new equilibrium level associated with the high salt intake during the course of the simulation despite marked and progressive increases in fluid volumes and cardiac output.

fluid volumes to affect BP lies at the heart of the RBFFM (Fig. 2), one can readily appreciate the importance of whole body autoregulation in allowing the RBFFM to regulate the long-term BP level in a highly effective manner without the need for the large changes in fluid volumes that would otherwise be necessary.

6. CONCLUSION

In this chapter we have touched on just two of the concepts that were developed and championed by Arthur C. Guyton, a founder of modern quantitative integrative physiology. Looking at the diagram of his large circulatory model (Fig. 1), it is hard to deny that his approach was integrative. However, perhaps the greatest evidence of his integrative approach was its achievements: concepts and principles (e.g., Table 1) critical to the understanding of the system, yet largely unexpected from previous studies of the individual components.

Perhaps what we can take most from Guyton's work is the understanding, even respect, for the multitude of control mechanisms that are simultaneously at work in physiological systems, with each mechanism providing a unique contribution to overall control. With this in mind, it is easy to understand Guyton's emphasis on the need to consider the behavior of entire systems (as systems do take on characteristics that are not apparent upon examination of their components) and the use of a quantitative approach at all levels of physiological organization (because not all control mechanisms are of equal importance, and because the importance of individual mechanisms changes with time and circumstance). We hope that such lessons will be helpful in advancing our understanding of physiological systems in the future.

REFERENCES

1. Brinson C. and Quinn J (1989) Arthur C. Guyton. His Life, His Family, His Achievements. *Oakdale Press,* Jackson, MI.
2. Hall, J.E., Cowley, A.W., Jr., Bishop, V.S., Granger, D.N., Navar, L.G. and Taylor, A.E. (2003) In Memoriam. Arthur C. Guyton (1919–2003). *Physiologist* **46,** 126–128.
3. Quinn, J. (1989) Dr. A.C. Guyton: Builder. *J. Miss. State Med. Assoc.* **30(8)**:255–258.
4. Guyton AC (1980) Circulatory Physiology III. Arterial Pressure and Hypertension. WB Saunders, Toronto.
5. Guyton, A.C., Coleman, T.G. and Granger, H.J. (1972) Circulation: Overall regulation. *Ann. Rev. Physiol.* 34:13–46.
6. Guyton, A.C., Jones, C.E. and Coleman, T.G. (1973) Circulatory Physiology: Cardiac Output and Its Regulation, 2nd edition. WB Saunders, Philadelphia.
7. Guyton, A.C., Taylor, A.E. and Granger, H.J. (1975) Circulatory Physiology II: Dynamics and Control of the Body Fluids. WB Saunders, Philadelphia.
8. Simanonok, K.E., Srinivasan, R.S., Myrick, E.E., Blomkalns, A.L. and Charles, J.B. (1994) A comprehensive Guyton model analysis of physiologic responses to preadapting the blood volume as a countermeasure to fluid shifts. *J. Clin. Pharmacol.* 34(5):440–453.
9. Srinivasan, R.S., Simanonok, K.E., Fortney, S.M. and Charles, J.B. (1993) Simulation of the fluid retention effects of a vasopressin analog using the Guyton model of circulation. *Physiologist* **36**(1 Suppl):S114–S115.
10. Selkurt, E.E. (1951) Effect of pulse pressure and mean arterial pressure on modification on renal haemodynamics and electrolyte water excretion. *Circulation* **4**:541–551.
11. Starling, E.H. and Verney, E.B. (1924–1925) The excretion of urine as studied in the isolated kidney. *Proc. R. Soc. Lond.* **97**:321–363.
12. Evans, R.G., Szenasi, G. and Anderson, W.P. (1995) Effects of N-nitro-l-arginine on pressure natriuresis in anesthetized rabbits. *Clin. Exp. Pharmacol. Physiol.* **22**:94–101.
13. Nafz, B., Ehmke, H., Wagner, C.D., Kirchheim, H.R. and Persson, P.B. (1998) Blood pressure variability and urine flow in the conscious dog. *Am. J. Physiol. Renal Physiol.* **274**:F680–F686.
14. Cowley, A.W., Jr. (1992) Long-term control of arterial blood pressure. *Physiol. Rev.* **72**:231–300.
15. Granger, J.P., Alexander, B.T. and Llinas, M (2002) Mechanisms of pressure natriuresis. *Curr. Hypertens. Rep.* **4**:152–159
16. Hall, J.E. (1991) The renin-angiotensin system: renal actions and blood pressure regulation. *Compr. Ther.* **17**:8–17.
17. Hall, J.E., Guyton, A.C. and Mizelle, H.L. (1990) Role of the renin-angiotensin system in control of sodium excretion and arterial pressure. *Acta Physiol. Scand. Suppl.* 1990;**591**:48–62.

18. Montani, J.P. and VanVliet, B.N. (2004) General Physiology and Pathophysiology of the Renin-Angiotensin System. In Handbook of Experimental Pharmacology, Volume **163/1**: Angiotensin (Thomas Unger and Bernward A. Schölkens, Eds.), Springer Verlag, Berlin, pp. 3–29.
19. Grisk, O. and Rettig, R. (2001) Renal transplantation studies in genetic hypertension. *News Physiol. Sci.* **16**:262–265.
20. Rettig, R. (1993) Does the kidney play a role in the aetiology of primary hypertension? Evidence from renal transplantation studies in rats and humans. *J. Hum. Hypertens.* **7**(2):177–180.
21. Rettig, R., Bandelow, N., Patschan, O., Kuttler, B., Frey, B. and Uber, A. (1996) The importance of the kidney in primary hypertension: insights from cross-transplantation. *J. Hum. Hypertens.* **10**(10):641–644.
22. Grisk, O., Klöting, I., Exner, J., et al. (2002) Long-term arterial pressure in spontaneously hypertensive rats is set by the kidney. *J. Hypertens.* **20**:131–138.
23. Botero-Velez, M., Curtis, J.J. and Warnock, D.G. (1994) Liddle's syndrome revisited—a disorder of sodium reabsorption in the distal tubule. *N. Eng. J. Med.* **330**:178–181.
24. Guidi, E., Menghetti, D., Milani, S., Montagnino, G., Palazzi, P. and Bianchi, G. (1996) Hypertension may be transplanted with the kidney in humans: a long-term historical prospective follow-up of recipients grafted with kidneys coming from donors with or without hypertension in their families. *J. Am. Soc. Nephrol.* **7**(8):1131–1138.
25. Churchill, P.C., Churchill, M.C., Bidani, A.K. and Kurtz, T.W. (2001) Kidney-specific chromosome transfer in genetic hypertension: The Dahl hypothesis revisited. *Kidney International* **60**:705–714.
26. Clemitson, J-R., Pratt, J.R., Frantz, S., Sacks, S. and Samani, N.J. (2002) Kidney specificity of rat chromosome 1 blood pressure quantitative trait locus region. *Hypertens.* **40**:292–297.
27. Strazzullo, P., Galletti, F. and Barba, G. (2003) Altered renal handling of sodium in human hypertension. Short review of the evidence. *Hypertens.* **41**:1000–1005.
28. Luft, F.C. (2000) Molecular genetics of human hypertension. *Curr. Opin. Nephrol. Hypertens.* **9**:259–266.
29. Meneton, P., Oh, Y.S. and Warnock, D.G. (2001) Genetic renal tubular disorders of renal ion channels and transporters. *Semin. Nephrol.* **21**:81–93.
30. Warnock, D.G. (2001) Genetic forms of human hypertension. *Curr. Opin. Nephrol. Hypertens.* **10**:493–499.
31. Reinhart, G.A., Lohmeier, T.E. and Hord, C.E., Jr. (1995) Hypertension induced by chronic renal adrenergic stimulation. Is angiotensin dependent. *Hypertens.* **25**:940–949.
32. Mattson, D.L., Lu, S., Nakanishi, K., Papanek, P.E. and Cowley, A.C. (1994) Effect of chronic renal medullary nitric oxide inhibition on blood pressure. *Am. J. Physiol. Heart Circ. Physiol.* **266**:H1918–H1926.
33. Smallegange, C., Kline,R.L. and Adams, M.A. (2003) Transplantation of enalapril-treated kidneys confers persistent lowering of arterial pressure in SHR. *Hypertens.* **42**(5):932–936.
34. Hall, J.E., Mizelle, H.L., Hildebrandt, D.A. and Brands, M.W. (1990) Abnormal pressure natriuresis. A cause or a consequence of hypertension? *Hypertens.* **15**:547–559.
35. Guyton, A.C. (1990) Long-term arterial pressure control: an analysis from animal experiments and computer and graphic models. *Am. J. Physiol.* **259** (*Regulatory Integrative Comp Physiol* 28):R865–R877.
36. Huang, M., Hester, R.L. and Guyton, A.C. (1992) Hemodynamic changes in rats after opening an arteriovenous fistula. *Am. J. Physiol.* **262**:H846–H851.
37. Huang, M., Hester, R.L., Guyton, A.C. and Norman, R.A., Jr. (1992) Hemodynamic studies in DOCA-salt hypertensive rats after opening of an arteriovenous fistula. *Am. J. Physiol. Heart Circ. Physiol.* **262**:H1802–H1808.
38. Guyton, A.C. (1988) Hypertenion. A neural disease? *Arch. Neurol.* **45**:178.
39. DiBona, G.F. and Kopp, U.C. (1997) Neural control of renal function. *Physiol. Rev.* **77**:75–197.

40. DiBona, G.F. (2002) Sympathetic nervous system and the kidney in hypertension. *Curr. Opin. Nephrol. Hypertens.* **11**:197–200.
41. Van Vliet, B.N., Smith, M.J. and Guyton, A.C. (1991) The time course of renal responses to greater splanchnic nerve stimulation. *Am. J. Physiol.* **260**:R894–R905.
42. Ehmke, H., Persson, P.B., Seyfarth, M. and Kirchheim, H.R. (1990) Neurogenic control of pressure natriuresis in conscious dogs. *Am. J. Physiol.* **259**:F466–F473.
43. Golin, R., Genovesi, S. and Castoldi, G., et al. (1999) Role of the renal nerves and angiotensin II in the renal function curve. *Archiv Ital. Biol.* **137**:289–297.
44. Steele, J.E., Koch, L.G. and Brand, P.H. (2000) State-dependent expression of pressure diuresis in conscious rats. *PSEBM* **224**:109–115.
45. Brand, P.H., Coyne, K.B., Kostrzewski, K.A., Shier, D., Metting, P.J. and Britton, S.L. (1991) Pressure diuresis and autonomic function in conscious dogs. *Am. J. Physiol.* **261**:R802–R810.
46. Peterson, T.V., Benjamin, B.A. and Hurst, N.L. (1988) Renal nerves and renal responses to volume expansion in conscious monkeys. *Am. J. Physiol.* **255**:R388–R394.
47. Peterson, T.V., Benjamin, B.A., Hurst, N.L. and Euler, C.G. (1991) Renal nerves and postprandial renal excretion in the conscious monkey. *Am. J. Physiol.* **261**:R1197–R1203.
48. Sadowski, J., Kurkus, J. and Gellert, R. (1979) Denervated and intact kidney responses to saline load in awake and anesthetized dogs. *Am. J. Physiol.* **237**:F262–F267.
49. Smith, F.G., Sato, T., McWeeny, O.L., Torres, L. and Robillard, J.E. (1989) Role of renal nerves in response to volume expansion in conscious newborn lambs. *Am. J. Physiol.* **257**:R1519–R1525.
50. Lohmeier, T.E., Hildebrandt, D.A. and Hood, W.A. (1999) Renal nerves promote sodium excretion during long-term increases in salt intake. *Hypertens.* **33**:487–492.
51. Lohmeier, T.E., Lohmeier, J.R., Reckelhoff, J.F. and Hildebrandt, D.A. (2001) Sustained influence of the renal nerves to attenuate sodium retention in angiotensin hypertension. *Am. J. Physiol. Regulatory Integrative Comp. Physiol.* **281**:R434–R443.
52. Mizelle, H.L., Hall, J.E. and Woods, L.L. (1988) Interactions between angiotensin II and renal nerves during chronic sodium deprivation. *Am. J. Physiol.* **255**, F823–F827.
53. Mizelle, H.L., Hall, J.E., Woods, L.L., Montani, J–.P., Dzielak, D.J. and Pan, Y.–J. (1987) Role of renal nerves in compensatory adaptation to chronic reductions in sodium intake. *Am. J. Physiol.* **252**, F291–F298.
54. Mizelle, H.L., Hall, J.E. and Montani, J–.P. (1989) Role of renal nerves in control of sodium excretion in chronic congestive heart failure. *Am. J. Physiol.* **256**, F1084–F1093.
55. Jacob. F., Ariza, P. and Osborn JW (2003) Renal denervation chronically lowers arterial pressure independent of dietary sodium intake in normal rats. *Am. J. Physiol. Heart. Circ. Physiol.* **284**:H2302–H2310.
56. Janssen, B.J., van Essen, H., Vervoort-Peters, L.H., et al. (1989) Effects of complete renal denervation and selective afferent renal denervation on the hypertension induced by intrarenal norepinephrine infusion in conscious rats. *J. Hypertens.* **7**:447–455.
57. Norman, R.A. Jr. and Dzielak, D.J. (1982) Role of renal nerves in onset and maintenance of spontaneous hypertension. *Am. J. Physiol.* **243**:H284–H288.
58. Säynävälammi, P., Vaalasti, A., Pyykönen, M.–L., Ylitalo, P. and Vapaatalo, H. (1982) The effect of renal sympathectomy on blood pressure and plasma renin activity in spontaneously hypertensive and normotensive rats. *Acta Physiol. Scand.* **115**:289–293.
59. Kassab, S., Kato, T., Wilkins, F.C., Chen, R., Hall, J.E. and Granger, J.P. (1995) Renal denervation attenuates the sodium retention and hypertension associated with obesity. *Hypertens.* **25**:893–897.
60. Ye, S., Gamburd, M., Mozayeni, P., Koss, M. and Campese, V.M. (1998) A limited renal injury may cause a permanent form of neurogenic hypertension. *Am. J. Hypertens.* **11**:723–728.
61. Ye, S., Ozgur, B. and Campese, V.M. (1997) Renal afferent impulses, the posterior hypothalamus, and hypertension in rats with chronic renal failure. *Kidney Int.* **51**(3):722–727.

62. Weinstock, M., Gorodetsky, E. and Kalman, R. (1996) Renal denervation prevents sodium retention and hypertension in salt-sensitive rabbits with genetic baroreflex impairment. *Clinical Science* **90**:287–293.

63. Osborn, J.L. and Camara, A.K.S. (1997) Renal neurogenic mediation of intracerebroventricular angiotensin II hypertension in rats raised on high sodium chloride diet. *Hypertens.* **30**(pt 1):331–336.

64. Osborn, J.L., Roman, R.J. and Ewens, J.D. (1988) Renal nerves and the development of Dahl salt-sensitive hypertension. *Hypertens.* **11**:523–528.

65. Wyss, J.M., Sripairojthikoon, W. and Oparil, S. (1987) Failure of renal denervation to attenuate hypertension in Dahl NaCl-sensitive rats. *Can. J. Physiol. Pharmacol.* **65**:2428–2432.

66. Katholi, R.E., Whitlow, P.L., Winternitz, S.R. and Oparil, S. (1982) Importance of the renal nerves in established two-kidney, one clip goldblatt hypertension. *Hypertens.* **4**(Suppl. 2):II–166 – II–174.

67. Katholi, R.E., Winternitz, S.R. and Oparil, S. (1981) Role of the renal nerves in the pathogenesis of one-kidney renal hypertension in the rat. *Hypertens.* **3**:404–409

68. Norman, R.A., Jr., Murphy, W.R., Dzielak, D.J., Khraibi, A.A. and Carroll, R.G. (1984) Role of the renal nerves in one-kidney, one clip hypertension in rats. *Hypertens.* **6**:622–626.

69. Villarreal, D., Freeman, R.H., Davis, J.O., Garoutte, G. and Sweet, W.D. (1984) Pathogenesis of one-kidney, one-clip hypertension in rats after renal denervation. *Am. J. Physiol.* **247**:H61–H66.

70. Wyss, J.M., Aboukarsh, N. and Oparil, S. (1986) Sensory denervation of the kidney attenuates renovascular hypertension in the rat. *Am. J. Physiol.* **250**:H82–H86.

71. Smithwick, R.H. (1951) Hypertensive cardiovascular disease: The effect of thoracolumbar splanchnicetomy upon mortality and survival rates. *JAMA* **147**:1611 – 1615.

72. Walker, J.R. and Guyton, A.C. (1967) Influence of blood oxygen saturation on pressure-flow curve of dog hindlimb. *Am. J. Physiol.* **212**:506–509.

73. Davis, M.J. and Hill, M.A. (1999) Signaling mechanisms underlying the vascular myogenic response. *Physiol. Rev.* **79**:387–423.

74. Borgstrom, P. and Gestrelius, S. (1987) Integrated myogenic and metabolic control of vascular tone in skeletal muscle during autoregulation of blood flow. *Microvasc. Res.* **33**:353–376.

75. Borgstrom, P., Grande, P.O. and Mellander, S. (1984) An evaluation of the metabolic interaction with myogenic vascular reactivity during blood flow autoregulation. *Acta Physiol. Scand.* **122**:275–284.

76. Morff, R.J. and Granger, H.J. (1982) Autoregulation of blood flow within individual arterioles in the rat cremaster muscle. *Circ. Res.* **51**(1):43–55. 77. Granger, H.J. and Guyton, A.C. (1969) Autoregulation of the total systemic circulation following destruction of the central nervous system. *Circ. Res.* **25**:379–388.

78. Hinojosa-Laborde, C., Greene, A.S. and Cowley, A.W., Jr. (1988) Autoregulation of the systemic circulation in conscious rats. *Hypertens.* **11**:685–691.

79. Liedtke, A.J., Urschel, C.W. and Kirk, E.S. (1973) Total systemic autoregulation in the dog and its inhibition by baroreceptor reflexes. *Circ. Res.* **32**:673–677.

80. Metting, P.J., Strader, J.R. and Britton, S.L. (1988) Evaluation of whole body autoregulation in conscious dogs. *Am. J. Physiol.* **255**:H44–H52.

81. Shepherd, A.P., Granger, H.J., Smith, E.E. and Guyton, A.C. (1973) Local control of tissue oxygen delivery and its contribution to the regulation of cardiac output. *Am. J. Physiol.* **225**:747–755.

82. Coleman, T.G. and Guyton, A.C. (1969) Hypertension caused by salt loading in the dog. III. Onset transients of cardiac output and other circulatory variables. *Circ. Res.* **25**:153–160.

83. Manning, R.D., Jr., Coleman, T.G., Guyton, A.C., Norman, R.A., Jr., McCaa, R.E. (1979) Essential role of mean circulatory filling pressure in salt-induced hypertension. *Am. J. Physiol.* **236**:R40–R47.

84. Patterson, G.C., Shepherd, J.T. and Whelan, R.F. (1957) The resistance to blood flow in the upper and lower limb vessels in patients with coarctation of the aorta. *Clin. Sci.* **16**:627–632.

85. Stanek, K.A., Coleman, T.G. and Murphy, W.R. (1987) Overall hemodynamic pattern in coarctation of the abdominal aorta in conscious rats. *Hypertens.* **9**:611–618.

86. Wakim, K.G., Slaughter, O. and Clagett, O.T. (1948) Studies of the blood flow in the extremities in cases of coarctation of the aorta: determinations before and after excision of the coarctate region. *Mayo Clin. Proc. Staff Meetings* **23**:347–351.

87. Struijker Boudier, H.A., le Noble, J.L., Messing, M.W., Huijberts, M.S., le Noble, F.A. and van Essen, H. (1992) The microcirculation and hypertension. *J. Hypertens. Suppl.* **10**:S147–S156.

88. Folkow, B., Gurevich, M., Hallback, M., Lundgren, Y. and Weiss, L. (1971) The hemodynamic consequences of regional hypotension in spontaneously hypertensive and normotensive rats. *Acta Physiol. Scand.* **83**:532–541.

89. Gangluli, M., Tobian, L. and Iwai, J. (1979) Cardiac output and peripheral resistance in strains of rats sensitive and resistant to NaCl hypertension. *Hyperten.* **1**:3–7.

90. Greene, A.S., Yu, Y., Roman, R.J. and Cowley, A.W. (1990) Role of blood volume expansion in Dahl rat model of hypertension. *Am. J. Physiol. (Heart Circ Physiol)* **258**:H508–H514.

91. Pfeffer, M.A., Pfeffer, J., Mirsky, I. and Iwai, J. (1984) Cardiac hypertrophy and performance of Dahl hypertensive rats on graded salt diets. *Hypertens.* **6**:475–481.

92. Simchon, S., Manger, W.M. and Brown, T.W. (1991) Dual hemodynamic mechanisms for salt-induced hypertension in Dahl salt-sensitive rats. *Hypertens.* **17**:1063–1071.

93. Krieger, J.E. and Cowley, A.W., Jr. (1990) Prevention of salt angiotensin II hypertension by servo control of body water. *Am. J. Physiol. Heart Circ. Physiol.* **258**:H994–H1003.

94. Krieger, J.E., Liard, J-F. and Cowley, A.W., Jr. (1990) Hemodynamics, fluid volume, and hormonal responses to chronic high salt intake in dogs. *Am. J. Physiol. Heart Circ. Physiol.* **259**:H1629–H1636.

95. Lund-Johansen, P. (1986) Hemodynamic patterns in the natural history of borderline hypertension. *J. Cardiovasc. Pharmacol.* **8** (Suppl 5):S8–S14.

96. Van Vliet, B.N., Lohmeier, T.E., Mizelle, L. and Hall, J.E. (1996) The Kidney. In: *Nervous Control of Blood Vessels. The Autonomic Nervous System, Vol. 8.* (T. Bennet and S. Gardiner, Eds.). Harwood Academic Publishers, London, pp. 371–433.

5

Neuroendocrine Networks

John A. Russell and Gareth Leng

1. INTRODUCTION

Neuroendocrine networks in the brain are interconnected circuits of neurons, forming the central components of neuroendocrine control systems, and, in some cases, extending into the spinal cord. Their activity organizes behaviors and desires, autonomic nervous action, and the activity of endocrine systems. *Neuroendocrinology* is the study of how the brain controls body function by regulating the secretion of hormones, and includes study of the actions of hormones on the brain. Almost all of the endocrine glands are controlled by the brain, and almost all hormones can affect (by "feedback") brain activity. The hypothalamus is the major neuroendocrine part of the brain, but it has important connections with brain regions concerned with monitoring the internal and external environments, and in the generation of emotion and behaviors.

Through neuroendocrine mechanisms, the brain controls water balance, all aspects of reproduction, diurnal rhythms, metabolism, growth, and responses to stress. These actions are exerted through discrete vitally important outlets from the brain: the posterior pituitary gland; the hypothalamic vascular connection to the anterior pituitary gland; the pineal gland; the sympathetic nervous outflow to the adrenal medulla; and the sympathetic/parasympathetic innervation of the pancreatic islets. The anterior pituitary hormones in turn regulate other endocrine glands.

Actions of hormones on the brain are important in development (thyroxine, steroid hormones), in reproduction (sex steroids and behavior), thirst (angiotensin), appetite (insulin, leptin), and mood (adrenocorticoids). Disorders of these mechanisms give rise to particular disease states (e.g., Cushing's disease), failure of normal function (infertility), or imbalance (obesity). Application of knowledge of neuroendocrinology enables choices to be exercised over otherwise automatic processes, as well as rational treatment of disorders.

1.1. Neurohormones

Hormones are substances produced by specialized cells to control activities of other cell groups, classically reaching target cells via the circulation. However, many substances described as "hormones" are also produced within the brain, by neurons, and function there as neurotransmitters, or as local neurohormones that diffuse more widely than the classical neurotransmitters that are confined to the synaptic cleft. Within the brain, these molecules

From: *Integrative Physiology in the Proteomics and Post-Genomics Age*
Edited by: W. Walz © Humana Press Inc., Totowa, NJ

are involved in the activity of the circuits or networks that are part of, or regulate, neuroendocrine systems. "Cross-talk" between the release of these neurohormones and the same (or very similar) circulating molecules is prevented by the blood–brain barrier (BBB). For peptide hormones, the tight junctions between endothelial cells of the blood capillaries in the brain that form the BBB prevent their entry. However, there are several sites, adjacent to the cerebral ventricles in the midline, where the BBB is open, and in these "circumventricular organs," peptide hormones can enter the brain, and penetrate for a short distance. For some peptide hormones there are transport mechanisms in the choroid plexus in the ventricles that transfer these peptides across the BBB into the cerebrospinal fluid, providing a route to the brain extracellular fluid.

1.2. Behavior

Because outputs from the hypothalamus include control of the autonomic nervous system, and because the hypothalamus is involved in regulation of behaviors, appropriate autonomic, behavioral and endocrine responses to external or internal disturbances can be organized. For optimal regulation of homeostasis, functions that are regulated by hormones often have appropriate associated behaviors. Examples include drinking as part of the regulation of water balance; feeding, as part of the regulation of metabolism; self-protective behaviors (escape or aggression) that are important mechanisms in surviving stressors; reproductive behaviors that are essential for procreation; and social behaviors that are important in rearing young or in maintaining cohesion in social groups. A useful postulate is that the actions of a "neuropeptide" in the brain are consonant with the actions of the same peptide when it is functioning as a hormone elsewhere in the body. For example, the actions of oxytocin on the uterus in parturition, and on the mammary gland in lactation, are clearly important for the survival of the offspring: in the brain oxytocin has actions in the initiation of maternal behavior, and in social behavior. By contrast, the actions of vasopressin as a hormone include contributing to the stimulation of adrenocorticotropic hormone (ACTH) secretion in response to stress, conserving body water when it is depleted, and maintaining blood pressure when blood is lost: in the brain vasopressin is involved in regulating aggressive behavior, which may be appropriate in situations where the systemic hormonal actions of vasopressin are needed.

1.3. Gene Expression

The classical description of neuroendocrine networks involves definition of the discrete brain regions that are essential for a particular homeostatic function, and of their interconnections, and identification of the transmitters, neurohormones, and intracellular cascades that are used in signaling. The availability of the signaling elements is modulated through the regulation of gene expression. At this level of describing neuroendocrine networks, gene expression is an important determinant of the activity of neuroendocrine systems. Thus, expression of receptor genes is necessary for all classes of neurohormones, altered expression of synthesizing or converting enzymes is important for steroids, and expression of genes for neuropeptide precursors is a major determinant of neuropeptide availability. Many neuropeptides have been identified, and their production is typically confined to relatively few neurons, often gathered together in discrete nuclei. Peptide neurons generally express the genes for more than one neuropeptide, and the pattern of expression is determined during development. However, there are also multiple, post- or nongenomic, levels of regulation of the activity of the neurohormone signaling mechanisms; for instance receptor internalization, phosphorylations of intracellular regulatory proteins, and alternative processing of neuropeptide precursors.

1.4. Feedback

Neuroendocrine networks are generally automatically regulated to perform a homeostatic function; typically, this involves monitoring an output, so that the secretion of hormone can be increased or decreased to produce an appropriate effect in the physiological context. Feedback signals act at multiple levels in neuroendocrine systems, and, in general, hormonal feedback signals act on the brain via receptors and postreceptor signaling mechanisms that are the same as, or very similar to, those mediating peripheral hormone actions. Thus, the genes for neurohormones, hormones, and their receptors are used parsimoniously in the control of body functions. Nonetheless, input from other sources can override negative feedback, and drive diversification of the responsiveness of endocrine systems to a changing environment.

1.5. Neuroendocrine Rhythms

Rhythmicity is characteristic of neuroendocrine networks, generating intermittent, pulsatile secretion of hormones from the anterior and posterior pituitary gland. There is a wide range of periodicity among the different neuroendocrine systems: individual vasopressin neurons discharge at intervals of a few seconds to a few minutes; oxytocin neurons synchronously fire bursts of action potentials every few minutes during birth or suckling; growth hormone-releasing hormone (GHRH) neurons secrete at 3– to 4-h intervals; corticotropin-releasing hormone CRH neurons are active under basal conditions only during the waking hours. These rhythms are generated by interactions between the inputs to the neurons and their intrinsic properties. In the case of CRH neurons, their diurnal rhythm is driven by the automatic circadian activity of suprachiasmatic neurons; the rhythmicity of these neurons is caused by a cascade of clock genes and their products.

1.6. Aim

Here, we focus first on magnocellular oxytocin and vasopressin neurons as classical representatives of peptide neurons, the essential functional unit in neuroendocrine networks. We illustrate the concept of *gene micronets* in relation to roles of oxytocin neurons in social behavior. We then discuss the organization and functions of the hypothalamic paraventricular nucleus (PVN) in the context of its role in the extended network that generates stress responses. We describe the mini-network that generates pulsatile growth hormone (GH) secretion, and how this is integral to a network that regulates metabolism and receives recently identified signals about metabolism that illuminate the problem of obesity. Finally, we outline a strategy for therapeutic developments from knowledge of neuroendocrine networks.

2. MAGNOCELLULAR OXYTOCIN AND VASOPRESSIN NEURONS

These prototypical neuroendocrine neurons are located in the supraoptic nuclei and the magnocellular subdivision of the paraventricular nuclei, and project to the posterior pituitary gland. These neurons function as sensory neurons, detecting osmolarity; and as effector neurosecretory neurons, secreting oxytocin or vasopressin; they secrete intermittently in pulses, so they are also peptide hormone "pattern generators." They have been extensively studied, first because their activity can be monitored by measuring circulating oxytocin and vaso-pressin, or their effects on peripheral targets; second because they can be studied in vivo and in vitro with electrophysiological and related techniques; and, third because they are so active in synthesizing and secreting peptide, to maintain effective circulating levels, that they have been a focus for investigation of the regulation of neuropeptide gene expression (*1*; Fig. 1A).

The cell bodies of magnocellular neurons have dendrites, and on these are many syn-apses, from several brain areas, and these use several chemical transmitters. Interactions between the presynaptic terminals and the postsynaptic dendrites creates a microenviron-ment, rich in chemical signals (*2*; Fig. 1B). Like other neurons, the activity of magnocellular neurons is influenced by many different inputs, which, through excitatory postsynaptic po-tentials (EPSPs) and inhibitory postsynaptic potentials (IPSPs), determine the firing activity (action potential/spike generation) of the cell bodies. Important excitatory transmitters for vasopressin neurons include glutamate (GLU), from rostral neurons in the lamina terminalis, mediating input from two circumventricular organs: subfornical organ and the organum vasculosum of the lamina terminalis, where circulating angiotensin and osmolarity are de-tected. Noradrenaline (NA) is in axon terminals that derive from neurons of the medulla oblongata, and these NA neurons mediate input from arterial baroreceptors, atrial volume receptors and chemoreceptors (*3*). Inhibitory transmitters include γ-aminobutyric acid (GABA), and this is synthesized in many afferent neurons, including some local neurons in the perinuclear zone around the magnocellular neurons. Oxytocin neurons in the rat have similar inputs, but the noradrenergic input from the medulla conveys signals from the con-tracting uterus at term, and, vitally, in lactation input from the nipples relays in the brainstem, and, via an as yet uncharacterized pathway, activates mechanisms intrinsic to oxytocin neu-rons that cause burst-firing (*4,5*). Magnocellular neurons promiscuously express genes for neurotransmitter and neuropeptide receptors, but the roles of many of these are not defined.

2.1. Dendritic Microenvironment

Signaling by synaptic inputs to magnocellular neurons is subject to powerful local modu-lation, as a result of secretory activity of the dendrites (*3*). This "talkback" has a key role in patterning the activity of the magnocellular neurons. The essential mechanism is that the dendrites release their neuropeptide(s) by exocytosis when cytoplasmic Ca^{2+} concentration increases as a consequence of transmitter action (*6*). For oxytocin neurons, dendritically released oxytocin has an essential role in the generation of their intermittent synchronised burst-firing during suckling. This oxytocin acts back on the dendrites, via oxytocin recep-tors, to stimulate further oxytocin release, and the generation of endocannabinoid (CB); this then acts on GLU terminals to inhibit GLU release and thus to inhibit the oxytocin neuron (*7*). These actions of oxytocin can be described as auto- or para-crine; only one oxytocin receptor gene has been identified, so these actions of oxytocin are via the same 7-transmem-brane domain G protein-coupled receptor that mediates peripheral oxytocin actions. When oxytocin occupies its receptor, phospholipase C is activated and inositol 1,4,5-trisphosphate (IP_3) and diacylglycerol (DAG) are generated, leading to increased cytoplasmic Ca^{2+} con-centration. α-melanocyte stimulating hormone (α-MSH), acting via melanocortin-4 recep-tors (MC4R), has the same effect on dendritic oxytocin release, and its action is confined to the dendrites as it does not excite the cell bodies (*8*).

In addition, nitric oxide is released by the dendrites when the neurons are activated, and this also acts presynaptically, increasing GABA release local neurons as well as directly inhibiting the oxytocin neurons. The interactions in this microcosm seem to limit excitation, but also during suckling *prime* the oxytocin neurons for an explosive burst of activity, synchronized among the magnocellular nuclei, leading to the secretion of a pulse of oxytocin to assist birth or for milk ejection *(5)*.

For vasopressin neurons, dendritic release of vasopressin and the co-produced opioid peptide, dynorphin, interact with the intrinsic properties of the membrane conductances of the neurons to terminate a burst of firing *(9)*. Here, it is important that the summated activity of vasopressin neurons is *not* pulsatile, as this would be inappropriate for the peripheral functions of vasopressin in the regulation of urine concentration and blood pressure.

2.2. Patterns of Electrical Excitation

Oxytocin and vasopressin neurons appear to be very similar, yet have quite separate functions, and they have evolved very different patterns of electrical activity to meet these functional demands. The phenotype of vasopressin neurons leads to firing in phasic bursts, each lasting for several seconds or a few minutes, with similar silences between bursts (Fig. 2). Oxytocin neurons fire continuously (about five action potentials per second), except during suckling or during parturition when they fire with a high frequency burst (about 40 action potentials per second for a few seconds, every few minutes), and all oxytocin neurons burst-fire at the same time *(5)*.

These different patterns are a consequence primarily of the intrinsic properties of the neurons, rather than of their inputs. The electrical properties of the oxytocin and vasopressin neurons that underlie these patterns reflect expression and activities of several membrane conductances. One of these conductances, which confers osmosensitivity, is a stretch-inactivated cationic channel *(10)*. The electrical activity of neurons reflects dynamic interactions between intrinsic membrane properties and perturbations by afferent inputs, and these interactions have been studied in stochastic computer models to successfully simulate the electrophysiological behavior of oxytocin and vasopressin neurons *(11)*. Such models have proven predictive value for hypothesis building for experimental testing of the mechanisms underlying the behavior of neurons in neuroendocrine networks *(12)*.

2.3. Production of Oxytocin and Vasopressin

Oxytocin and vasopressin are closely related nonapeptides, cleaved from distinct precursors, each the product of a separate gene, which are closely related. The genes have evolved from a common ancestor by gene duplication and mutation, and are on adjacent DNA strands, transcribed in opposite directions. While expression of the oxytocin and vasopressin genes is upregulated when the neurons are strongly stimulated, as in lactation and dehydration, the mechanisms controlling expression of the oxytocin and vasopressin genes are not yet clear *(1)*.

As oxytocin has essential roles in reproductive processes, possible positive regulation of gene expression by sex steroid hormones has been repeatedly examined, generally without finding strong effects. Steroids may however have effects that are not mediated by direct regulation through steroid receptor interactions with the oxytocin gene: the gene lacks an estrogen response element *(13)*, and in rats and mice oxytocin neurons lack estrogen receptor (ER)α, although they do express ERβ (*see* ref. *14*). Studies on mice with targeted ERβ gene inactivation show that ERβ can mediate a modest stimulation of oxytocin gene expression by estrogen *(15)*. Oxytocin neurons also lack cytoplasmic androgen and progesterone receptors, so genomic actions of testosterone and progesterone through these classical

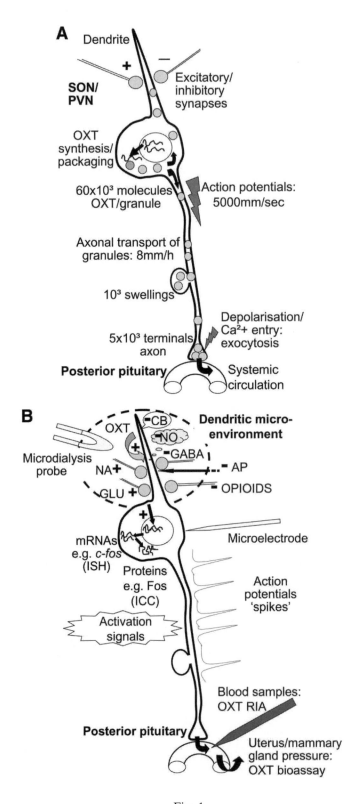

Fig. 1

mechanisms are not likely. However, actions of estrogen and progesterone on plasma membrane receptors could indirectly affect gene expression. The best described actions at this level are on electrical activity; thus, oestrogen has excitatory actions, which may be mediated by cAMP. Progesterone is inhibitory, but only after local enzymatic conversion involving 5α-reductase to allopregnanolone (AP), which is an allosteric modifier of $GABA_A$ receptors. AP may act by binding to a subunit of the $GABA_A$ receptors on oxytocin neurons, and when GABA is also bound to the receptor the Cl^- channel opening time is prolonged *(16)*. The extended duration of IPSCs enhances the inhibitory tone on the neurons, so they are less responsive to excitatory inputs. As progesterone secretion is greatly increased in pregnancy, this mechanism evidently contributes to preventing premature activation of oxytocin secretion.

Neuropeptide synthesized and processed in the cell bodies is slowly transported by axonal transport to the posterior pituitary (Fig, 1A). Secretion of oxytocin or vasopressin from the posterior pituitary occurs when the axon terminals that store the peptide are depolarized by the arrival of action potentials from the cell body, propagated along the axon. This depolarization triggers Ca^{2+} entry and exocytosis, and the released hormone enters the adjacent capillaries for transport in blood around the body to target tissue(s).

This stimulus–secretion coupling process is modulated locally. The secretion rate of vasopressin or oxytocin increases disproportionately with the frequency of arrival of action potentials at the axon terminals, so the arrival of a burst of action potentials favors the secretion of a high-amplitude pulse of peptide. This frequency facilitation is probably due to summation of the entry of Ca^{2+} into the terminals that is triggered by each spike, which enhances exocytosis. This exemplifies how the secretory terminals of a peptide neuron convert the digitally coded pattern of action potentials into an analog signal—the secretion of a pulse of neuropeptide characterized by its amplitude and duration.

There are also actions of modulators released along with the peptide; these include inhibitory actions of opioid peptides and facilitatory actions of adenosine triphosphate, which are followed by inhibitory actions after conversion to adenosine *(17)*.

Fig. 1. *(continued from facing page)* A typical neuroendocrine neuron: neuropeptide production and activation in a magnocellular oxytocin (OXT) neuron. **(A)** Summary of stages in oxytocin production: from gene transcription through translation, processing, packaging, and transport in vesicles into dendrites and in the axon to nerve terminals in the posterior pituitary for release by Ca^{2+}-dependent exocytosis, stimulated by action potentials arriving from the cell body in the suproptic (SON) or paraventricular (PVN) nuclei. **(B)** Electrical excitation: action potentials arise when the cell body depolarizes through suprathreshold excitatory synaptic activity. Excitatory transmitters include glutamate (GLU), noradrenaline (NA), opposed by inhibitory transmitters or neuropeptides including γ-aminobutyric acid (GABA) and opioids. The dendrites release oxytocin when excited, which has auto- or paracrine excitatory actions, whereas inhibitory nitric oxide (NO) and endocannabinoids (CB) are also released by dendrites; these signals from dendrites act partly by modulating synaptic input (talkback). Allopregnanolone (AP) acts on $GABA_A$ receptors to augment GABA inhibition. Release of chemical signals by presynaptic terminals and dendrites into the dendritic microenvironment can be measured with microdialysis; electrical activity is measured by single cell microelectrode recording; responses of populations of neurons to stimuli is evaluated by postmortem *in situ* hybridization (ISH) or immunocytochemistry (ICC) for, respectively, *c-fos* mRNA or Fos protein expression; or by measurement of oxytocin secretion by assay of blood concentrations.

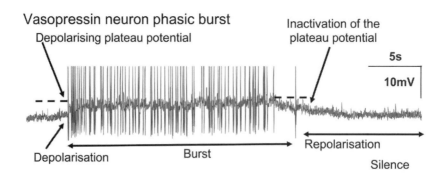

Fig 2. Magnocellular vasopressin neuron burst firing. Phasic action potential firing of vasopressin neurons is due to intrinsic properties: excitatory synaptic input triggers firing, which continues because of maintained depolarization (plateau potential), which eventually activates hyperpolarising mechanisms that stop the burst; in the following silence they slowly become more excitable, until triggered into activity again. Vasopressin release varies with the proportion of time each neuron is burst-firing because the neurons fire *asynchronously* with each other. Synaptic input modulates burst frequency and duration. Strong stimulation prolongs bursts and shortens silences, so more neurons are firing at any time and more vasopressin is secreted from the posterior pituitary gland. (Intracellular recording—supraoptic nucleus in vitro; note the action potentials have been truncated. Illustration courtesy of Dr. C. H. Brown.)

The burst-firing behavior of vasopressin and oxytocin neurons serves as a working model for the generation of pulses of secretion of hypothalamic-releasing hormones, and hence of pulsatile secretion of anterior pituitary hormone secretion.

2.4. Oxytocin and Behavior

The general hypothesis that actions of oxytocin released in the brain are complementary to its peripheral actions, on the uterus in parturition and on the mammary gland in lactation, is supported by a range of studies *(4)*. These have used central administration of oxytocin or oxytocin antagonist to study effects on behavior and on neurotransmitter release in discrete brain regions; lesioning; immediate-early gene expression to locate activated neurons; and manipulations of expression of the genes for oxytocin, the oxytocin receptor, and estrogen receptors.

The sources of oxytocin released in the brain that acts in networks involved in maternal and social behavior may be either the dendrites of magnocellular neurons, or centrally projecting parvocellular neurons in the ventromedial PVN, with actions in the limbic brain.

These studies have revealed that oxytocin has actions in the initiation of maternal behavior, in the expression of social affiliative behavior, and in social memory (recognizing "friends"). Identifying this cluster of behaviors in which oxytocin is involved has led to studies to seek abnormalities in oxytocin function in human disorders, such as autism, in which social interaction is disturbed *(18)*. To date, no naturally occurring disabling mutations in the oxytocin peptide or receptor genes have been described, which contrasts with the multiple mutations known for the vasopressin gene. However, the phenotype for failure of the vasopressin system is diabetes insipidus, which is overt, but survivable by greatly increased drinking. Mice with genetically engineered inactivation of the oxytocin peptide or receptor genes show only one overt phenotypic characteristic, and this is complete failure of milk ejection, so the young die unless the mothers are injected with oxytocin. Evidently, inheritance of a natural, disabling mutation in the oxytocin peptide gene would not be possible, except for species like humans that can cross-foster *(4)*.

However, mice with engineered oxytocin gene inactivation do show impairment of social behavior when appropriately tested: they do not learn to recognize by smell unfamiliar mice they have just met *(19,20)*. Mice with knockout of the ERα or ERβ genes show a similar phenotype. This is attributable to the action of oestrogen through ERβ in driving oxytocin gene expression in centrally projecting PVN neurons, and to the dependence on estrogen action through ERα of oxytocin-receptor gene expression in the amygdala, where oxytocin from the PVN projection interacts with olfactory processing. Hence, the genes for oxytocin peptide and receptor, and for ERα and ERβ are described as comprising a "4-gene micronet" within the PVN–amygdala–olfactory neuroendocrine network involved in social memory *(20)*.

3. PARAVENTRICULAR NUCLEUS AND STRESS

The roles of the PVN include, through its populations of magnocellular neurons, contributions to the secretion of oxytocin and vasopressin by the posterior pituitary. However, the PVN has additional neuroendocrine functions. Thyrotropin-releasing hormone (TRH) neurons in the PVN control the secretion of thyroid-stimulating hormone from the anterior pituitary to regulate the activity of the thyroid gland; and the PVN also regulates the anterior pituitary–adrenal cortex axis (together described as the hypothalamic–pituitary–adrenal [HPA] axis). Central outputs from the PVN also regulate the autonomic nervous system, especially the sympathoadrenal medullary (SAM) system, and regulate reproductive, social and eating behaviors, and emotionality.

Different sets of neurons in the PVN are responsible for these different functions. The primary organization of the PVN is into groups of neurons, or subdivisions, that can be distinguished morphologically, by the neuropeptides they produce, by their projections, by their inputs, and thus the stimuli that control them, and by their functions *(21,22)*. These subdivisions comprise (a) the lateral magnocellular oxytocin and vasopressin neurons that project to the posterior pituitary gland; (b) the dorsomedial subdivision, where parvocellular neurons producing CRH and some vasopressin are gathered, projecting to the median eminence; (c) medial and periventricular parvocellular subdivisions, containing hypophysiotropic TRH neurons; (d) the ventromedial group of centrally projecting neurons, which produce oxytocin and nociceptin (orphanin-FQ, OFQ); and (e) the dorsal and posterior group of neurons, producing oxytocin or CRH, projecting to the brainstem and spinal cord. Around the lateral periphery of the PVN are GABA neurons that project axons into the nucleus. The magnocellular subdivision is further organized so there is a central core of vasopressin neurons surrounded by a shell of oxytocin neurons. This arrangement may provide for economical contacting by axon terminals from distant inputs, and for efficient autocrine or paracrine interactions and regulation, as discussed earlier.

In terms of function, or physiology, the task is to identify the roles of the different elements of the PVN in performing and coordinating its different functions, and how the elements receive and integrate information that signal homeostatic disturbances. This means that it is necessary to characterize the origin of neural input; the chemical coding used by distant and local synaptic inputs; humoral factors, which may act after crossing into the brain from the blood, or be locally generated either within the PVN or elsewhere in the brain, reaching the PVN by volume transmission; intracellular postreceptor signaling and process-

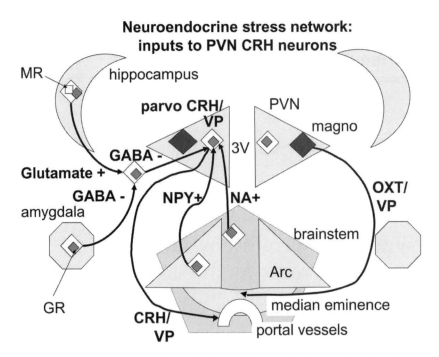

Neuroendocrine stress network: inputs to PVN CRH neurons

Fig 3. Neuroendocrine stress network and the hypothalamo–pituitary–adrenal axis: paraventricular nucleus (PVN) corticotropin-releasing hormone (CRH) neurons. Dorsomedial parvocellular (parvo) PVN CRH neurons, co-expressing vasopressin (VP), project to the hypothalamo–hypophysial portal capillaries in the median eminence. The general properties of these neuropeptide neurons are similar to those of magnocellular oxytocin or vasopressin neurons (Fig 1). Magnocellular (magno) oxytocin (OXT) and vasopressin (VP) neurons in the lateral PVN project to the posterior pituitary via the median eminence. Parvo PVN CRH neurons receive input from: the limbic system (hippocampus, amygdala via the bed nucleus of stria terminalis), using glutamate and γ-aminobutyric acid (GABA) interneurons, and encoding information about emotional stressors; the arcuate nuclei (Arc), signaling metabolic stress, including projections from excitatory neuropeptide Y (NPY) neurons; noradrenergic (NA) neurons in the brainstem, transmitting signals from immune stimulation, cardiorespiratory and visceral disturbances. Glucocorticoid feedback acts at multiple sites via glucocorticoid receptors (GR; small diamonds), and at basal levels of secretion via mineralocorticoid receptors (MR; square) expressed in the hippocampus. Processing of signals regulating parvo PVN CRH neurons involves interactions in their dendritic microenvironment like that of the magnocellular neurons (Fig 1B). Diagram indicates organization in the coronal plane, with different rostro-caudal levels represented in one plane. 3V, third ventricle.

ing; the coding of output as frequency, pattern and chemical signals; and the targets for output from the PVN. This account will focus on the role of the PVN in neuroendocrine stress responses.

4. HYPOTHALAMIC–PITUITARY–ADRENAL AXIS

The best characterized of the roles of parvocellular PVN neurons is in the control of the HPA axis. The effector neurons are cells in the dorsomedial part of the PVN. These neurons project to the median eminence and release CRH, a 41 amino acid peptide, together with varying amounts of vasopressin, from their terminals on the portal vessels (Fig. 3). Acting on CRH type 1 receptors on anterior pituitary corticotrophs, CRH then stimulates ACTH secretion, and increases expression of the pro-opiomelanocortin (POMC) gene. ACTH is a 39 amino acid peptide cleavage product of the precursor, POMC. POMC is also produced by neurons in the arcuate nucleus, and here it is processed to α-MSH (involved in appetite regulation, see below) and β-endorphin (an opioid peptide). ACTH secreted by the anterior pituitary acts in turn on adrenal cortex melanocortin-2 receptors that are coupled to adenylyl cyclase, and stimulates the synthesis and secretion of glucocorticoid (corticosterone in rodents; cortisol in humans). Vasopressin can directly stimulate ACTH secretion, via V1b receptors; but its most powerful role is to synergistically augment the action of CRH: CRH and vasopressin act on pituitary corticotrophs by separate postreceptor signaling pathways, respectively involving protein kinase A and protein kinase C activation—leading to increased intracellular $[Ca^{2+}]$ and stimulation of exocytosis. ACTH secretion can be inhibited at the level of the anterior pituitary by atrial natriuretic peptide, secreted by the hypothalamus, and by feedback actions of glucocorticoids on the anterior pituitary gland.

4.1. CRH Neurons

The electrophysiological properties of the PVN CRH neurons are more difficult to study than magnocellular oxytocin or vasopressin neurons, because in vivo anesthesia confounds study of their responses to stressors. Nonetheless, in vitro studies, monitoring PVN CRH neuron activity or CRH release, have given details about the regulation of their activity. Local regulation of CRH neurons involves similar mechanisms to those locally regulating magnocellular neurons *(23)*. In vivo studies of CRH release into hypothalamic–hypophysial portal blood are not feasible without anaesthesia in rodents, and CRH is not detectable in the general circulation, but changes in ACTH concentrations in the circulation can be monitored as an indicator of CRH secretion. Fortunately, also, elements of the neuroendocrine stress network show rapid stimulation of the expression of relevant genes during stress. For example, PVN CRH neurons show increased expression of a cascade of genes over several hours after stress that can be evaluated with *in situ* hybridization, and these changes correspond to increased ACTH secretion *(24)*. Thus, there are changes in expression of the *c-fos*, nerve growth factor inducible protein B, CRH, vasopressin, and CRH1 receptor genes, and an increase in any of these at an appropriate time after stress is evidence of CRH neuron activation, and increased CRH secretion can probably be assumed. As CRH increases the transcription rate of the POMC gene, measurement of this may also suggest increased CRH secretion. Increased CRH and POMC gene expression serve to replenish the stores of CRH or ACTH after depletion by stress.

4.2. Stress

Mammals generally observe a circadian cycle of activity linked to the photoperiod—thus rats, as nocturnal animals, increase their activity at the onset of darkness, whereas humans naturally wake up in the morning. The HPA axis is activated at the onset of the daily active period through mainly indirect connections from the suprachiasmatic nucleus to the PVN CRH neurons. Stressors additionally activate these CRH neurons by different pathways, projecting from many brain areas because of the diverse nature of stressful stimuli. By definition, stress is *a state of real or perceived threat to survival or homeostasis*, resulting in activation of the HPA axis. Other neuroendocrine systems also respond (e.g., adrenaline is secreted from the adrenal medulla; vasopressin or oxytocin secretion increases, depending on the species, as does GH and prolactin; gonadotropin pulses are suppressed), and there are changes in autonomic nervous system activity. Behavioral responses are driven by fear and anxiety, or despair, with loss of appetite; this emotional context is generated within the limbic system, especially involving the amygdala.

4.2.1. Stress Processing

The common pattern of responses to stress is a consequence of convergence of different types of stressor onto a neuroendocrine network in which the PVN is a key processing component, and the organizer of the major responses.

There are many different types of stressor, but in terms of the neural circuits that process stressful stimuli, it is useful to categorize these as emotional (psychogenic) or physical. Emotional stressors (e.g., social conflict, novel environment) are processed, after perception through the special or somatic senses, in the limbic circuits, which project to the PVN, mainly indirectly to the CRH neurons, including via the bed nucleus of the stria terminalis *(25)*. Physical stressors may arise externally and produce pain, which is processed through nociceptive mechanisms that activate brainstem projections; pain from internal causes acts similarly *(26)*. Physical stressors also arise internally, as signals indicating threats to homeostasis; in order to understand the processing pathways, such signals are usefully categorized as related to metabolic disturbance (e.g., hypoglycemia), circulatory or respiratory malfunction (e.g., hemorrhage or hypercapnia), or inflammation. The processing of signals indicating metabolic stress and the chemical coding in projections to the PVN are discussed here. Baro- and chemoreceptor and visceral vagus input pathways via the brainstem to the hypothalamus are well described *(3)*.

4.2.2. Immunoneuroendocrine Signaling

Immunoneuroendocrine signaling involves circulating cytokines produced by activated immune cells (especially interleukin [IL]-1β) acting on IL receptors on brain vascular endothelial cells, and triggering the local generation of prostaglandin E2 via cyclooxygenase-2 *(27)*. In the nucleus tractus solitarii (NTS) in the medulla oblongata, prostaglandin produced in this way diffuses into the brain, and acts on prostanoid receptors on noradrenergic neurons, exciting them. The NTS noradrenergic (A2) neurons project directly to PVN CRH neurons, which in turn are stimulated via their α1-adrenoreceptors *(28)*.

4.2.3. Emotionality

Importantly, both for the generation of emotional change and cognitive effects of stress, and for the modulation or adaptation of stress responses, there are reciprocal connections among the brain regions involved in stress responses—the limbic brain, the PVN, and the brainstem. In one respect, this leads to "memory of the context" of an emotional stress, and its future avoidance, and in another respect it leads to the emotional context of illness owing to infection. Thus, systemic IL-1β activates neurons in the amygdala as well as the PVN *(29)*.

The evident evolutionary principle of using the same, or closely related, neuropeptides to perform related functions in different places applies to CRH and vasopressin. There are CRH neurons and CRH1 receptors in the amygdala, where CRH acts to increase anxiety. Similarly, vasopressin is released in the limbic brain during stress and increases anxiety where it acts via V1a receptors, which are coupled to phospholipase C and IP$_3$ and DAG generation leading to increased cytoplasmic Ca^{2+} concentration *(30)*. Accordingly, CRH1 and V1a receptor antagonists are anxiolytic *(31,32)*, and mice with engineered inactivation of the CRH1 receptor gene show reduced anxiety *(33)*.

4.3. Terminating Stress Responses

Glucocorticoid actions in stress responses involve catabolism, mobilizing energy, redistributing fat, cardiovascular strain, immune suppression, heightened emotion, and sharpened cognition. These actions ensure prompt and focused coping responses to a threatening stressor, but the effects accumulate deleteriously as "allostatic load" with repeated stress *(34)*. Clearly, it is advantageous that stress responses are terminated rapidly when the stressor is removed, or, for an emotional stressor, is recognised as harmless, and that the activity of the neuroendocrine network can adapt during repeated stress. Both mechanisms involve glucocorticoid feedback, and intrinsic mechanisms in the network.

Failure or suppression of these mechanisms predisposes to stress-related physical (including obesity and its sequelae) and mental (anxiety and depression) illness.

4.3.1. Glucocorticoid Feedback

Under basal conditions, with a low-circulating concentration of corticosterone, only mineralocorticoid receptors (MR), which have a high affinity for corticosterone and are expressed in the hippocampus, are fully occupied *(35)*. This accords with the concept that hippocampal inputs to the PVN regulate basal HPA axis activity. The glucocorticoid receptors (GR) are occupied when corticosterone secretion is increased during stress. GR are expressed widely in the brain, including the hippocampus, brainstem, amygdala, and PVN CRH neurons, and in the corticotrophs. Thus, through genomic mechanisms, corticosterone can alter activity in the neuroendocrine stress network. The net effect is to reduce HPA axis activity, and to hasten recovery to basal activity after stress. There are also rapid actions of glucocorticoid on neuronal membrane channels *(36)*, and stress levels of glucocorticoid act, via a genomic mechanism, to inhibit ACTH secretion from corticotrophs by hyperpolarization through enabling the activity of large-conductance potassium channels *(37)*. Glucocorticoid also rapidly inhibits CRH neurons by a presynaptic action on GLU terminals mediated by CB *(23)*.

The inactive metabolite of corticosterone, 11-dehydrocorticosterone, is converted back to corticosterone by 11β-hydroxysteroid dehydrogenase type 1 (11β-HSD1), which is expressed in glucocorticoid feedback sites and effectively amplifies the feedback signal. Mice with targeted deletion of the 11β-HSD1 gene have elevated basal HPA axis activity and greater stress responses, and show reduced corticosterone feedback during stress *(38)*.

4.3.2. Intrinsic Network Feedback

Within the PVN, CRH acts on CRH1 receptors to enhance HPA axis stress responses; notably, CRH1 receptor gene expression in the PVN is rapidly increased by stress. Several other neuropeptides modulate HPA axis recovery from stress *(39)*. These include urocortins I, II, and III which are related to CRH, and act on the CRH1 (CRH; urocortin I) or CRH2 receptor (urocortins I, II, III) *(40)*. Whereas the CRH1 receptor in the neuroendocrine stress network mediates expression of stress responses *(33)*, urocortins II and III actions via CRH2 receptors have the opposite effect *(40)*. Activation of this neuropeptide urocortin–CRH2 receptor mechanism during stress is likely to be an important attenuator or terminator of stress responses.

4.3.3. Multistage Adaptations

In general, every signaling component of the neuroendocrine stress network shows adaptations with chronic stress, or as a result of past experience of stress (as in posttraumatic stress disorder, or perinatal stress resulting in programming). These adaptations are detected as changes in gene expression, and they alter the set point, so far as HPA axis regulation is concerned, and the responsiveness to stress, affecting the emotional reactions as well as neuroendocrine responses. HPA axis responsiveness to stress can be an inherited trait, along with emotionality, and in one selectively bred rat line is evidently linked to a single nucleotide polymorphism (SNP) in regulatory elements of the vasopressin gene *(30)*.

Some of these changes are driven by glucocorticoid action on the brain. These changes include GR downregulation in the hippocampus, CRH gene expression upregulation in the amygdala, and downregulation in the PVN—where with chronic stress vasopressin gene expression in the CRH neurons is upregulated *(41)*. Although the consequences of any one of these changes are predictable, the emergent properties of the network with these multiple changes are not. However, their identification offers leads for therapeutic intervention, and for molecular genetic analysis of underlying causes of inherited traits.

Manipulating stress responses thus involves specific targeting of harmful peripheral effects (interfering with intracellular glucocorticoid metabolism or action); and intercepting the central signaling mechanisms that engender the disturbed emotional state accompanying chronic stress, or that attenuate neuroendocrine stress responses. These candidate targets for therapeutic intervention come from identifying the neuropeptides and their receptors used as intercellular signaling molecules in the neuroendocrine network that organizes stress responses, and their possible contributions to changes in chronic stress states. Strong leads come, for example, from observing that mice with targeted deletion of the CRH1 receptor gene are less anxious, but mice with CRH2 receptor gene deletion are highly anxious *(33,40)*. The strategy for these approaches is outlined here.

4.4. Programming

Prenatal exposure to stress or to synthetic glucocorticoid "programs" the offspring HPA axis to be more active later in life, both under basal and stress condition, and has a serious impact on physical and mental health and longevity *(42)*. A poor quality of mothering has similar effects *(43)*. Mechanisms of programming involve altered serotonin mechanisms in the hippocampus and reduced GR expression *(44)*.

Although placental 11β-HSD2 inactivates glucocorticoid, and thus reduces transfer of glucocorticoid across the placenta from the mother that might otherwise adversely program the fetal HPA axis *(42)*, there is another protective mechanism. The responsiveness of the maternal HPA axis to a wide range of stressors is attenuated in pregnancy, and this involves inhibitory actions of endogenous opioid on the PVN CRH neurons, or their inputs *(45)* and AP *(46)*. The actions of AP, which is produced in the brain in larger amounts in pregnancy from the greatly increased availability of progesterone evidently overwhelm the stimulatory actions of oestrogen on CRH neuron responses to stress *(47)*.

4.5. Metabolic Regulation

The PVN is a second- or third-order component of the network in the hypothalamus that controls metabolism, and receives synaptic input from the arcuate nucleus, brainstem, and lateral hypothalamus (see later). The roles of the PVN in metabolic regulation include its neuroendocrine function in controlling the activity of the HPA axis, its pivotal position in determining the activity of the autonomic nervous system, especially the SAM system, and

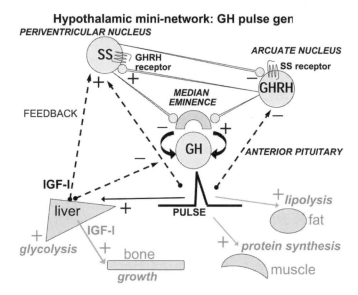

Fig 4. Hypothalamic mini-network generating growth hormone (GH) pulses. A pulse of GH is secreted by the anterior pituitary every 3 to 4 h as a result of GHRH release into the hypothalamo–hypophysial portal vessels. GH acts directly on fat, muscle and liver. Insulin-like growth factor (IGF)-I secretion by the liver is stimulated, which acts on bone and as a feedback signal inhibiting GH secretion by stimulating somatostatin (SS) neurons. Also, GHRH neuron projections might stimulate SS neurons, and GH has feedback effects, respectively slowly excitatory and inhibitory, on SS and GHRH neurons. Then SS secreted into portal blood inhibits GH secretion, and released at synapses on GHRH neurons SS inhibits GHRH neurons, thus ending the GH pulse. Slow decline in the GH and IGF-I feedback signals on SS neurons, owing to decreased GH secretion, leads to reduced activity of the SS neurons. This allows GHRH neurons to become active and another GH pulse is generated. Other signals act through this mini-network to modulate GH secretion (e.g., ghrelin, exercise, hypoglycaemia).

its actions in regulating appetite. Both oxytocin and CRH from PVN neurons have anorexigenic actions *(48)*.

5. GROWTH HORMONE: GENERATION OF PULSES BY A NEUROENDOCRINE MINI–NETWORK

GH is an anterior pituitary hormone important for growth in childhood, and for regulating metabolism and body composition in adults; it acts to increase availability of energy (in opposition to insulin), and to increase protein synthesis. Secretion of GH is regulated by the brain, and it is secreted in large pulses, every 3 to 4 h. Its peripheral actions are partly direct (e.g., via its receptor, a member of the cytokine receptor family, on hepatocytes, adipocytes, skeletal myocytes) and partly indirect (e.g., on bone growth) via insulin-like growth factor (IGF)-I. IGF-I is secreted by the liver when stimulated by GH, and IGF-I not only mediates some GH actions but also acts as a negative feedback signal to the neuroendocrine hypothalamus (Fig. 4). The heterotetrameric IGF-I receptor has intrinsic tyrosine kinase domains, autophosphorylating and phosphorylating other proteins when the receptor is occupied, and is expressed in the basal hypothalamus.

5.1. Hypothalamic-Releasing Factor Neurons

The hypothalamus is the essential regulator of GH secretion—the origin of the pulsatile pattern and the integrator of various feedback signals related to metabolism (Fig. 4). The hypothalamus regulates GH secretion through neurons that produce a stimulatory hypothalamic-releasing factor (GHRH, a 44 amino acid peptide), and a different set of neurons that produce an inhibitory factor (somatostatin, a 14 amino acid peptide). GHRH neurons are in the arcuate (tuberal) nucleus of the hypothalamus—close to the primary capillaries of the hypothalamic–hypophysial portal system, and accessible to circulating peptides *(49)*. Somatostatin neurons are found alongside the third ventricle, ventromedial to the PVN, and like GHRH neurons they secrete into the portal system. The GHRH and somatostatin receptors are 7-transmembrane domain, G protein-coupled receptors: the GHRH receptor is positively coupled by G protein and adenylyl cyclase *(50)* and the somatostatin receptor (of which there are five subtypes, of which sst2 is prevalent in somatotrophs *[51]*) is negatively coupled via G_i protein.

5.2. GHRH, Somatostatin, and GH Pulses

Each pulse of GH is the result of stimulation by a pulse of GHRH. Between pulses, GH secretion is suppressed by inhibiting actions of somatostatin both at the anterior pituitary and at the GHRH neurons (Fig. 4 *[52]*). Thus, GHRH neurons receive somatostatin synapses and express somatostatin receptors (sst1 and 2 *[51]*). Somatostatin neurons are stimulated by GH itself and by IGF-I, secreted by the liver in response to GH. Thus, somatostatin neurons are stimulated as an indirect result of the burst of activity of GHRH neurons that produces a GH pulse, and they may also be more directly stimulated by GHR. This stimulation of somatostatin neurons reimposes inhibition of GHRH neurons, terminating the GH pulse. Stimulation of the somatostatin neurons by GH and IGF-I slowly wanes as their levels decrease, and consequently somatostatin inhibition is withdrawn, permitting another GHRH and GH pulse. Pulses in females are less clear, and smaller than in males, perhaps the result of sexually dimorphic behavior of the somatostatin neurons. The evidence for the generation of a GH pulse from this neuroendocrine network comes from studies using systemic infusions of the secretagogues and antisera, and from electrophysiological experiments *(52, 53)*.

This framework provides for the generation of basal pulsatile GH secretion, upon which many signals related to metabolism act to modify GH secretion; these include stimulation by hypoglycaemia or fasting, aerobic exercise, but also by sleep, and inhibition by increased blood glucose and fatty acids. Inputs to GHRH and somatostatin neurons include cholinergic, serotoninergic, opioidergic, and noradrenergic projections. A part of the effects of fasting is mediated by circulating ghrelin, secreted by an empty stomach. Ghrelin acts directly on somatotrophs, through distinct receptors (GH secretagogue receptor [GHSR]) to stimulate GH secretion, but it also acts in the hypothalamus on GHRH neurons *(54)*, which express the GHSR. Synthetic compounds acting as agonists on the GHSR may be useful in the treatment of GH deficiency where the somatotrophs are responsive, including in old age *(55)*.

6. OBESITY: A HYPOTHALAMIC NEUROENDOCRINE NETWORK PROBLEM

The regulation of both the HPA axis and GH secretion contributes to the control of metabolism. The control of food intake is clearly a major element in metabolic regulation, and excess energy storage in particular, and consequently in the imbalance in this control as seen in obesity. The neuroendocrine network regulating metabolism is being defined through the discovery of peptides signaling about metabolic state from the periphery to the brain. This network includes the PVN, the GH mini-network, the arcuate nucleus, the hypothalamic ventromedial nucleus, and the dorsomedial and lateral hypothalamic areas.

6.1. The Scale of Obesity

The human species is getting fatter and fatter; obesity is a growing global health problem *(56)*. In the United States, the 1999–2000 National Health and Nutrition Examination Survey (NHANES) *(see* ref. *57)* reported that 64% of US adults (and 15% of children aged 6–11) are overweight, with a body mass index (BMI) exceeding 25 kg/m^2. According to The World Health Organization, more than 1 billion people living today are overweight, and at least 300 million of these are clinically obese.

A BMI above 40 kg/m^2 carries a 90% risk of developing type 2 diabetes, and in the United States, one out of every five adults is now insulin-resistant. Obesity and diabetes account for between 5 and 8% of all health care expenditures in the United States, and for 39 million workdays lost annually. In Europe, the number of adults with diabetes is predicted to rise from 26.5 million in 2000 to 32.9 million by 2010. Many obese individuals die prematurely from coronary and cerebrovascular disease, have increased risk of some cancers including especially large-bowel cancers, and suffer reduced quality of life from musculoskeletal and gastrointestinal disease, depression, social isolation, and poor mobility. Obese people spend more on health care and on medications than non-obese people; they earn less, are more likely to be unemployed, are less likely to be promoted in their jobs, and are at greater risk of marital breakdown and clinical depression. The Rand Institute reported that obesity is more strongly linked to chronic diseases than poverty, smoking, or drinking alcohol, and this report equated being obese with *aging by 20 yr.*

Until recently, obesity was widely regarded as a "disease of choice." The common presumption was that, with better education, sound nutritional advice, and a little willpower, encouraged by environmental and lifestyle initiatives, obesity could be controlled. However, neither dieting nor lifestyle changes have proved effective in achieving sustained weight reduction in more than a small minority of subjects.

Powerful forces defend body fat mass against imposition of negative energy balance. With chronic overweight, the brain interprets energy restriction as a threat to survival. If energy intake is reduced, then hypothalamic control systems respond by reducing metabolic rate, and by diverting energy utilization into maintenance of body composition. The obese subject on enforced calorie restriction is, physiologically, in a state of energy starvation. Metabolic rate slows, and the immune system is depressed; diet-induced weight loss results in increased vulnerability to infection, in feeling cold, lethargic and depressed, and in a powerful drive to eat that ultimately is difficult to resist because the drive reflects a powerful perceived physiological imbalance.

Current pharmacological treatments are scarcely more effective than dieting. Some therapies target the noradrenergic and serotonergic systems of the brain, which are widely distributed, diffuse systems that influence many different physiological functions. These drugs are limited in efficacy, and, perhaps predictably considering the targets, have many side effects. Other potentially more selective candidates have poorly sustained efficacy: the neuronal networks that govern appetite and energy balance have multiple homeostatic control mechanisms—thus induced imbalance often results in compensatory counterregulation within the networks.

Yet these problems may not be intractable. Although body weight is powerfully defended, in some physiological circumstances the set point of the control mechanisms can change. For instance, pregnancy is associated with a reversible increase in adiposity, and in seasonal animals such as Siberian hamsters, appetite and body weight follow a seasonal pattern. Moreover, for severely obese patients, the last resort of bariatric surgery (such as gastric bypass) is very effective in most cases at inducing a sustained weight loss accompanied by a normalization of appetite. Because the loss of weight and appetite are sustained, the surgery evidently does indeed re-set the set point for energy balance. Finally, there is overwhelming evidence that human obesity is influenced by variations in the structure and expression of genes that encode molecules involved in the control of energy homeostasis.

Recent research has identified many new factors that play specific roles in the regulation of energy homeostasis, and many of these signals converge on specific neuronal networks in the hypothalamus. Although it has long been known that the hypothalamus includes populations of neurons that respond to glucose availability, we now know that many circulating factors signal aspects of energy balance. Exactly how these signals from the stomach, fat stores and pancreas are regulated, and how they interact, is unclear. The role of the many newly identified ligands that are specifically expressed in discrete subsets of hypothalamic neurons is also unclear, as are the pathways by which the hypothalamus regulates energy balance and adipocyte function. However, the discrete expression of new candidate drug targets and the specificity of their physiological roles invites optimism. From an understanding of these regulatory networks, and from understanding the mechanisms of action of these specific factors, guided by discoveries of linkage between genes and body composition in humans and by studies of mutagenesis in animal models, we may find new targets that will enable the pharmacological manipulation of the set point for energy homeostasis.

6.2. Present State of Knowledge

Since the discovery of leptin in 1994 *(58)* and ghrelin in 1999 *(59)* there has been a wavefront of published data on these and other systems reflecting a large number of independent studies conducted in diverse laboratories around the world. This account aims to describe the scheme for the neuroendocrine network regulating metabolism as presently indicated by these findings.

6.2.1. Leptin and Adipose Signaling to the Hypothalamus

Information about fat mass is signaled to the hypothalamus by leptin, an adipose-derived peptide that is present in the circulation at concentrations proportional to the total body fat mass. Leptin is transported across the BBB into the brain by a dedicated, saturable transport system, and acts at specific receptors expressed on discrete populations of central neurons, including cell groups in the hypothalamus and brainstem. Leptin was discovered by positional cloning of the *obesity* (*ob*) locus in 1994, and the leptin receptor was cloned the following year. Leptin receptors are products of a single gene that occur as several variants; a short form of the leptin receptor is involved in transport of leptin into the brain at the choroid plexus and possibly elsewhere, whereas the long forms act as membrane receptors. Leptin is not the only signal that reaches the brain at levels proportional to fat mass, the hypothalamus also contains receptors for insulin, which is produced by pancreatic β-cells. Insulin secretion is influenced by the amount of stored fat, with the consequence that total daily insulin secretion is proportional to stored fat. However, insulin secretion is much more dynamic than that of leptin, and fluctuations of insulin concentrations, with food intake and exercise, provide the brain with minute-by-minute information about glucose availability. When administered directly into the brain, both leptin and insulin are effective in reducing food intake. Leptin is also effective when given systemically, but insulin given systemically produces a hypoglycaemia, which stimulates appetite.

Congenital leptin deficiency leads to severe obesity in humans *(60)* and recombinant leptin is a highly successful treatment for affected children. Deficiencies in leptin itself or its receptor are, however, very rare in humans—indeed, obese humans are generally characterised by high-circulating concentrations of leptin, leading to the current theory that obesity reflects a state of leptin resistance, that may be considered analogous to insulin resistance as a cause of type 2 diabetes.

Studies with leptin and other circulating hormones, in particular insulin, have thus unmasked a neuroendocrine network that intercommunicates with peripheral organs as well as governing central circuits controlling energy balance. Tellingly, insulin and leptin signals converge on a discrete population of neurons in the hypothalamus—and in those neurons insulin and leptin converge on *the same intracellular signaling cascades,* indicating that there may be a basis at this level for the analogy between insulin resistance and leptin resistance.

Among the neurons that express receptors for leptin are several populations of neurons in the hypothalamus, including populations previously linked with the regulation of appetite. Experiments involving discrete lesions of hypothalamic areas, performed mainly in the 1950s, produced changes in feeding behavior that led to the characterization of the ventromedial nucleus (VMN) as a "satiety center," and the lateral hypothalamic area (LHA) as a "feeding center." Two other areas of the hypothalamus are now conspicuously implicated in appetite regulation: the arcuate nucleus and the PVN, which was previously discussed.

6.2.2. The Arcuate Nucleus of the Hypothalamus

Apart from containing GHRH neurons, the arcuate nucleus also contains several neuronal populations with different roles in the control of feeding. The ventromedial portion of this nucleus contains a large population of centrally projecting neurons that co-synthesize neuropeptide Y (NPY), agouti-related peptide (AGRP), and the inhibitory neurotransmitter GABA. These neurons express receptors for both leptin and ghrelin (Fig. 5A). NPY is a highly potent orexigen when administered centrally, and NPY circuits are widely believed to play an important role in appetite regulation; however NPY knockout mice show a normal body composition phenotype indicating redundancy in the control mechanisms.

The more dorsolateral portion of the arcuate nucleus contain neurons that synthesize POMC derived peptides, and one of these, α-MSH, is a highly potent inhibitor of appetite, and believed to mediate satiety (Fig. 5B *[61]*). This action of α-MSH is mediated primarily through MC4Rs (two of the five cloned MRs, MC3 and MC4, show significant expression in brain). The MC4R is currently the best-established link between genetic variance and severe obesity in man *(62)*. Central administration of α-MSH or of specific MC4 agonists suppresses food intake and produces weight loss in rodents, and transgenic mice with global disruption of the MC4R display increased food intake and body weight. The expression of POMC messenger RNA (mRNA) in the arcuate nucleus is reduced in fasted rats, and centrally administered leptin and insulin both increase POMC mRNA expression in the arcuate nucleus, whereas

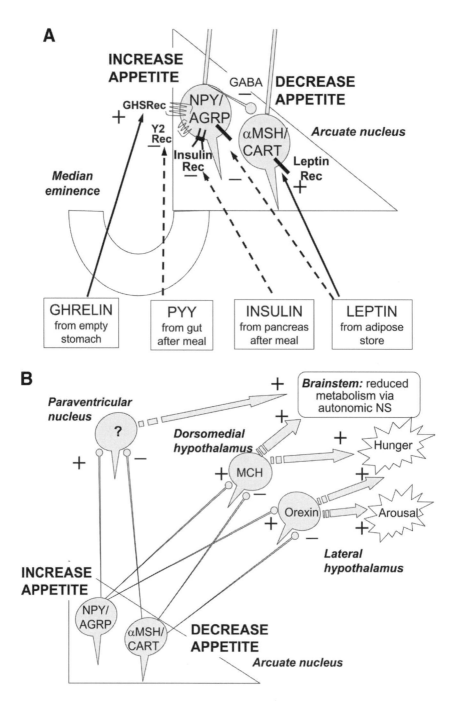

Fig. 5. Metabolic signaling and processing in the hypothalamus. (**A**) Systemic signals to the arcuate nucleus. Neurons producing neuroptide Y (NPY) and agouti-related peptide (AGRP), which both increase appetite, are *stimulated* by ghrelin acting on growth hormone secretagogue (GHS) receptors. NPY/AGRP neurons are *inhibited* by peptide YY (PYY; via Y2 receptors), insulin and leptin, signaling replenished energy reserves. Neurons producing α-melanocyte stimulating hormone (α-MSH) and cocaine and amphetamine-regulated transcript (CART), which both decrease appetite, are *stimulated* by leptin, but *inhibited* by NPY/AGRP neurons via their γ-aminobutyric acid (GABA)-producing axon

studies in vitro have shown that local application of leptin increases the electrical activity of arcuate POMC neurons *(63)*. Thus, it is currently hypothesized that leptin and insulin reduce food intake as a result of increased central release of α-MSH, signaling through central MC4Rs in the neuroendocrine network regulating metabolism *(see* refs. *48,64,65)*

The catabolic actions of both leptin and insulin can be ameliorated by pretreatment with MC4R antagonists. Interestingly, AGRP, co-produced in NPY neurons of the arcuate nucleus, also acts at the MC4R, where it is a highly potent competitive antagonist of the actions of α-MSH. In fasted rats, and in leptin-deficient mice, AGRP mRNA expression is increased. Central administration of AGRP increases food intake, and this response can last up to 6 d. However, genetic disruption of AGRP results in little or no change in food-intake and weight-gain parameters, again indicating redundancy. Like POMC neurons, the AGRP–NPY neurons in the arcuate nucleus are direct targets of leptin; in this case, the actions of leptin are inhibitory.

6.2.3. Other Hypothalamic Sites Implicated in Appetite Regulation

Dorsal to the arcuate nucleus is a band of neurons that express ghrelin *(66)*—an important factor secreted from the gastrointestinal tract, but also expressed in this discrete neuronal location. Above this band lies the VMN, which contains subpopulations of neurons that are sensitive to leptin, ghrelin, glucose, and insulin, as well as to the centrally released neuropeptides. Dorsal to the VMN, the dorsomedial hypothalamus includes neurons that synthesise melanin-concentrating hormone, and the adjacent LHA contains a population of neurons that express orexins, peptides named for their conspicuous effects on feeding. Lesions of the VMN induce hyperphagia, whereas lesions of the LHA produce hypophagia.

Outputs from the PVN include several pathways involved in energy regulation. The PVN is a mediator of autonomic regulation of metabolism, as well as acting via its control of the HPA axis. The PVN receives inputs from the arcuate nucleus, and activation of CRH neurons and the HPA axis by hunger signals (hypoglycaemia, intrahypothalamic NPY, orexin, ghrelin) is consistent with the metabolic actions of glucocorticoid. Centrally projecting neurons originating in the PVN include a population of neurons that project to the brainstem to control gastric and intestinal motility. Other paraventricular neuroendocrine neurons regulate the secretion of thyroid-stimulating hormone from the anterior pituitary gland, to regulate basal metabolic rate.

Fig. 5 *(continued from facing page)* collaterals. The relative activity of the orexigenic (NPY/AGRP) and anorexigenic (α-MSH/CART) neurons results from the net effects of the systemic signals. In addition, ghrelin is also produced by dorsal arcuate neurons, and stimulates GHRH neurons, so increasing GH secretion *(see* Fig 4). (**B**) Arcuate nucleus signaling in the neuroendocrine network regulating eating. Arcuate neurons project: to the PVN, regulating metabolism via autonomic outflow and influencing appetite (possibly involving CRH and oxytocin); and to melanocyte concentrating (MCH) and orexin neurons, which both increase eating (through incompletely defined routes), whereas MCH reduces metabolism, and orexin arouses activity. NPY excites MCH and orexin neurons, and reduces metabolism. α-MSH and CART have opposite actions to NPY. Notably, α-MSH acts via melanocortin-4 receptors (MC4R), and AGRP is a MC4R antagonist.

6.2.4. Ghrelin, and Other Signals From the Gastrointestinal Tract

Hunger and satiety are driven by endocrine factors from the stomach and gastrointestinal tract, most notably ghrelin and peptide YY (PYY [3-36]) and by insulin, released from the pancreas in response to food ingestion. Ghrelin and PYY(3-36) act on specific receptors expressed discretely on populations of hypothalamic neurons, interacting with leptin and insulin-responsive circuits.

The origins of ghrelin research derive from studies of synthetic secretagogues for GH (now known to be ghrelin mimetics), which act on the somatotrophs but not through the GHRH receptor. These compounds also have potent, selective and direct actions on the hypothalamus *(67)*. In 1996, the receptor for these compounds was cloned, and in 1999, an endogenous ligand for this receptor, ghrelin, was discovered *(59)*, and found to be secreted from the stomach in very high concentrations during food restriction. Ghrelin was then found to be a powerful orexigenic agent—the most potent such agent to date *(68)*—and it is also a potent stimulator of adiposity. It is possible that decreased ghrelin section may contribute to the success of bariatric surgery (e.g., gastric bypass). Another gut-derived hormone PYY(3-36), secreted by the intestine after a meal, seems to act in a counterregulatory manner to ghrelin: PYY(3-36) powerfully inhibits feeding by acting on the hypothalamic Y2 receptor, which is an inhibitory autoreceptor for NPY located on NPY neurons *(69)*.

6.2.5. Other Signals: Exercise and IL-6

Exercise and energy utilization also signal back to the hypothalamus to regulate energy homeostasis. These pathways are poorly understood, but the current best candidate for mediating this feedback is the muscle-derived cytokine IL-6, circulating levels of which are increased 100-fold during exercise. Central administration of IL-6 leads to increased energy expenditure and transgenic mice deficient in IL-6 develop obesity *(70)*.

7. STRATEGIES FOR THE FUTURE: WHERE WILL OUR NEW DRUGS COME FROM?

7.1. Neuroendocrine Networks and Obesity

The Human Genome Project has provided vast raw material for identifying genes that contribute to malfunctioning of neuroendocrine networks regulating stress responses and energy balance. Taking obesity as an example, candidate genes for future investigation may be chosen based on their known biological functions, and on their involvement in processes regulating energy balance, but, increasingly, data mining activities are exploiting the vast genomic, expression, and positional information now available to explore genes with more subtle links to obesity. Rational choice of genes to study allows emphasis to be placed on genes of potential pharmaceutical relevance—those with discrete sites of expression, specific functional relevance, and potentially accessible to pharmacological manipulation. Human genetic studies now include high-throughput molecular scanning studies in extreme phenotypes (cohorts of both lean and obese individuals), and large case-control SNP-based association studies, and also allow the study of gene–environment interaction.

The objective of genetics studies may be said to be identification of candidate *molecular drug targets* that is, molecular sites of potential therapeutic intervention in future treatments for obesity. Such targets may be, for instance, molecules involved in the transmission of information relating to energy usage and expenditure from the periphery to the brain (such as ghrelin, leptin, and IL-6), signaling molecules involved in the integration of this information in neuronal networks (such as NPY, AGRP, and α-MSH); or molecules involved in the regulation of energy expenditure (such as glucocorticoids and thyroid hormone). Receptors for these molecules include *nuclear* and *membrane* receptors, or molecules involved in intracellular signaling pathways used to transduce this information at specific sites. *Target identification* may involve (a) human *molecular genetics* approaches, to identify gene mutations associated with relevant body composition phenotypes (e.g., ref. *62*); and (b) *mutagenic techniques* in rodents to generate mutants with alterations in body fat mass. The outcome of mutagenesis is a specified gene mutation with a demonstrated association to an obesity-related phenotype. This association can be tested experimentally by targeted mouse mutagenesis (conditional, cell-specific gene disruption or over-expression).

Few genes that have a specific association to obesity are likely to be suitable candidates for intervention. Some genes may influence phenotype through actions only in early development, others may have widespread effects associated with weak specificity of expression or function, and others may have specific effects on phenotype but which are associated with dysregulation of other physiologically important functions.

7.1.1. Target Validation

This involves establishing how a gene mutation gives rise to altered body composition. Specifically, this involves establishing the *site of expression* of the target molecules, and their *functional role* within pathways that regulate energy balance. For targets expressed in the hypothalamus, target validation would involve mapping the distribution of mRNA expression, relating this to the pathways known to be involved in regulation of energy balance, and establishing the functional role of the encoded proteins at a cellular level.

Because of the neuroanatomical complexity of the hypothalamus, patterns of co-expression of several signaling molecules must be established using techniques for simultaneous visualization of multiple signals either through immunocytochemistry or through *in situ* hybridization. Recently, this has been greatly facilitated by the generation of transgenic mice expressing reporter constructs that allow direct visualization sites of specific mRNA expression, for example using green fluorescent protein to visualize neurons that express the MC4R *(71)*.

Even full biochemical characterization is insufficient, because to relate neuronal populations to other known populations requires combining biochemical identification with anatomical specification. Currently this is usually achieved by the use of retrograde-tracing molecules such as Fluorogold or fluorescent-labeled microspheres injected into discrete brain nuclei. The former diffuse further, so are useful for widespread projection sites, and the latter label more focal projections.

Studying the functional role of specific populations of neurons presents still greater challenges. The approaches include the following:

1. Using transgenic rodent models of under- or overexpression of the target molecule to study effects on phenotype (e.g., refs. *70,72*).
2. Using immediate-early gene mapping to establish the functional involvement of a given pathway in a particular physiological state. For example, neurons in the arcuate nucleus that express NPY are activated by systemically administered ghrelin, as revealed by the induction of expression of the immediate early, transcription factor gene, *c-fos* in these neurons. This is measured as increased c-*fos* mRNA or Fos protein expression by *in situ* hybridization or immunocytochemistry respectively (e.g., ref. *73*)

3. Measuring changes in mRNA expression of the target molecule in defined physiological states. Depending on how restricted the expression of the gene is, the specificity of changes in the target gene might be usefully studied with a *cDNA microarray*. However, there are concerns about quantification and lack of statistical power of this approach, while there is due skepticism about this intrinsically unfocussed strategy, and its problems in validation, reproducibility, and interpretation; but it is potentially useable as an initial screen for genes of possible interest, in the absence of clear leads.

As novel players emerge in the hypothalamic orchestra of energy balance homeostasis, it will be important to establish how expression of these agents is regulated. Laser capture microdissection is a potentially important recent technical development. This allows microdissection of single cells or groups of cells of interest from a hypothalamic section and measurement of expression of specific mRNAs by polymerase chain reaction (PCR).

There are many other penetrating avenues of experimentation available, notably the application of single-cell electrophysiological techniques in vivo and in vitro. These approaches can be very effective at resolving clearly defined questions, such as exactly what information is carried by neurons that project from the arcuate nucleus to the PVN, and how is it encoded and processed?

The activity of single hypothalamic neurons can be recorded either in vivo or in vitro (in, e.g., hypothalamic slices). The in vivo approach (e.g., ref. *67*) offers the advantage of recording the effects of systemic injection or local microdialysis administration of hormones, compounds, and candidate drugs on the firing activity of single neurons, including neuroendocrine cells identified prospectively by electrophysiological criteria according to their projection sites and responses to specific stimuli.

In vitro studies offer the advantage of determining whether single neurons receive and integrate information directly or indirectly from signaling molecules, and these studies can involve analysis of membrane actions using intracellular or patch-clamp recording techniques. Presently recorded neurons are phenotyped retrospectively (by immunocytochemistry or PCR), but prospective definition will be possible as animals with visualisable cell-specific reporters (green fluorescent protein) are generated.

Broadly, in vivo studies will tell about relevance of drug action, and in vitro studies report cellular mechanisms.

8. CONCLUSION

Neuroendocrine networks perform complex regulatory functions that involve processing and modulation of signaling at every point of communication between and within the neurons that comprise the network. Identifying functionally specific and potent signaling molecules in a network, the regulation of their expression, and their mechanisms of action is an aim in developing new drugs to prevent or reverse obesity and its sequelae, and similarly to limit the mental and physical consequences of unremitting stress.

REFERENCES

1. Burbach, J.P.H., Luckman, S.M., Murphy, D. and Gainer, H. (2001) Gene regulation in the magnocellular hypothalamo-neurohypophysial system. *Physiol. Rev.* **81**:1197–1267.
2. Ludwig, M. and Pittman, Q.J. (2003) Talking back: Dendritic neurotransmitter release. *Trends Neurosci.* **26**:255–261.
3. Leng, G., Brown, C.H. and Russell, J.A. (1999) Physiological pathways regulating the activity of magnocellular neurosecretory cells. *Prog. Neurobiol.* **57**:625–655.

4. Russell, J.A. and Leng, G. (1998) Sex, parturition and motherhood without oxytocin? *J. Endocrinol.* **157**:343–359.
5. Russell, J.A., Leng, G. and Douglas, A. J. (2003) The magnocellular oxytocin system, the fount of maternity: Adaptations in pregnancy. *Front. Neuroendocrinol.* **24**:27–61.
6. Ludwig, M., Sabatier N., Bull, P. M., Landgraf, R., Dayanithi, G., Leng ,G. (2002) Intracellular calcium stores regulate activity-dependent neuropeptide release from dendrites. *Nature* **418**, 85–89.
7. Hirasawa, M., Schwab, Y., Sharkey, K.A. and Pittman, Q.J. (2003) Dendritic oxytocin release induces and amplifies endocannabinoid-mediated retrograde transmission in the supraoptic nucleus. *Soc. Neurosci. Abstr.* **29**:709.2.
8. Sabatier, N., Caquineau, C., Dayanithi, G., et al. (2003) α-Melanocyte-stimulating hormone stimulates oxytocin release from the dendrites of hypothalamic neurons while inhibiting oxytocin release from their terminals in the neurohypophysis. *J. Neurosci.* **23**:10351–10358.
9. Brown, C.H., Ludwig, M. and Leng, G. (2004) Temporal dissociation of the feedback effects of dendritically co-released peptides on rhythmogenesis in vasopressin cells. *Neuroscience* **124**:105–111.
10. Bourque, C.W., Voisin, D.L. and Chakfe, Y. (2002) Stretch-inactivated cation channels: Cellular targets for modulation of osmosensitivity in supraoptic neurons. *Prog. Brain Res.* **139**:85–94.
11. Leng, G., Brown, C.H., Bull, P.M., et al. (2001) Responses of magnocellular neurons to osmotic stimulation involves coactivation of excitatory and inhibitory input: An experimental and theoretical analysis. *J. Neurosci.* **21**:6967–6977.
12. Leng, G., Reiff-Marganiec, A., Ludwig, M. and Sabatier, N. (2004) Generating quantitatively accurate, but computationally concise, models of single neurons. In *Computational Neuroscience* (J. Feng, Ed.), Chapman & Hall/CRC Press, London, UK, pp.185–214.
13. Stedronsky, K., Telgmann, R., Tillmann, G., Walther, N. and Ivell, R. (2002) The affinity and activity of the multiple hormone response element in the proximal promoter of the human oxytocin gene. *J. Neuroendocrinol.* **14**:472–485.
14. Somponpun, S., Holmes, M.C., Seckl, J.R. and Russell, J.A. (2004) Modulation of oestrogen receptor-β mRNA expression in rat paraventricular and supraoptic nucleus neurones following adrenal steroid manipulation and hyperosmotic stimulation. *J. Neuroendocrinol.* **109**:472–482.
15. Nomura, M., McKenna, E., Korach, K. S., Pfaff, D. W. and Ogawa, S. (2002) Estrogen receptor-beta regulates transcript levels for oxytocin and arginine vasopressin in the hypothalamic paraventricular nucleus of male mice. *Mol. Brain Res.* **109**:84–94.
16. Koksma, J.-J., Van Kesteren, R. E., Rosahl, T. W., et al. (2003) Oxytocin regulates neurosteroid modulation of GABA$_A$ receptors in supraoptic nucleus around parturition. *J. Neurosci.* **23**:788–797.
17. Lemos, J.R. and Wang, G. (2000) Excitatory versus inhibitory modulation by ATP of neurohypophysial terminal activity in the rat. *Exp. Physiol.* **85(Suppl)**: 67S–74S.
18. Green, L.A., Fein , D., Modahl, C., Feinstein, C., Waterhouse, L. and Morris, M. (2001) Oxytocin and autistic disorder: Alterations in peptide forms. *Biol. Psychiatry.* **50**:609–613.
19. Winslow, J.T. and Insel, T.R. (2002) The social deficits of the oxytocin knockout mouse. *Neuropeptides* **36**:221–229.
20. Choleris, E., Gustafsson, J.-A., Korach, K.S., Muglia, L.J., Pfaff, D.W. and Ogawa, S. (2003) An estrogen-dependent four-gene micronet regulating social recognition: A study with oxytocin and estrogen receptor-α and -β knockout mice. *Proc. Natl. Acad. Sci.U.S.A.* **100**:6192–6197.
21. Swanson, L.W. and Kuypers, H.G. (1980) The paraventricular nucleus of the hypothalamus: cytoarchitectonic subdivisions and organization of projections to the pituitary, dorsal vagal complex, and spinal cord as demonstrated by retrograde fluorescence double-labeling methods. *J. Comp. Neurol.* **194**:555–570.
22. Sawchenko, P.E., Brown, E.R., Chan, R. K., et al. (1996) The paraventricular nucleus of the hypothalamus and the functional neuroanatomy of visceromotor responses to stress. *Prog. Brain Res.* **107**:201–222.

23. Di. S., Malcher-Lopes, R., Halmos, K.C. and Tasker, J.G. (2003) Nongenomic glucocorticoid inhibition via endocannabinoid release in the hypothalamus: A fast feedback mechanism. *J. Neurosci.* **23**:4850–4857.

24. Kovacs, K.J and Sawchenko, P.E (1996) Sequence of stress-induced alterations in indices of synaptic and transcriptional activation in parvocellular neurosecretory neurons *J. Neurosci.* **16**:262–273.

25. Herman, J.P., Figueiredo, H., Mueller, N.K., et al. (2003) Central mechanisms of stress integration: Hierarchical circuitry controlling hypothalamo-pituitary-adrenocortical responsiveness. *Front. Neuroendocrinol.* **24**:151–180.

26. Blackburn-Munro, G. and Blackburn-Munro, R.E. (2001) Chronic pain, chronic stress and depression: Coincidence or consequence? *J. Neuroendocrinol.* **13**:1009–1023.

27. Rivest, S. (2001) How circulating cytokines trigger the neural circuits that control the hypothalamic–pituitary–adrenal axis. *Psychoneuroendocrinology* **26**:761–788.

28. Ericsson, A., Arias, C. and Sawchenko, P.E. (1997) Evidence for an intramedullary prostaglandin-dependent mechanism in the activation of stress-related neuroendocrine circuitry by intravenous interleukin-1. *J. Neurosci.* **17**:7166–7179.

29. Buller, K.M., Crane, J.W. and Day, T. A. (2001) The central nucleus of the amygdala: A conduit for modulation of HPA axis responses to an immune challenge? *Stress* **4**:277–287.

30. Wigger, A., Sanchez, M.M., Mathys, K.C., et al. (2004) Alterations in central neuropeptide expression, release, and receptor binding in rats bred for high anxiety: critical role of vasopressin. *Neuropsychopharmacology* **29**:1–14.

31. Serradeil-Le Gal, C., Derick, S., Brossard, G., et al. (2003) Functional and pharmacological characterization of the first specific agonist and antagonist for the V1b receptor in mammals. *Stress* **6**:199–206.

32. Habib, K.E., Weld, K.P., Rice, K.C., et al. (2000) Oral administration of a corticotropin-releasing hormone receptor antagonist significantly attenuates behavioral, neuroendocrine, and autonomic responses to stress in primates. *Proc. Natl. Acad. Sci. U.S.A.* **97**:6079–6084.

33. Muller, M.B., Uhr, M., Holsboer, F. and Keck, M.E. (2004) Hypothalamic-pituitary-adrenocortical system and mood disorders: highlights from mutant mice. *Neuroendocrinology* **79**:1–12.

34. Goldstein, D.S. and McEwen, B. (2002) Allostasis, homeostats, and the nature of stress. *Stress* **5**:55–58.

35. De Kloet, E. R. (2003) Hormones, brain and stress. *Endocr. Regul.* **37**: 51–68.

36. Joels, M. (2001) Corticosteroid actions in the hippocampus. *J. Neuroendocrinol.* **13**:657–669.

37. Lim, M. C., Shipston, M.J. and Antoni, F. A. (2002) Posttranslational modulation of glucocorticoid feedback inhibition at the pituitary level. *Endocrinology* **143**:3796–3801.

38. Harris, H.J., Kotelevtsev, Y., Mullins, J. J., Seckl, J.R. and Holmes, M.C. (2001) Intracellular regeneration of glucocorticoids by 11beta-hydroxysteroid dehydrogenase (11β-HSD)-1 plays a key role in regulation of the hypothalamic-pituitary-adrenal axis: Analysis of 11β-HSD-1-deficient mice. *Endocrinology* **142**:114–120.

39. Jessop, D.S. (1999) Central non-glucocorticoid inhibitors of the hypothalamo-pituitary- adrenal axis. *J.Endocrinol.* **160**:169–180.

40. Bale, T.L. and Vale, W.W. (2004) CRF and CRF Receptors: role in stress responsivity and other behaviors. *Annu. Rev. Pharmacol. Toxicol.* **44**:525–557.

41. Ma, X.-M. and Lightman, S. L. (1998) The arginine vasopressin and corticotrophin-releasing hormone gene transcription responses to varied frequencies of repeated stress in rats. *J . Physiol.* **510**:605–614.

42. Welberg, L.A.M. and Seckl J.R. (2001) Prenatal stress, glucocorticoids and the programming of the brain. *J. Neuroendocrinol.* **13**, 113–128.

43. Liu, D., Caldji, C., Sharma, S., Plotsky, P.M. and Meaney, M.J. (2000) Influence of neonatal rearing conditions on stress-induced adrenocorticotropin responses and norepinepherine release in the hypothalamic paraventricular nucleus. *J. Neuroendocrinol.* **12**:5–12.

44. Andrews, M.H. and Matthews, S.G. (2004) Programming of the hypothalamo-pituitary-adrenal axis: serotonergic involvement. *Stress* **7**:15–27.

45. Douglas, A.J., Johnstone, H.A., Wigger, A., Landgraf, R., Russell, J.A. and Neumann I. D. (1998) The role of endogenous opioids in neurohypophysial and hypothalamo-pituitary-adrenal axis hormone secretory responses to stress in pregnant rats. *J. Endocrinol.* **158**:285–293.

46. Brunton, P.J., Harrison, C.E.L. and Russell, J.A. (2004) Allopregnanolone is involved in reduced HPA axis responses to immune challenge in late pregnancy. *Endocr. Abstr.* **7**, OC1

47. Lund, T.D., Munson, D.J., Haldy, M.E. and Handa, R.J. (2004) Androgen inhibits, while oestrogen enhances, restraint-induced activation of neuropeptide neurons in the paraventricular nucleus of the hypothalamus *J. Neuroendocrinol.* **16**:272–278.

48. Schwartz, M.W., Woods, S.C., Porte, D. Jr., Seeley, R.J. and Baskin, D.G. (2000) Central nervous system control of food intake. *Nature* **404**:661–671.

49. Balthasar, N., Mery, P.-F., Magoulas, C.B., et al. (2003) Growth hormone-releasing hormone (GHRH) neurons in GHRH-enhanced green fluorescent protein transgenic mice: A ventral hypothalamic network. *Endocrinology* **144**:2728–2740.

50. Gaylinn, B.D. (2002) Growth hormone releasing hormone receptor. *Receptors & Channels* **8**:155–162.

51. Zhang, W.-H., Beaudet, A. and Tannenbaum, G. S. (1999) Sexually dimorphic expression of sst1 and sst2 somatostatin receptor subtypes in the arcuate nucleus and anterior pituitary of adult rats. *J. Neuroendocrinol.* **11**:129–136.

52. Robinson, I.C.A.F., Jeffery, S. and Clark, R.G. (1990) Somatostatin and its physiological significance in regulating the episodic secretion of growth hormone in the rat. *Acta Paediatr. Scand. Suppl.* **79**:87–92.

53. Dickson, S.L., Leng, G. and Robinson, I.C.A.F. (1994) Electrical stimulation of the rat periventricular nucleus influences the activity of hypothalamic arcuate neurones. *J. Neuroendocrinol.* **6**:359–367.

54. Tannenbaum, G.S., Epelbaum, J. and Bowers, C.Y. (2003) Interrelationship between the novel peptide ghrelin and somatostatin/growth hormone-releasing hormone in regulation of pulsatile growth hormone secretion. *Endocrinology* **144**:967–974.

55. Merriam, G. R., Schwartz, R.S. and Vitiello, M.V. (2003) Growth hormone-releasing hormone and growth hormone secretagogues in normal aging. *Endocrine.* **22**:41–48.

56. Hill, J.O., Wyatt, H.R., Reed, G.W. and Peters, J.C. (2003) Obesity and the environment: Where do we go from here? *Science* **299**:853–855.

57. U.S. Department of Health and Human Services (2001). *The Surgeon General's Call To Action To Prevent and Decrease Overweight and Obesity* (Rockville, MD). <http://www.surgeongeneral.gov/topics/obesity/calltoaction/toc.htm>

58. Zhang, Y.Y., Proenca, R., Maffei, M., Barone, M., Leopold, L., and Friedman, J.M. (1994) Positional cloning of the mouse obese gene and its human homolog. *Nature* **372**:425–432.

59. Kojima, M., Hosoda, H., Date, Y., Nakazato, M., Matsuo, H., and Kangawa, K. (1999) Ghrelin is a growth-hormone-releasing acylated peptide from stomach. *Nature* **402**:656–660.

60. Montague, C.T., Farooqi, I.S., Whitehead, J.P., et al. (1997) Congenital leptin deficiency is associated with severe early-onset obesity in humans. *Nature* **387**:903–908.

61. Fan, W., Boston, B., Kesterson, R., Hruby, V. and Cone, R. Role of melanocortinergic neurons in feeding and the agouti obesity syndrome. *Nature* **385**:165–168 (1997).

62. Farooqi, I.S., Keogh, J.M., Yeo, G.S.H., Lank, E., Cheetham, C.H., and O'Rahilly, S. (2003) Clinical spectrum of obesity and mutations in the melanocortin 4 receptor gene. *N. Engl. J. Med.* **348**:1083–1093.

63. Cowley, M.A., Smart, J.L., Rubinstein, M., et al. (2001) Leptin activates anorexigenic POMC neurons through a neural network in the arcuate nucleus. *Nature* **411**:480–484.

64. Berthoud, H.R. (2002) Multiple neural systems controlling food intake and body weight. *Neurosci. Biobehav. Rev.* **26**:393–428.

65. Grill, H.J. and Kaplan, J.M. (2002) The neuroanatomical axis for control of energy balance. *Front. Neuroendocrinol.* **23**:2–40.

66. Cowley, M.A., Smith, R.G., Diano, S., et al. (2003) The distribution and mechanism of action of ghrelin in the CNS demonstrates a novel hypothalamic circuit regulating energy homeostasis. *Neuron* **37**:550–553.

67. Dickson, S.L., Leng, G., and Robinson, I.C.A.F. (1993) Systemic administration of growth hormone-releasing peptide activates hypothalamic arcuate neurons. *Neuroscience* **53**:303-306.

68. Tschöp, M., Smiley, D.L. and Heiman, M.L. (2000) Ghrelin induces adiposity in rodents. *Nature* **407**:908–913.

69. Batterham, R.L., Cowley, M.A., Small, C.J., et al. (2002) Gut hormone PYY3-36 physiologically inhibits food intake. *Nature* **418**:650–654.

70. Wallenius, V., Wallenius, K., Ahren, B., et al. (2002) Interleukin-6-deficient mice develop mature-onset obesity. *Nat. Med.* **8**:75–79.

71. Liu, H., Kishi, T., Roseberry, A.G., et al. (2003) Transgenic mice expressing green fluorescent protein under the control of the melanocortin-4 receptor promoter. *J. Neurosci.* **23**:7143–7154.

72. Huszar, D., Lynch, C. A., Fairchild-Huntress, V., et al. (1997) Targeted disruption of the melanocortin-4 receptor results in obesity in mice. *Cell* **88**:131–141.

73. Hewson, A.K., Tung, L.Y.C., Connell, D.W., Tookman, L. and Dickson, S. L. (2002) The rat arcuate nucleus integrates peripheral signals provided by leptin, insulin, and a ghrelin mimetic. *Diabetes* **51**:3412–3419.

6
Physiology and Behavior

Energy Balance

Michel Cabanac

1. A LIVING ORGANISM IS AN OPEN SYSTEM

Living organisms are not closed systems but are open to the environment. At each instant they receive energy and matter from the environment, and lose to the environment an equal flow of the same. The very process of life entails this exchange with the environment. Without these in-and-out energy and matter flows, life is not possible. The water tank of Fig. 1 represents any of these flows that traverse the living body.

In Fig. 1 a tank containing water permanently receives and looses in-and-out equal flows of water. The tank in the figure is a model of these flows that traverse the body and its subcompartments, consisting of water, energy, or mass of any of the constituents of the body, such as carbon, nitrogen, sodium, calcium, and so on. The inflow demonstrates the rate of intake of these elements, and the outflow depicts the rate of output leaving the body. In the model, the mass of water in the tank is analogous to the amount of heat, water, nitrogen, glucose, sodium, and the like, received from the environment and stored in the body, and h, the level of the water in the tank, indicates the tension reached by the variable in question (water, nitrogen, glucose, sodium, etc.). Thus, the level of the water in the tank is analogous to the concentration of solutes, the temperature, the pressure, and so on, achieved in the body or in its subcompartments.

A proper functioning of the water tank implies a modulation of the inflow and/or outflow faucet as, for example, achieved in Fig. 2.

In Fig. 2, the water tank is also equipped with sensors of the water level, the floats, a negative feedback loop controlling the inflow, and a positive feed-forward loop controlling the outflow (In this chapter the words "control" and "regulation" are in no way synonymous. They are used according to Brobeck's definitions [1]). Any rise of h will tend to retroact negatively and reduce the inflow (negative feedback), or to anteact positively and increase the outflow (positive feed-forward). Reciprocally, any drop of h will retroact negatively and increase the inflow or will anteact positively and decrease the outflow. The water tank thus regulates its level and this regulation is achieved through the control of the interface with its environment.

From: *Integrative Physiology in the Proteomics and Post-Genomics Age*
Edited by: W. Walz © Humana Press Inc., Totowa, NJ

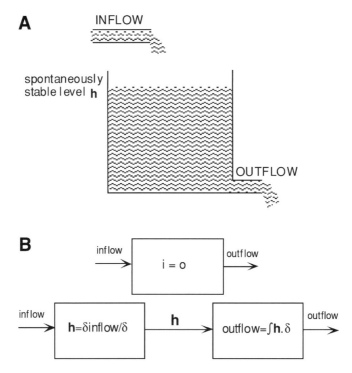

Fig. 1. Living animals are analogous to the above tank. They are open systems in a steady state receiving a continuous inflow of matter and energy, and loosing an equal outflow of matter and energy. The black-box block diagram describes the system and analyzes the system with its input, and output faucets related by the level h, which is a derivative of inflow and an integral of outflow. (From ref. *84*.)

The water tank is a strict analogue of the body. The various regulations control directly or indirectly the various inflows and outflows entering and leaving the body, and maintain the *milieu intérieur* nearly constant. It is through the control of the interface between the body and the environment that these regulations are achieved. The control of these flows is essentially behavioral because it involves the environment; it is physiological, but not in the classical acceptation of the word, which implies ''beneath the skin.'' This chapter deals with the exchange of energy between body and environment.

The control of behavioral exchange of energy between the body and its environment takes place both in the short term, as heat, and in the long term, as chemical potential energy.

2. BEHAVIORAL SHORT-TERM CONTROL OF EXCHANGES: TEMPERATURE REGULATION

In the short term, the energy exchanged with the environment is heat and the regulated variable of the *milieu intérieur* is core body temperature. Heats flows, both inward and outward, are tightly controlled through behavior. Several types of behavior are used in the animal kingdom to achieve the balance between in-and-out heat flows.

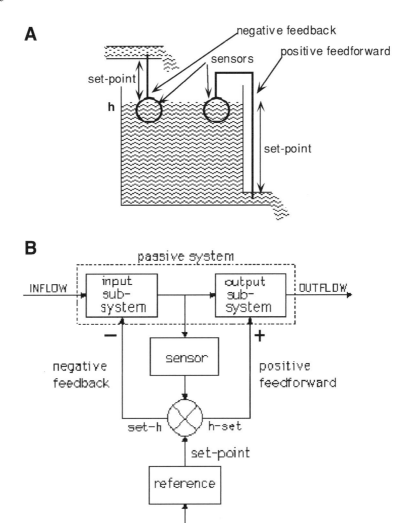

Fig. 2. (**A**) The steady state becomes a regulation, when equipped with a sensor (float) of the regulated variable (h), a negative feedback modulating the input flow, and a positive feed-forward modulating the output flow. The set-point of the system is a built-in property incorporated in the length of the shafts between float and input and output faucets. (**B**) The new diagram proposed for regulation. This underlines the flow of energy and matter through the system, that exists already in the steady state. In addition, it shows that the negative feedback controlling the input is anatomically different from the positive feed-forward on the output. Both tend to keep steady the regulated variable h. (From ref. *84*.)

2.1. The Various Behavioral Responses to Ambient Temperature

2.1.1. Postural Adaptation

Among the simplest behaviors, postural adaptation is already quite efficacious when applied to the flux of energy of solar radiation. All terrestrial animals position their body

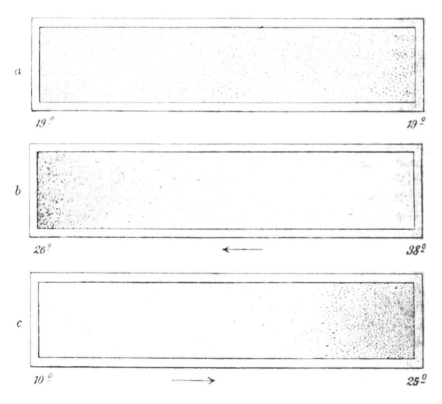

Fig. 3. (**A**) A population of *Paramecium aurelia* placed in a water bath at an even temperature of 19°C, occupies the whole space. (**B**) However, when a temperature gradient is present they move to avoid heat (38°C) and (**C**) cold (10°C below). The animals' thermopreferendum is at 25–26°C. The arrows, present in the original figure, indicate the Paramecia migrations. (Adapted from ref. *3*.)

parallel to the solar beams when their body temperature is high, and perpendicular when their body temperature is low. In a dry climate, the sun irradiates about 1 kW.m^{-2}. Such a heat flux, when applied to the wings of a butterfly, is sufficient to warm the animal in seconds. Even when the mass–area ratio is larger, as in big mammals, simple postural adaptation is used to modulate radiative heat gain (e.g., in humans the irradiated area changes by a factor of 3 when a subject crouches or faces the sun standing with spread limbs).

2.1.2. Migration

Migration can be understood in a classical sense, a seasonal continental displacement by flocks of birds and herds or groups of mammals that migrate to seek a more favorable environment with approaching winter. Examples are the migrations of passerine birds, geese, and caribou. Migration, however, can also be understood as more modest displacement when animals seek a barrier or a shelter to protect them from a cold wind or from solar beams *(2)*. In all cases of migration locomotion is a behavioral means to satisfy a physiological need. Modest crawling in a temperature gradient *(3)* is a type of migration (Fig. 3). The difference between unicellular thermotropism and goose behavior lies in the magnitude of the behavior, not in the principle that body temperature is threatened and that locomotion satisfies a physiological need.

 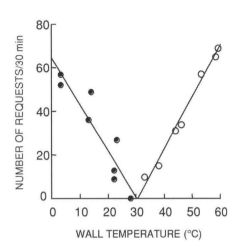

Fig. 4. A dog's thermoregulatory operant behavior. Left, the apparatus: the dog lies in a small climatic chamber with two light beams and photocells in front of its nose. When it cuts the left light beam, the dog obtains a burst of cool air; when it cuts the right light beam, the dog obtains a few seconds of infrared heat. Right, the resulting behavior is plotted against ambient temperature. It can be seen that the dog requests heat or cold as soon as the ambient temperature differs from neutral approx 30°C.

2.1.3. The Building of Microenvironments

The building of microclimates saves locomotion and, at the same time, satisfies physiological needs. This is commonplace in the human species, which lives in a totally artificial environment. In our species, the natural environment is artifactual. The human species has been able to penetrate all continents by building thermal microclimates. Heated houses of course, provide favorable temperatures, but clothes *(4)* and beds *(5,6)* also slow body heat loss and provide warmer surface temperatures. As a result, even under extreme latitudes, humans live permanently in the tropical conditions of their origins.

One can find many examples of artificial microclimates in animals. Nests, burrows, and dens are usually seen as protection against predators, but they also provide a favorable thermal microclimate. This behavior is not limited to homeotherms because social insects also regulate their environmental temperature. The oxygen consumption within the beehive increases, thus maintaining a safe internal temperature, when the outside ambient temperature decreases *(7)*. Reciprocally, on warm days the bees behaviorally increase the internal convection of the hive and bring water inside for evaporation *(8)*.

2.1.4. Operant Behavior

A variation of building microclimates is the use of operant behavior, which experimentally modifies the subject's immediate environment. Most often, the behavior made available to the animal is lever pressing to obtain a puff of cool air or a few seconds of infra-red heat as first described by Weiss and Laties *(9)*. This method has been used extensively for experimental purposes because the quantification of the behavior displayed by the animal is easily obtained. Figure 4 gives an example of such behavioral responses. The dog corrects its environmental strain with infrared heat or bursts of cool air. The response is a thermoregulatory behavior, proportional to the thermal need on both sides of a neutral environmental temperature, near 30°C for the small dog.

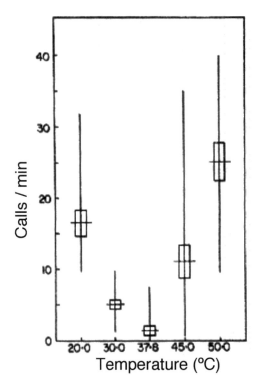

Fig. 5. Number of calls by white pelican chicks, still in their eggs before hatching. Each dot represents a mean of 15 eggs exposed for 10 min to various temperatures (abscissa). Horizontal lines indicate means, rectangles ±s.err., and vertical lines extreme values. It can be seen that chicks call for help proportionally to their temperature from 37.8°C. (Adapted from ref. *14*.)

2.1.5. Parental Behavior

Another behavioral adjustment of the environment is social. When parents provide their offspring with thermal protection, they behaviorally ensure the environmental conditions for survival. Under extreme latitudes, parental thermoregulatory behavior is especially vital, for example, in white bears *(10)* and emperor penguins *(11)*. Sometimes the protection consists in providing to the offspring oxygen *(12,13)* that will enable them to raise their own heat production. Figure 5 provides such an example where the corrective regulatory response to the piping egg should be parental *(14)*. As can be seen, the call for help is proportional to the chick's thermal needs, and is thus strictly regulatory. In the human species, thermoregulatory behavior in neonates consists of an audible request for parental protection, even by premature infants *(15)*.

2.1.6. Behavioral Self-Adjustment of the Subject

Behavior may be focused not only on the modulation of the environment, but also on the subject itself. In that case, behavior is quite similar to an autonomic response because the response lies within the body, and a separation of behavior from autonomic response may be not easy. An example of such a response is thermoregulatory heat production from muscular exercise when a subject deliberately works his or her muscles to warm him or herself *(16)*. Hibernation may be considered as a reflex but is accompanied by a whole sequence of behaviors that render hibernation possible and safe for the animal. Sleep enters into the category of behavioral self-adjustment of the subject although the result of sleep on temperature is still remote.

An interesting response of some endothermic species to hypoxia and low oxygen supply is the seeking of cooler environments to lower their body temperature and, in turn, their metabolism *(17)*. This behavior seems to be common in ectothermic vertebrates. Such a behavior can be described as behavioral anapyrexia *(18)* .

2.2. Behavioral Temperature Regulation

Temperature regulation provides the best example of short-term precisely quantitated behavioral adjustments to physiological needs *(19)* . Here again, Figs. 4 and 5 provide such examples of temperature regulation with the behavioral response proportional to the deviation of ambient temperature from its neutral point (i.e., near 30 for the small dog and 37.8°C for the piping egg). The autonomic response to such a need is aroused mainly from signals in the thermal core. Is this also true for the behavioral response? The behaviors that oppose ambient changes actually also defend the stability of deep core temperature. The main signal for thermoregulatory behavior is core temperature, mainly hypothalamic (Fig. 6). The equations describing the thermoregulatory corrective responses as functions of the *body temperatures*, where sensors have been identified, are strikingly similar when the response is autonomic and behavioral:

BEHAVIORAL RESPONSE $R = a \,(T_{hypothalamus} - T_{set})$ *(20)*

AUTONOMIC RESPONSE $R = a \,(T_{body} - T_{set})$ *(21)*

Thus, the behavioral response is especially well adapted to the thermoregulatory needs.

Figure 6 shows that a change in hypothalamic temperature is sufficient to trigger a behavior that opposes the stimulus. *How* behavior is triggered to achieve such a performance still requires an explanation.

2.3. The Mental Signal

In humans, and presumably other mammals also, the behavioral response is a conscious phenomenon (i.e., behavior takes place to satisfy a motivation to behave, thermal comfort/ discomfort, which sits in the mental space). In the case of temperature regulation, the answer to the question of the adaptation of behavior to the body's thermoregulatory needs is to be found in sensation, which is the mental object that describes the temperature signal from the skin. Both skin and core temperatures contribute to this mental signal *(22)*. The thermoregulatory signal in temperature sensation is its pleasure or displeasure.

The hedonic dimension of sensation is narrowly dependent on signals from the thermal core (Fig. 7). The actual signal from the thermal core is the algebraic difference between core temperature and set temperature. A given stimulus arouses pleasure when it corrects a core temperature problem and displeasure when it contributes to worsen the internal trouble *(23)*. For example, a 44°C skin is pleasurable in a hypothermic subject, or a feverish patient, and unpleasant in a hyperthermic subject. The word "alliesthesia" describes the fact that pleasure is contingent on signals arising from the core temperature and that pleasure is aroused by stimuli that are useful to restore homeothermia. Two models have been proposed to predict preferred skin temperature (Tp):

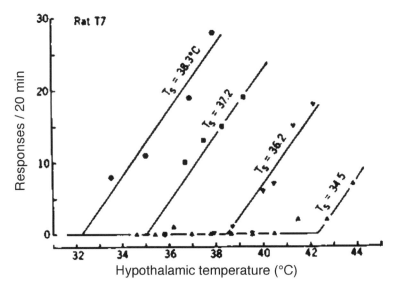

Fig. 6. Frequency of bar presses by a rat plotted against hypothalamic temperature. Each bar press lowers local hypothalamic temperature to 38°C, for 15 s. It can be seen that the rat's behavior was proportional to hypothalamic temperature. In addition, a signal from skin temperature lowered the hypothalamic threshold: the cooler the skin temperature, the warmer the hypothalamic threshold. (Adapted form ref. *20*.)

$$Tp = a(Tc-b)Tmean\ sk+c \quad (24)$$

$$Tp = a+bTc+cTmean\ sk \quad (25)$$

and one predicting subjective assessment (SA):

$$SA = aTc + bdTc/dt + cTmean\ sk + edTmean\ sk/dt + fS + g \quad (26)$$

where a, b, c, e, f, and g are constant parameters, Tc is core temperature, Tmean sk is mean skin temperature, and S is a shivering factor (0 or 1). One model was proposed to predict alliesthesia:

$$a = f(Tc-Tset)Tmean\ sk, \quad (26)$$

where a, is a measure of thermal alliesthesia, Tc core temperature, and Tset the set-point temperature of the biological thermostat. It follows from the variable Tset, that during fever, when the set-point is raised by pyrogens or by emotion, that alliesthesia defends this higher set-point. Alliesthesia will affect different locations on the skin's surface simultaneously *(27)*, a phenomenon that accounts for comfort and discomfort in various circumstances. For example, comfort persists when one side of the body is cooled and the other side warmed, or when the body receives asymmetrical radiation producing a difference of up to 13°C between front and back skin temperatures *(28)*.

   If pleasure is a mental signal for an adapted behavior, it follows that this signal is necessarily transient because as soon as the pleasant stimulus has corrected the internal trouble, pleasure disappears because the stimulus is no longer useful. In turn, thermal comfort must be defined therefore as the stated indifference to the environmental microclimate *(18)*.

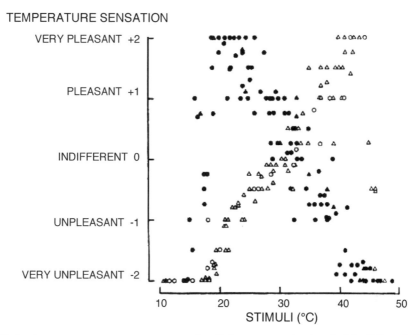

Fig. 7. Hedonic responses to temperature stimuli between pain thresholds (15 and 45°C) of a subject whose hand has been stimulated for 30 s, and whose body has been immersed in a well-stirred bath. Each dot is the response to a stimulus. The subject is immersed to the chin in a water bath, in order to control mean skin temperature. Triangles, cold bath; circles, warm bath; hypothermic subject, open dots; hyperthermic subject, solid dots. It can be seen that core temperature determined the hedonic experience; for example, to the hyperthermic subject, cold stimuli felt pleasant, and warm stimuli unpleasant. (Adapted from ref. *23*.)

2.4. Conclusion: Block Diagram of Behavioral Thermostat

Figure 8 is an adaptation of Fig. 2 to thermoregulatory behavior.

Figure 8 emphasizes that behavior is the most potent way to control inflow and outflow of heat, first into the body, then into the environment. The control of these exchange flows allows core temperature to be regulated. The time constant of the energy balance achieved through temperature regulation is of the order of minutes to hours. Yet, behavior plays also a fundamental role in energy balance for longer periods, such as hours, days, and months. The energy flow thus controlled is under chemical form.

3. BEHAVIORAL CONTROL OF LONG-TERM EXCHANGES: BODY WEIGHT REGULATION

The inflow and outflow rates of Figs. 1 and 2 apply as well to long-term as to short-term energy exchanges and balance. In the case of the long term, the energy received from the environment is under the form of potential chemical energy. Behavior also plays a major role here. The body receives an intermittent flow of energy under chemical form through food. Such intake is thus totally and exclusively behavioral. Therefore, everything that enters

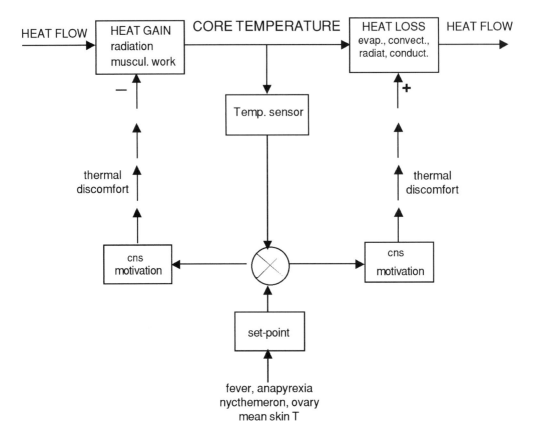

Fig. 8. Tentative block diagram of behavioral temperature regulation. Full arrows indicate signals. Interrupted arrows indicate behavior. A + sign, indicates a controlling action in the same direction as the change in the regulated variable (glucocorticoid concentration). A – sign, indicates a controlling action in the direction opposite to the change in the regulated variable. cns, central nervous system.

the body is under behavioral control. The body returns an equal flow of energy to the environment in a current that is both under autonomic and behavioral modulation. It also returns metabolic water to the environment through sweat, that is autonomically controlled for short-term energy balance, and in respiratory gas that is usually not controlled behaviorally but can occasionally become so, and in urine with behavioral relaxation-type modulated micturition. Finally, the body returns unabsorbed chemical energy in faeces and catabolites in urine, both under behavioral control, carbon dioxide in respiratory gas under occasional behavioral control, and various body secretions such as milk or sperm, mostly under behavioral control. The loss of blood and skin exfoliates from skin and phaneres is almost independent from behavior.

3.1. The Various Alimentary Behaviors

3.1.1. Food Intake

From the environment, the body receives intermittent inflows of chemical potential energy, vitamins, metabolites, oligo-elements, and water. All of those enter into the body through the mouth and digestive tract (i.e., are behaviorally controlled).

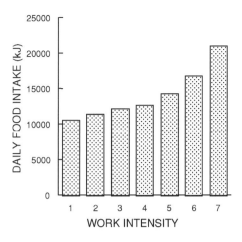
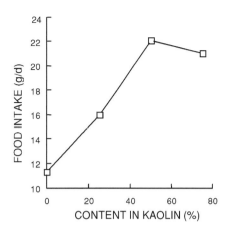

Fig. 9. Left: Amount of energy *spontaneously* ingested daily by people with the following work: 1. sedentary, 2. clerk, 3. taxi driver, 4. restaurant helper, 5. dustman, 6. smith, 7. coal miner (from various publications). Right: Mass of food ingested by rats when an inert substance (kaolin) is mixed with their regular chow. It can be seen that behavior compensates for the deficit in food and doubles with dilution. Thus the rats' needs are still covered. It is only when dilution is extreme that behavior cannot cover the need. (Adapted from ref. *85*.)

The intake is permanently adjusted according to both the body needs and the substrate availability. Food intake rises when energy expenditure rises and when energy density in food drops (Fig. 9). In addition, the lowering of ambient temperature increases the need for energy, as temperature regulation often raises heat production to match heat loss; in that case, food intake is also raised.

The examples provided in Fig. 9 show quantitative optimization of behavior, (i.e., in the long term the energy balance is nil). Such an optimization takes place not only with food intake but also with each step in the sequence of actions in the procurement of food: search, procurement, and handling *(29,30)*.

On the life-span term, food intake is extremely precise as shown by Hervey, who estimated that the average English woman ingests 20 tons of food between the ages of 25 and 65 while she gains 11 kg. Yet, during this period, the slight but gradual 11-kg upward weight change of her body corresponds to an average error of only 0.1 g per meal over 40 yr *(31)*. The application of behavior to long-term physiological regulations has been developed by Mrosovsky *(32)*. Over a month, a year, or a life, long-term adjustments of the set-point take place *(33)*. This long-term resetting has been called *homeorhesis (34)*, and *rheostasis (32)*.

3.1.2. Hoarding

The hoarding of food *(35)* in the nest, the den, the cache, the hive, or the home is also a long-term enrichment of the immediate environment in response to an anticipated physiological need of nutrients and energy. Hoarding behavior is so tightly linked to the need for energy that this response may be used to probe the underlying physiological function; several examples are given here. The amount of food stored by a rat is proportional to the decrease of its body weight below set-point *(36)* (Fig. 10). It follows that the amount hoarded, may be used as an argument in favor of body weight regulation, as is discussed later.

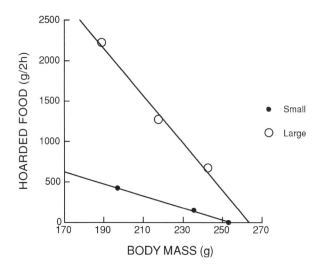

Fig. 10. Mass of food hoarded by a rat in repeated 2-h sessions plotted against the animal's body mass. Each dot stands for a different day. Sessions alternated with small and large pellets. (From Charron and Cabanac, unpublished.)

3.1.3. Hibernation

Some behaviors satisfy both short- and long-term survival. Hibernation, as seen previously, allows some animals to save energy by resetting their thermostat at a lower temperature set-point during winter. At the same time, hibernation saves food by allowing the body weight to drop. Hibernation is an example of circannual *homeorhesis/rheostasis* regarding body weight regulation.

3.1.4. Parental Behavior

In many animal species, the young are too immature to be able to feed themselves. In mammals the nursing mother provides behaviorally the needed energy through her milk. Parental feeding is almost universal in birds, in altrichial species of course, but also in species whose chicks are able to feed themselves. Many insects and other arthropods also provide behaviorally a nutritional environment favorable to their offspring' s survival.

3.1.5 Auto Modification

The behaviors listed here modify the subjects' internal state or their nutritional environment, but they are not necessarily accomplished with the satisfaction of a physiological need in mind. The subjects just satisfy their motivation to behave, reported as conscious hunger in humans, and the result is improved physiology. But, when human subjects purposely jog frequently and raise their aerobic heat production in order to lower their blood cholesterol or their body weight, they intend to behaviorally modify their energy content. These efforts are somewhat different from other behaviors because they do not fulfil a physiological need, but rather oppose the subjects' normal physiological functioning.

3.2. Behavioral Regulation of Body Weight

To return to Hervey's calculations (*see* Section 3.1.1.), if, as suggested by the extremely small error of 0.1 g per meal, body weight is regulated, then the life drift of 11 kg in the English woman's body weight, is owing to a rise of her body weight set-point with aging, and the behavioral response may be considered as perfect *(31)*. This implies, of course, that body weight is regulated.

3.2.1. Body Weight: Regulation vs Steady State

Following his calculations, Hervey hypothesized that body mass, or a variable closely correlated to body mass, such as body fat content, is constant over the adult life span because it is regulated. However, the hypothesis has been the object of a debate, because the concept of set-point appears as a circular explanation to many. As an alternative, Wirtshafter and Davis *(37,38)* hypothesized that body weight remains constant at or near a "settling point," the adjustment taking place passively without a regulation and a set-point. Examples from the literature show that this point of view is shared by other physiologists. Le Magnen considered the stability of body weight as resulting from the equilibrium between filling and emptying the adipose reservoir *(39)*. The constancy of body weight in obesity *(40)* and in anorexia nervosa was also described as simply resulting from imperfect energy balance *(41)*, insufficient energy expenditure resulting in elevated body weight, or insufficient food intake resulting in lowered body weight.

This approach considers the stability of body weight as resulting from the equilibrium between filling and emptying the adipose reservoir, as in Fig. 1. If the stability of body weight were produced by a steady state, then it would be the simple result of the balance of food intake with energy expenditure, as suggested by the concept of a settling point. According to the steady-state hypothesis, obesity would be the result of excessive food intake and/or insufficient energy expenditure. Similarly, leanness would be the result of insufficient food intake and/or excessive energy expenditure. The alternative hypothesis is regulation and set-point resetting.

A regulated system is a steady state equipped with regulatory loops. Of course, if body weight is to be constant, equal inflow and outflow of energy must also occur in the regulated system of Fig. 2, as well as in the steady state of Fig. 1, and transient unequal inflow and outflow of energy must occur, if any change in body weight is to be observed. However, the concept of regulation implies that equal flow is not the final cause, but rather only a means to achieve the goal of body weight equal to its set-point.

One way to differentiate a regulated system from a simple steady-state system consists in measuring how the system responds to a global perturbation. A response opposed to the perturbation indicates that we are dealing with a regulated system. That was clearly the case with temperature regulation, above. Now, the mere fact that food intake increases, after a body weight decrease resulting from starvation, then returns to control value when body weight has recovered, is an indication that the stability of body weight is not passive but is the result of a regulatory process.

In addition, all the available evidence points to a resistance to body weight changes in health as well as obesity:

1. During hypocaloric diet, the basal metabolism is depressed *(42)*. After ending their diet, subjects become hyperphagic and regain their initial body weight.
2. The weight of animals with a circannual cycle is defended by the animals' food intake *(33)*.
3. After middle as well as lateral hypothalamic lesions, the body weight is defended by the animals' food intake and heat production *(43,44)*.
4. Periods of limited access to food are followed by compensatory overfeeding when body weight is below set-point but not when body weight is above set-point *(45)*.
5. Gavage is accompanied by a lower food intake and a raised thermogenesis *(46)*.

6. After ending a hypercaloric diet, subjects spontaneously reduce their intake and return to their initial body weight *(47)*.

Such responses are pathognomonic of regulation. According to the regulation hypothesis, obesity is the result of an elevated set-point (reached through transient increased food intake and/or decreased energy expenditure). Similarly, leanness is the result of a lowered set-point (reached through lowered food intake and/or excessive energy expenditure). All the available evidence demonstrates that this takes place.

In open-loop situations, regulatory responses are prevented from adjusting body weight. Such open-loop responses have also been used to study the regulation of body weight. The experimental results also point toward a regulated body weight (see review in ref. *48*) [48]: (a) satiety in humans, and disgust mimics in rats tend to be delayed when body weight is below set-point, in healthy as well as obese humans and rats; (b) saliva secretion after alimentary stimuli respond similarly; (c) the hoarding of food by rats takes place only when body weight is below set-point in control, this goes as well for obese rats; (d) female rats hoard food at a lower body weight set-point during pre- than post-ovulation periods *(49)*.

3.2.2. What is the Regulated Variable?

Despite all the overwhelming evidence accumulated, there is very little chance that body weight *sensu stricto* is regulated. It is not that a simple steady state would achieve the stability of body weight or that no underlying regulation would be at work, but there is another reason. The weight of the body is stable but weight reflects the mass, which is not a tensive variable; strictly speaking, therefore, body weight cannot be regulated. Another variable closely correlated to body weight must first be regulated. So far, this variable to which body weight is correlated, has not yet been identified. In his 1969 article, Hervey postulated that the likely mechanism by which body weight was regulated, was blood steroid concentration. Steroids being soluble in fat, he argued that circulating concentration would be reduced by any increase in body fat. If this is the signal for food intake/satiety, then the subject will eat less. Any decrease in body fat will tend to raise circulating steroid concentration and in turn food intake. This hypothesis led to the proposal that the variable, whose regulation results in body weight stability, is the hypothalamic concentration of corticotropin releasing hormone (CRH).

Several experimental results seem to lead to that conclusion (see review in ref. *50*): (a) ablation of the adrenal gland is followed by a decreased body weight set-point (lowering of the rat's hoarding threshold); (b) glucocorticoid administration produces opposite results; (c) direct intracerebroventricular injection of CRH lowers the hoarding threshold; (d) various stresses transiently lower the hoarding threshold: muscular exercise, gentle handling and emotional fever, and surgery; (e) fenfluramine, a serotonin agonist, likely involves CRH and, was shown to lower the set-point.

3.3. The Mental Signal

We saw in Fig. 7 that a given temperature can arouse a pleasant or an unpleasant sensation, depending on the subject's internal deep core temperature. A phenomenon we called alliesthesia. Alliesthesia is not limited to temperature sensation but takes place also with food sensations. Food tastes and odors are pleasant in fasting subjects and turn unpleasant with satiation. Sensory pleasure defends not only the body weight stability but, as we just saw above, defends the set-point of the ponderostat *(45)*. The mechanism is thus identical to what takes place on the short term with alliesthesia and temperature set-point.

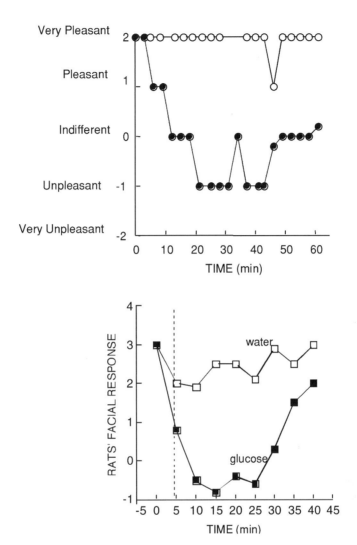

Fig. 11 Top: Hedonic responses given by a human subject when tasting, then repeatedly spiting out small samples of sweet water. Dark symbols: the subject receives intragastrically a load of concentrated glucose. A pleasant sensation turns into an unpleasant one, then the subject recovers. Open symbols, the subject receives no gastric load. Bottom: The same experiment with a rat that receives minute sweet samples intra-orally (5 µL). Instead of collecting verbal ratings as above, the experiment consists in rating the behavioral signs of pleasure/displeasure in the rat. In the control session, the rat received an intragastric load of water. (Adapted from ref. *86.*) It can be seen that the patterns are strikingly similar in humans and in rats.

3.4. Conclusion: Block Diagram of Behavioral Ponderostat

Figure 12 presents a tentative block diagram integrating the hypothalamic–pituitary–adrenocortical axis and the ponderostat. In the model, the regulated variables are the concentrations of glucocorticoid and leptin in the blood and the main variables controlled are the inflow of food intake and hoarding, and the outflow of energy expenditure. The set-point of glucocorticoid concentration and body weight is determined by CRH hypothalamic concentration. The regulation ensures a rate of secretion by the adrenal glands so as to keep a constant blood concentration. In addition, a specific catabolism of glucocorticoids by adipocytes cannot be ruled out. Thus, for a given constant adrenal secretion rate, the larger the fat stores, the lower the hormonal concentration in the blood. According to Fig. 12 the concentration of circulating glucocorticoids would be the variable whose regulation leads to body weight stability, and the set-point would tend to keep that variable constant. A short feedback loop regulates this concentration via adrenocorticotropic hormone and CRH messages. Two other loops translate the concentration of CRH in the brain into behavioral responses. The first is a negative feedback that reduces food intake and food hoarding when CRH increases. The second loop is a positive feed-forward that increases energy expenditure when CRH increases. Constancy of body weight is thus achieved by control in flowing energy (food intake) and out flowing energy (metabolism, muscular work).

Body weight set-point regulation does not originate in the adipocytes but rather in the hypothalamus, as suggested by others. This signal could represent a balance between activation of the hypothalamic CRH production by leptin, and inhibition of the hypothalamic CRH production by glucocorticoids.

The direct involvement of CRH in the regulation of body weight was suggested before, in the cases of anorexia nervosa *(51)* and obesity *(52–61)*. Obesity would result from a drop of hypothalamic CRH and a corresponding increase in the body weight set-point. Accordingly, the higher concentration of circulating glucocorticoids observed in obese subjects might be a causal factor. Conversely, anorexia nervosa would result from a rise in hypothalamic CRH and in a corresponding decrease in the body weight set-point; the higher circulating glucocorticoid concentration in anorexia nervosa would simply reflect higher CRH activity. Such effects closely parallel those of intracerebral CRH, as found in rats with hoarding, and also with food intake.

The hypotheses developed here are presented in Fig. 12, as a block-diagram extension of Fig. 2.

4. COMPLEMENTARITY OF AUTONOMIC AND BEHAVIORAL RESPONSES: THE GAIN OF FREEDOM

Behavior appeared above as the prevalent response to achieve stability in the energy content of the body. Of course, the need to oppose deviation from set-point does not rely on behavior only. Autonomic responses are present also and can be powerful in higher vertebrates, mammals and birds.

Figure 13 provides an example of the complementarity of autonomic and behavioral responses. In this experiment, pigeons used their operant response to counterbalance behaviorally the ambient temperature stress and maintain both their body temperature stable, and their evaporative heat loss low.

Complementarity takes place both ways to save either autonomic or behavioral response, according to the environment and the subject's priority.

4.1. Autonomic Compensation of Behavioral Strain

Some species have developed particular autonomic capacities that make them far better performers in a given environment than other species or breeds. For example, the black Bedouin goat is able to store water in its rumen, maintain its urine flow extremely low, recycle in its gut up to 90% of its urea production, and to digest fibers that are indigestible to other goats *(62)*. All these capacities make the black Bedouin goat better suited than other goats and most other mammals to survive in desert environment. The camel also thrives in

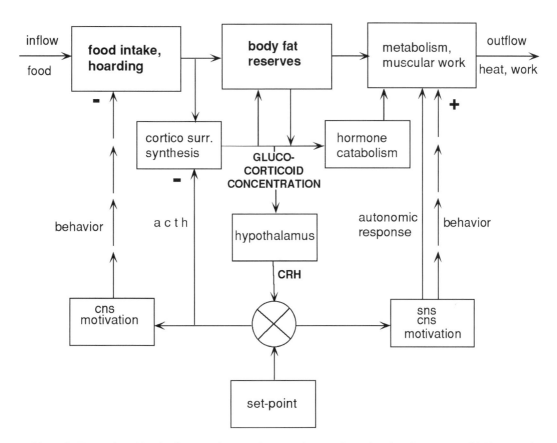

Fig. 12. Tentative block diagram integrating corticotropin releasing hormone (CRH) and the ponderostat. Interrupted arrows indicate a behavioral pathway. Bold characters indicate the factors involved in the present chapter. A + sign, indicates a controlling action in the same direction as the change in the regulated variable (glucocorticoid concentration). A – sign, indicates a controlling action in the direction opposite to the change in the regulated variable. CNS, central nervous system; SNS, sympathetic nervous system.

the desert because it recycles its urea, can depress its thyroid and insulin activity, tolerates high trunk temperature, glycemia, and osmotic pressure when needed, and is able to rehydrate within minutes *(63)*. The mole rat tolerates high concentrations of carbon dioxide, and low oxygen concentration in its respiratory environment, the underground collective burrows; thus it is able to maintain its normal activity and metabolism *(64)*. The emperor penguin is able to store 40% of its body weight as fat, which allows for almost 4 mo of fasting during the reproductive season in the Antarctic continent, 120 km from the sea *(11)*.

These extraordinary adaptations of autonomic functions free the behavior and allow these species to dwell in areas where their predators cannot follow them. One may wonder what first started, the autonomic properties that free behavior and allow long stays in hostile environments such as deserts, underground burrows rich in carbon dioxide, and the Antarctic, or the behavioral trend to be free from predators in environments that demand a particular autonomic adaptation. The behavioral trend to seek security from predators, and the extreme efficacy of autonomic responses to resist environmental constraints, probably result from co-evolution of autonomic and behavioral traits. There must be a cost to be paid for these

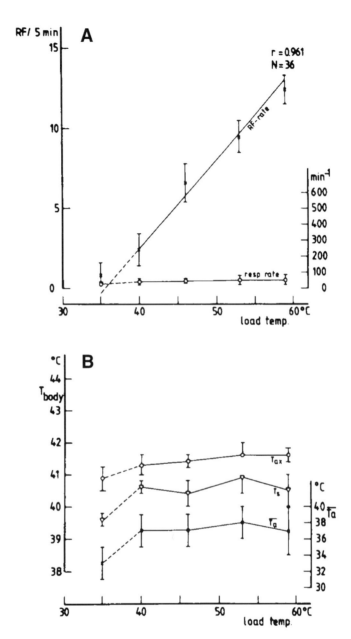

Fig. 13. A pigeon stands in a climatic chamber the temperature of which (load temperature) is imposed by the experimenter. The bird can peck at a key and thus obtain a burst of cool air. The figure gives the mean resulting body temperatures and behaviors of three pigeons performing in 43 sessions. (**A**) rate of pecking: RF, respiratory frequency: resp. rate. This shows that the animals' behavior was proportional to ambient temperature and therefore directly thermoregulatory. As a result, the birds saved their evaporative heat loss and did not hyperventilate while maintaining stable body temperatures (**B**); Tax, axillary temperature; Ts, dorsal skin temperature; Ta, temporal mean ambient temperature. (Adapted from ref. *87*.)

exceptional autonomic adaptations, if not by the individual, at least by the species that is able to survive in these extreme environments. The cost is probably to be found in low reproductive rates owing to scarcity of food.

4.2. Behavioral Compensation of Autonomic Strain

Such compensation must take place in ectothermic animals, because they only rely on their behavior to regulate their body temperature. This is also true for some species of mammals. The mole rat, a poikilothermic mammal, does not possess the autonomic defense against cold stress and must use its behavior to prevent hypothermia when the ambient temperature drops *(65)*. Symmetrically, the pig does not sweat, is a poor panter, and its only autonomic response to heat stress is skin vasomotion. It must, therefore, use its behavior to prevent hyperthermia in a warm environment *(66)*. This allows pigs to be covered with wet mud to escape the heat of the summer.

4.3. Pathological Situations

The complementarity of behavioral and autonomic functions can be found in pathological situations. When the autonomic response to environmental stress is hindered by disease or by the experimenter, the behavioral response takes over the regulatory process. This may result in another cost, behavioral this time, which must be paid. Psychological or behavioral aggressions have somatic impacts. Indeed, psychological stimuli are among the most potent of all stimuli to affect the pituitary adrenal cortical system and they lead to a stress reaction only if the subject shows an emotional response *(67)*. Stress is now viewed mainly as a general biological response to environmental demands *(68)*, and experimental studies of stress always consist of behavioral hindrance.

5. CONFLICTS OF MOTIVATION: SHORT- VS LONG-TERM NEEDS

The complementarity of autonomic and behavioral responses gives one degree of liberty to those animals that are equipped with both. Such a degree of liberty can be used in situations of conflict where short- and long-term energy needs clash.

5.1. Food Intake vs Air Temperature

Food intake ranks high among the various conflicting physiological needs, especially when the environmental temperature is not too aggressive *(69)*. However, when the environmental conditions of temperature and humidity become tropical, the pigs' rectal temperatures rise (1–2°C) and their food intake drops. The heavier the animal, the more profound the temperature influence. If their owners provide the animals with a properly equipped environment, pigs will seek shade and wet themselves *(70)*. Paradoxically the enrichment of their regime in lipids improves the pigs' nutrition, especially in lactating sows *(71)*.

When food is scarce, pigeons behaviorally compensate food scarcity by selecting a higher ambient temperature *(72)*. This demonstrates that pigeons can substitute one type of thermoregulatory response, here from autonomic to behavioral, as seen in Fig. 13. The opposite is also true. Pigeons can use their autonomic response to free their behavior for another purpose (e.g., flight from a predator). The gain in freedom is larger when adaptation improves the autonomic performance, and especially in the case of cross Adaptation *(73)*. The upward drift of body weight of aging humans, might be an adaptation of a somatic nature to the declining behavioral capacity to forage. The example of behavioral flexibility in response to a conflict of motivations can be found, when the need to feed clashes with the need to thermoregulate. Experiments showed that not only mammals but also reptiles were able to solve that problem and to optimize their response to it.

5.1.1. The Rat

Rats suffer more from high than from low ambient temperature, which modulates their food intake *(74)*. Specific exploration of the conflict between cold vs food intake, showed that their behavior was especially flexible to face both their environmental and nutritional needs *(75)*. Rats were fed in an air temperature of −15°C at distances 1, 2, 4, 8, or 16 m from a thermoneutral refuge. As the distance between the feeder and thermoneutrality increased, the number of excursions to feed decreased from 37 to 7 during the 2-h sessions; concomitantly, the meal duration increased from 1.2 to 5.2 min. The rate of feeding and the total feeding time were the same at all distances. The mass of food ingested was also constant except for a slight decline at 16 m. Meal duration was strongly correlated (R = 0.9) with the time taken to reach the feeder at each distance, whereas estimated cost–benefit of feeding episodes increased with distance. Estimates of body temperature indicated that significant drops in skin temperature occur even over short distances/durations, whereas over greater distances core temperature probably also decreases. In the range of distances studied, rats accorded their food drive a higher priority than temperature preference, and chose to feed while tolerating greater thermal disturbance.

5.1.2. The Lizard

Ectotherms are especially vulnearble to ambient temperature changes, as they do not possess autonomic responses to temperature stress, and must only rely on their behavior to thermoregulate. When their body temperature rises to the optimal approx 35°C their digestive and locomotor performances improve *(76)*. Yet, they also display a capacity to adjust behaviorally to conflicts between short- and long-term energy needs *(77)*.

Juvenile *Tupinambis teguixin* were placed in a conflicting situation: feeding vs cold environment. To feed they had to leave a warm refuge (ambient temperature = 44–45°C) and go 1.5 m to where food was presented at an ambient temperature varying from 25 to 0°C. When ambient temperature was decreased, the lizards managed to ingest a constant amount of food by modifying their behavior. They shortened the duration and increased the number of meals, by returning to the warm refuge between meals. In the cold, they left the food when their cloacal temperature dropped to about 32°C. After satiation, they maintained their cloacal temperature behaviorally between 34 and 38°C. The attempt to increase the lizards' drive for food by increasing the duration of a fast preceding their access to food from 1 to 17 d, did not result in any behavioral change during feeding. The only modification was a decrease in the amount ingested, when the fast was shorter than 3 d. In a warm environment, when the intervals between feeding increased from 1 to 17 d, the lizards main response was not an increase in food intake, but rather, a decrease in the growth rate and sloughing frequency (Fig. 14).

5.2. Sensory Pleasure, the Optimizer of Behavior

Sections 2.3. and 3.3. showed that the mental signal that controlled adapted behavior to temperature and food needs was sensory pleasure. Maximization of pleasure would optimize energy balance in the short term, as well as in the long term, when one of these motivations was present alone. This is also true in animals as well as in humans, when they are simultaneously present and when they clash.

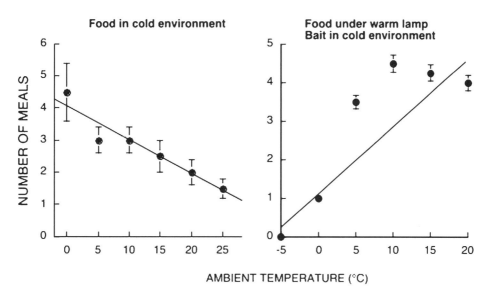

Fig. 14. Left. Mean number of foraging trips (meals) of lizards placed in a situation of conflict : the food is located in a cold environment, away from the heated corner of their terrarium. (Adapted from ref. *77*.) Right. Mean number of foraging trips (meals) of iguanae placed in a situation of conflict : the palatable bait (lettuce) is located in a cold environment, away from the heated corner of their terrarium where regular iguana chow is available. (Adapted from ref. *80*.)

Thus, when there is no alternative, reptiles are able to forage in the cold. The rising number of meals with decreasing ambient temperature, indicates that the duration of meals was shorter in a cold environment, thus preventing hypothermia. However, when the reptiles are not forced to venture into the cold because regular food is available in the warm corner, they decrease the number of trips toward the palatable bait and, eventually renounce it. The opposite patterns obtained in the two situations, likely indicate that on the right, the conflict was between palatability and cold discomfort, two motivations indicative, of a mental space in reptiles.

5.2.1. Rat Palatability vs Ambient Temperature

Rats were trained to feed each day from 10 AM to 12 noon. Once a week, in an environment of −15°C, additional food was made available 16 m from a thermoneutral refuge *(78)*. The additional food offered was either shortcake, meat paté, peanut butter, Coca-cola®, all of these ("cafeteria"), or laboratory chow. Although laboratory chow was also always available in their thermoneutral refuge, rats invariably ran in the cold to the feeder, especially so when the food offered was highly palatable. With such foods, rats took as much as half their nutrient intake in the cold. For less palatable food, rats went only once or twice to the feeder, and there ingested smaller quantities. The attractiveness of the various baits was ranked similarly; by the amounts ingested, the number of excursions to the feeder, and the time spent feeding in the cold. Meal duration and speed of running to the food were not influenced by palatability. For the whole group of rats, the preference was shortcake, Coca-cola®, meat paté, peanut butter, and chow. There was considerable variation between rats in their attraction to different foods. Feeding behavior in a situation of conflict could be used to measure palatability. One explanation for the individual differences in taste preference might be the extent of tastes and flavors met by the young at the time of weaning: a richer experience when starting with solid food, tends to mask the palatability of future baits *(79)*.

5.2.2. Lizard Palatability vs Ambient Temperature

Juvenile green iguanas were placed in a situation of conflict between two motivations: a thermoregulatory drive and the attraction of a palatable bait *(80)* . To be able to reach the bait (lettuce), they had to leave a warm refuge, provided with standard food, and venture into a cold environment. In experiment 1 the time interval between sessions with bait ranging from 1 to 8 d had no effect on the duration of stay on the bait. This result shows that the lettuce was not a necessary food, deprivation of which would have had to be compensated for. In experiment 2, as the ambient temperature at the bait decreased the lizards spent less time feeding on lettuce and they visited the bait less frequently. This result shows that the lizards traded off the palatability of the bait with the disadvantage of the cold. These findings support the hypothesis that a common currency makes it possible for lizards to compare two sensory modalities.

5.3. Conflicts of Motivation in Humans

The information about sensory pleasure provided by animal experiments is clear. Nevertheless, it is preferable to have also human data that lead to the same conclusions regarding mental signals. Several experiments attempted to verify whether pleasure is the common currency in conflicts of motivations. The pleasure of eating palatable food was pitted against the displeasure of giving money away in a situation where subjects had to spend money to buy their food. Ten healthy subjects taken individually had lunch in the laboratory on 4 different days. During the first session, they rated the palatability of small sandwiches of 10 different varieties. In the three following sessions, they were asked to eat the same number of sandwiches as in the first session and they had to buy each sandwich at a price that increased with palatability. The rate of the price increase varied in the three sessions. In view of the price increase, the subjects moved their preference to the less palatable sandwiches. The subjects' actual behavior was predictable from the quantitative relationship of ratings and prices. This result supports the hypothesis according to which behavior tends to maximise multidimensional pleasure experiences.

Six men were placed in a situation of physiological conflict, of fatigue vs cold discomfort. Dressed in swimsuits and shoes they walked at 3 km/h on a treadmill placed in a climatic chamber. The slope of the treadmill varied from 0 to 24% and the ambient temperature from 25 to 5°C. The subjects could choose temperature when slope was imposed or the converse. They rated pleasure and displeasure of temperature and exercise. Deep body temperature and heart rate were monitored. The results show that the subjects adjusted their behavior to maintain approximate steady deep body temperature and to limit heart rate below 120 beats per minute. The physiological compromise was thus correlated to the drive for the algebraic sum of maximal pleasure–minimal displeasure in the two sensory dimensions fatigue and discomfort.

6. CONCLUSION: THE RANKING OF PRIORITIES

Almost the total exchange of the body with its environment is, therefore, under behavioral control. These behavioral responses modify either the environment or the subject itself. Adjustment of energy balance is but one example of optimization taking place behaviorally in animals and humans. All motivations follow the same pattern and the complementarity of behavioral and autonomous responses is universal. At the present time, the concept that behavior belongs as much to physiology as to psychology or zoology (ethology) is not yet commonplace, but behavioral research still finds a place in several physiological journals. The notion of emergence has become accepted by physiologists, and even mentation is of interest to them *(81,82)*.

The examples of behaviors adapted to physiological aims, provided in the sections here, were all similar to autonomic responses in that they controlled the inflow or outflow of matter and energy entering and leaving the body. Autonomic and behavioral responses are therefore complementary and can be substituted for one another. Thus, an organism may have several possible responses to a given environmental challenge, and may be able to play with possible substitutions. This flexibility is useful when several physiological functions compete for a given response, and also when several motivations compete for the "behavioral final common path" *(83)*. The common currency, that allows the trade-offs between short- and long-term energy needs for access to behavioral satisfaction, is most likely to be sensory pleasure. This psychological mechanism is rooted in biology, as its antiquity shows, since it most likely emerged with reptiles.

REFERENCES

1. Brobeck, J.R. (1965) Exchange, control, and regulation. In *Physiological Controls and Regulations* (Yamamoto, W.S. and Brobeck, J.R., eds.), Saunders, Philadelphia, PA, pp. 1–14.
2. Berbigier, P. and Christon, R. (1983) Efecto de la sombra y la aspersion sobre las temperaturas rectal y cutanea, fresuencia respiratoria y tasa de crecimiento de cerdos creole jovenes en Guadalupe (Antillas francesas). *Rev. Salud. Anim.,* 785–792.
3. Mendelsohn, M. (1895) Über den Thermotropismus sinzelliger Organismen. *Arch. Gesamte Physiol.* 60:14.
4. Scholander, P.F., Anderson, N., Krog, J., Lorentzen, F.V. and Steen, J. (1957) Critical temperature in Lapps. *J. Appl. Physiol.* 10:231–234.
5. Goldsmith, R., Hampton, R. and Hampton, I.F.G. (1968) Nocturnal microclimate of man. *J. Physiol. London.* 194:32–33.
6. Candas, V., Libert, J.P., Vogt, J.J., Ehrhart, J. and Muzet, A. (1978) *Proc. Internat. Indoor Climate Sympos., København.*
7. Lindauer, M. (1951) Temperaturregulierung der Bienen bei Stocküberhitzung. *Naturwissenschaften.* 38:308–309.
8. Hazelhoff, E.H. (1954) Ventilation in a bee-hive during summer. *Physiologia Comparata et Œcologia.* 3:343–364.
9. Weiss, B. and Laties, V.G. (1961) Behavioral thermoregulation. *Science.* 20:1338–1344.
10. Blix, A.S. and Steen, J.B. (1979) Temperature regulation in newborn polar homeotherms. *Physiol. Rev.* 59:285–304.
11. LeMaho, Y. (1977) The emperor penguin: a strategy to live and breed in the cold. *Am. Sci.* 65:680–693.
12. Courtenay, S.C. and Keeleyside, M.H.A. (1983) Wriggler-hanging: a response to hypoxia by brood-rearing *Herotilapia multispinosa* (Teleostei, Cichlidae). *Behaviour.* 85:183–197.
13. VanIersel, J.J.A. (1953) An analysis of the parental behavior of the male three-spined stickleback (*Gastrosteus aculeatus* L.). *Behav. Suppl.* 3:1–159.
14. Evans, R.M. (1990) Vocal regulation of temperature by avian embryos: a laboratory study with pipped eggs of American white pelican. *Anim. Behav.* 40:968–979.
15. Brück, K. (1968) Which environmental temperature does the premature infant prefer? *Pediatrics.* 41:1027–1030.
16. Cabanac, M. and LeBlanc, J. (1983) Physiological conflict in humans: fatigue *vs* cold discomfort. *Am. J. Physiol.* 244:R621–R628.
17. Wood, S.C. (1991) Interactions between hypoxia and hypothermia. *Ann. Rev. Physiol.* 53:71–85.

18. (IUPS), C. f. t. p. (1987) Glossary of terms for thermal physiology. *Pflüg. Arch.* 410:567–587.

19. Cabanac, M. (1979) Le comportement thermorégulateur. *J. Physiol. Paris* 75:115–178.

20. Corbit, J.D. (1973) Voluntary control of hypothalamic temperature. *J. Comp. Physiol. Psychol.* 83:394–411.

21. Hardy, J.D. (1965) The "set-point" concept in physiological temperature regulation. In *Physiological Controls and Regulations* (Yamamoto, W.S. and Brobeck, J.R., eds.), Saunders , Philadelphia, PA, pp. 98–116..

22. Frank, S., Raja, S., Bulcao, C. and Goldstein, D. (1999) Relative contribution of core and cutaneous temperatures to thermal comfort and autonomic responses in humans. *J. Appl. Physiol.* 86:1588–1593.

23. Cabanac, M. (1969) Plaisir ou déplaisir de la sensation thermique et homéothermie. *Physiol. Behav.* 4:359–364.

24. Cabanac, M., Massonnet, B. and Belaiche, R. (1972) Preferred hand temperature as a function of internal and mean skin temperatures. *J. Appl. Physiol.* 33:699–703.

25. Bleichert, A., Behling, K., Scarperi, M. and Scarperi, S. (1973) Thermoregulatory behavior of man during rest and exercise. *Pflüg. Arch.* 338:303–312.

26. Attia, M. and Engel, P. (1982) Thermal pleasantness sensation: an indicator of thermal stress. *Eur. J. Appl. Physiol.* 50:55–70.

27. Attia, M. and Engel, P. (1981) Thermal alliesthesial response in Man is independent of skin location stimulated. *Physiol. Behav.* 27:439–444.

28. Olesen, S., Bassing, J.J. and Fanger, P.O. (1972) Physiological comfort conditions at sixteen combinations of activity, clothing, air velocity, and ambient temperature. *ASHRAE Trans.* 78:199–206.

29. Collier, G. and Rovee-Collier, C.K. (1981) A comparative analysis of optimal foraging behavior: laboratory simulations. In *Foraging Behavior* (Kamil, A.C. and Sargent, T.D., eds.), Garland STPM, New York, NY, pp. 39–76..

30. Collier, G.H., Johnson, D.F., Naveira, J. and Cybulski, K.A. (1989) Ambient temperature and food costs: effects on behavior patterns in rats. *Am. J. Physiol.* 257:R1328–R1334.

31. Hervey, G.R. (1969) Regulation of energy balance. *Nature.* 223:629–631.

32. Mrosovsky, N. (1990) *Rheostasis, the physiology of change.* Oxford University Press, New York, NY.

33. Mrosovsky, N. and Fisher, K.C. (1970) Sliding set-points for body weight in ground squirrels during the hibernation season. *Can. J. Zool.* 48:241–247.

34. Nicolaïdis, S. (1977) Physiologie du comportement alimentaire. In *Physiologie humaine* (Meyer, P., ed.), Flammarion, Paris, pp. 908–922..

35. VanderWall, S.B. (1990) *Food hoarding in animals.* University of Chicago Press, Chicago, IL.

36. Fantino, M. and Cabanac, M. (1980) Body weight regulation with a proportional hoarding response in the rat. *Physiol. Behav.* 24:939–942.

37. Wirtshafter, D. and Davis, J. (1977) Set-points, settling points, and the control of body weight. *Physiol. Behav.* 19:75–78.

38. Davis, J. and Wirtshafter, D. (1978) Set-points or settling points for body weight?: A reply to Mrosovsky and Powley. *Behav. Biol.* 24:405–411.

39. LeMagnen, J. (1984) Is regulation of body weight elucidated? *Neurosci. Biobehav. Rev.* 8:515–522.

40. Himms-Hagen, J. (1984) Thermogenesis in brown adipose tissue as an energy buffer. *N. Engl. J. Med.* 311:1549–1558.

41. Bernstein, I.L. and Borson, S. (1986) Learned food aversion: A component of anorexia syndromes. *Psychol. Rev.* 93:462–472.

42. Apfelbaum, M. and Bostsarron, J. (1969) Le bilan d'énergie de l'obèse soumis à un régime restrictif. *Presse Med.* 77:1941–1943.

43. Keesey, R.E. and Powley, T.L. (1975) Hypothalamic regulation of body weight. *Am. Scientist.* 63:558–565.
44. Keesey, R.E. and Hirvonen, M.D. (1997) Body weight set-points: Determination and adjustment. *J. Nutr.* 127:S1875–S1883.
45. Fantino, M. (1984) Role of sensory input in the control of food intake. *J. Auton. Nerv. Syst.* 10:326–347.
46. Lavine, J.A., Eberhardt, N.L. and Jensen, M.D. (1999) Role of nonexercise activity thermogenesis in resistance to fat gain in humans. *Science.* 283:212–214.
47. Pasquet, P. and Apfelbaum, M. (1994) Recovery of initial body weight and composition after long-term massive overfeeding in men. *Am. J. Clin. Nutr.* 60:861–863.
48. Cabanac, M. (1991) Open-loop methods for studying the ponderostat. In *Appetite and Nutrition* (Friedman, M.I., Tordoff, M.G. and Kare, M.R., eds.), Marcel Dekker Inc., New York, NY pp. 149–170.
49. Fantino, M. and Brinnel, H. (1986) Body weight set-point changes during the ovarian cycle: experimental study of rats during hoarding behavior. *Physiol. Behav.* 36:991–996.
50. Cabanac, M., Michel, C. and Gosselin, C. (2000) Corticotropin releasing hormone and body weight regulation: the behavioral approach. *Nutrit. Neurosci.* 2:385–401.
51. Cavagnini, F., Invitti, C., Passamonti, M. and Polli, E.E. (1986) Impaired ACTH And cortisol response to CRH in patients with anorexia nervosa. In *Disorders of Eating Behaviour, a Psychoneuroendocrine Approach* (Ferrari, E. and Brambilla, P., eds.), Pergamon Press, Oxford, pp. 229–233.
52. York, D.A. (1992) Central regulation of appetite and autonomic activity by CRH, glucocorticoids and stress. *Progr. NeuroendocrinImmunol.* 5:153–165.
53. Guillaume-Gentil, C., Rohner-Jeanrenaud, F., Abramo, F., Bestetti, G.E., Rossi, G.L. and Jeanrenaud, B. (1990) Abnormal regulation of the hypothalomo–pituitary–adrenal axis in the genetically 0bese FA FA rat. *Endocrinol.* 126:1873–1879.
54. McGinnis, R., Walker, J., Margules, D., Aird, F. and Redei, E. (1992) Dysregulation of the hypothalamo–pituitary–adrenal axis in male and female genetically obese (ob/ob) mice. *J. Neuroendocrinol.* 4:765–771.
55. Plotsky, P.M., Thrivikraman, K.V., Watts, A.G. and Hauger, R.L. (1992) Hypothalamic–pituitary–adrenal axis function in the Zucker obese rat. *Endocrinol.* 130:1931–1941.
56. Mortola, J.F., Rasmussen, D.D. and Yen, S.S.C. (1989) Alteration of the adrenocorticotropin–cortisol axis in normal weight bulimic women: evidence for a central mechanism. *J. Clin Endocrinol. Metab.* 68:517–522.
57. DeVos, P., Saladin, R., Auwerx, J. and Staels, B. (1995) Induction of *ob* gene expression by corticosteroids is accompanied by body weight loss and reduced food intake. *J. Biol. Chem.* 270:15958–15961.
58. Fukagawa, F., Knight, D.S., Price, H.V., Sakata, T. and Tso, P. (1996) Transplantation of lean fetal hypothalamus restaures hypothalamic function in Zucker rats. *Am. J. Physiol.* 40:R55–R65.
59. Costa, A., Poma, A., Martignoni, E., Nappi, G., Ur, E. and Grossman, A. (1997) Stimulation of cortico-releasing hormone release by the obese (ob) gene product, leptin, from hypothalamus explants. *Neuroendocrinology.* 8:1131–1134.
60. Huang, Q.L., Rivest, R. and Richard, D. (1998) Effects of leptin on corticotropin-releasing factor (CRF) synthesis and CRF neuron activation in the paraventricular hypothalamic nucleus of obese (ob/ob) mice. *Endocrinology.* 139:1524–1532.
61. Debons, A.F., Zurek, L.D., Tse, C.S. and Abrahamsen, S. (1986) Central nervous system control of hyperphagia in hypothalamic obesity: dependence on adrenal glucocorticoids. *Endocrinology.* 118:1678–1681.
62. Shkolnik, A. (1992) The black Bedouin goat. *Bielefelder ökol. Beiträge.* 6:53–60.
63. Yagil, R. (1985) *The Desert Camel. Comparative Physiology Adaptation.* Comparative Animal Nutrition Series N°5, Krager.

64. Arieli, R., Ar, A. and Shkolnik, A. (1977) Metabolic responses of a fossorial rodent (*Spalax ehrenbergi*) to simulated burrow conditions. *Physiol. Zool.* 50:61–75.

65. Yahav, S. and Buffenstein, R. (1991) Huddling behavior facilitates homeothermy in the naked mole rat *Heterocephalus glaber*. *Physiol. Zool.* 64:871–884.

66. Christon, R. (1983) Effet d'un ombrage et du taux protéique de la ration sur la croissance du porc local en Guadeloupe. *Revue d'Élevage et Médecine VÉtérinaire en Pays Tropical.* 36:191–196.

67. Mason, J.W., Maher, J.T., Hartley, L.H., Mougey, E.H., Perlow, M.J. and Jones, G.J. (1976) Selectivity of corticosteroid and catecholamine respons to various natural stimuli. In *Psychopathology of Human Adaptation* (Sarban, G., ed.). Plenum, New York, NY.

68. Dantzer, R. and Kelley, K.W. (1989) Stress and immunity: an integrated view of relationship between the brain and the immune system. *Life Sci.* 44:1995–2008.

69. Ingram, D.L. and Legge, K.F. (1970) The thermoregulatory behavior of young pigs in a natural environment.. 5:981–987.

70. Christon, R. (1988) The effect of tropical ambient temperature on growth and metabolism in pigs. *Journal of Animal Science.* 66:3112–3123.

71. Christon, R., Saminadin, G., Lionet, H. and Racon, B. (1999) Dietary fat and climate alter food intake, performance of lactating sows and their litters and fatty acid composition of milk. *Animal Science.* 69:353–365.

72. Ostheim, J. (1992) Coping with food-limited conditions: feeding behavior, température preference, and nocturnal hypothermia in pigeons. *Physiol. Behav.* 51:353–361.

73. LeBlanc, J. (1992) Interactions between adaptation to cold and to altitude. In *High Altitude Medicine* (Ueda, G., ed.), Shinshu University Press, Matsumoto, pp. 475–481.

74. Christon, R., LeDividich, J., Seve, B. and Aumaitre, A. (1984) Influence de la température ambiante sur l'utilisation métabolique de l'énergie et de l'azote alimentaire chez le rat en croissance. *Reprod. Nutr. Develop.* 24:327–341.

75. Johnson, K.G. and Cabanac, M. (1982) Homeostatic competition in rats fed at varying distances from a thermoneutral refuge. *Physiol. Behav.* 29:715–720.

76. Chen, X.-J., Xu, X.-F. and Ji, X. (2003) Influence of body temperature on food assimilation and locomotor performance in white-striped grass lizards, *Takydromus wolteri* (Lacertidae). *Journal of Thermal Biology.* 28:385–391.

77. Cabanac, M. (1985) Strategies adopted by juvenile lizards foraging in a cold environment. *Physiol. Zool.* 58:262–271.

78. Cabanac, M. and Johnson, K.G. (1983) Analysis of a conflict between palatability and cold exposure in rats. *Physiol. Behav.* 31:249–253.

79. Stasiak, M. (2002) The development of food preferences in cats: the new direction. *Nutritional Neuroscience.* 5:221–228.

80. Balaskó, M. and Cabanac, M. (1998) Behavior of juvenile lizards (*Iguana iguana*) in a conflict between temperature regulation and palatable food. *Brain Behavior and Evolution.* 52:257–262.

81. Bunge, M. (1989) From neuron to mind. *NIPS.* 4:206–209.

82. Bunge, M. (2003) *Emergence and Convergence.* University of Toronto Press, Toronto.

83. McFarland, D.J. and Sibly, R.M. (1975) The behavioural final common path. *Philos. Trans. roy. Soc. London.* 270:265–293.

84. Cabanac, M. and Russek, M. (1982) *Régulation et contrôle en biologie.* Presses Université Laval, Québec.

85. Kennedy, G.C. (1950) The hypothalamic control of food intake in rats. *Proc. Roy. Soc. London.* 140 B:, 535–548.

86. Cabanac, M. and Lafrance, L. (1990) Postingestive alliesthesia: The rat tells the same story. *Physiol. Behav.* 47:539–543.

87. Schmidt, I. (1978) Interaction of behavioral and autonomic thermoregulation in heat-stressed pigeons. *Pflügers Arch.* 374:47–55.

7

Neural Circuits and Behavior

Theodore H. Bullock

1. INTRODUCTION

The hypothesis has become widely accepted that behavior is the output of neural circuits; that a one-to-one relation exists between behavior and the spatiotemporal pattern of neuronal activity. One way to look at the physiology of the nervous system, sense organs, and effectors (chiefly muscles and glands) is to ask whether the observed behavior can be "explained" in terms of these systems. By "explain," we usually mean describing what goes on at the next or another lower integrative level (organs, parts of organs, cells, parts of cells, molecules, ions, and processes including movements, modulations, reactions, triggering, blocking, gating, exciting, or inhibiting, etc.) leading to propagated signals that initiate or prevent organized and time-patterned activity of subsets, sets, and assemblies of sets of cells. The aims of this chapter are, first, to show that the hypothesis just stated can go very far in satisfying this definition of explanation (but *see* ref. *1*, Chapter 6, "How Far Does Connectivity Get Us?") and, second, to show that this "explanation" of behavior is quite inadequate and needs new terminology, new models, discovery of principles, phenomena and dependencies (which is to say, descriptive natural history), and the subsequent reductionist explanations of them.

2. ADEQUACY OF CONNECTIVITY

The last several decades have seen an exciting upswing in optimism that we will be able to explain a significant fraction of behavior because we have learned how to trace connectivity in the nervous system, with new techniques in *hodology* (the logos of pathways) and microelectrode recording of neuronal activity. The great simplifying concept we owe to the 19th century and Santiago Ramon y Cajal was the neuron doctrine—that the activity of the brain and the rest of the nervous system is essentially the activity of nerve cells (neurons). This concept was extended in the 20th century to the further simplifying belief that the relevant activity of neurons is all-or-none firing of propagated *nerve impulses* and their consequences, the *synaptic responses* of the target cells to which the active axons are connected. These two forms of activity can be recorded from neurons, extra- and intracellularly, confirming in general the pathways deduced from anatomical hodology. This optimism was greatly enhanced by the successes of invertebrate neurophysiology in finding *"identifiable"*

From: *Integrative Physiology in the Proteomics and Post-Genomics Age*
Edited by: W. Walz © Humana Press Inc., Totowa, NJ

neurons in many species, orders, even phyla of animals—unique cells that can be demonstrated in every individual of the species, with consistent responses to a preferred form of stimulation and consistent consequences of their activity, bespeaking consistent and characteristic connections, input and output *(2)*. Being able to find the same cell in each successive specimen studied—of a snail, a leech or polychaete annelid, an insect, crustacean or a nematode, makes it much easier to work out the connections (called the circuit) for given pieces of behavior and we now know the circuits of many such examples—for feeding, startle, flight or walking, cell by cell *(3–8)*. In other taxa, especially for vertebrates, circuits have been worked out with types of cells, rather than identifiable single cells. Outstanding examples are the circuits involved in production and recognition of bird song *(9)* and in postural reflexes *(10)* and associated eye movements *(11)*. Perhaps the best known piece of behavior in neural terms is the "jamming avoidance response" (JAR) in certain kinds of tropical freshwater, weakly electric fishes *(12,13)*.

3. EXAMPLES OF WELL-KNOWN BEHAVIOR EXPLAINED THROUGH 14 ORDERS OF NEURONS

The behavior (JAR) observed is that individuals of certain species among the gymnotiforms that maintain a relatively high-frequency electric organ discharge (EOD) extremely steadily, day after day, can shift the frequency very predictably when another fish comes within range and happens to have a slightly different EOD frequency. Both fish shift but in opposite directions, so as to increase the difference and avoid jamming, which would interfere with the object-locating function that enables these fish to detect and localize anything of a slightly different electrical impedance from that of the water, by the distortion of the fish's own EOD field. We have learned what is the adequate stimulus to initiate the JAR behavior and how that sensory input is processed in the brain, from two sets of electroreceptors widely distributed in the skin, one that follows the time course of the field amplitude of the mixed signals and their beat frequency, the other that follows the phase advance and retard through each beat cycle. When the phase advance accompanies the amplitude increase, cells that sample both sets of receptors stimulate the pacemaker cells to speed up their commands to the electromotor cells in the spinal cord, and vice-versa when the phase advance accompanies the amplitude decrease. The circuits also have to satisfy another requirement, by comparing input from converging subsets—namely receptors from different parts of the body. Another fish, whether detected by its EOD or as a passive island of different impedance, will cause the receptor population in the skin of a receiving fish to signal distortion unequally in the left and right or dorsal and ventral or rostral and caudal subsets and thus distinguish which EOD in the mixture is foreign and which is its own, at the same time indicating the direction, proximity, and character of the neighbor. The circuits also provide for other forms of modulation of the EOD, whether spontaneous or event-related social situations, explaining "chirps," silent periods, and the like. We have only begun to reveal the pathways and roles of connections from the pallium (areas of the surface of the cerebrum), subpallium, chemoreceptive regions, and hypothalamus) or to localize and characterize such "higher functions" as responses to novelty, different states of arousal or attention, expectation, and learned events. But we are confident, by modest extrapolation, that circuit analysis will explain a great deal of behavior

Circuit tracing and uncovering the dynamics of successive stages have turned up emergents and the corresponding questions about meaning, and mechanism, such as the convergence already referred to, where we compare two sources of inputs or combining submodalities. The prevalence of *reciprocal connections* has been as much a source of wonder as a help in explaining. How nervous systems deal with signal detection in the presence of large, *self-generated noise* has been investigated and explained, especially in several species of electroreceptive fish that use distinct principles *(14,15)*.

4. MULTIPLE MAPS OF THE SAME SENSORY SURFACE

The electrosensory skin of these fish *(16)* are perhaps the first, phylogenetically, apart from retinal projections on the midbrain tectum. The question of what differences there may be in the several maps or submodalities so mapped has been partly answered by the dynamics and best stimuli, but perhaps not completely. *Command cells* that determine the exact spatiotemporal sequence of discharge in the motor neuron pool can be simple rhythmic *pacemakers* or generators of relatively patterned trains of impulses, arising in a circuit or in a neuron. We do not know how much elaboration of brain function can be attributed basically to greater *numbers of neurons or synapses*, but this author is convinced that it plays a limited, although significant role and at least the major grades of complexity of central nervous function involve emergent properties in the realm of dynamics. This means that we are severly limited in the *utility of models* if they depend on extrapolating from relatively simplified networks.

5. MAJOR EMERGENTS

A class of major emergents is *induced rhythms (17)*, relatively regular discharging of nerve impulses or of nonimpulsive, subthreshold fluctuations of for example, dendritic origin. Another class is marked by synchrony of discharge of a population of neurons, whether of spikes or of subthreshold slow waves. One subset of these synchronized rhythms are the γ waves (25–50 or more Hz) of the human electroencephalogram (EEG) believed to be involved in some cognitive processes. Rhythms are prominent in many human EEGs, although we do not yet have any clear role, significance, or mechanism of their influence. It may be meaningful that such rhythms are very inconspicuous in most invertebrate brains and even in mammalian and human EEGs much if not most of the time and, when present, adding a fraction to the total power, except for special states such as seizures.

6. INTEGRATIVE VARIABLES

Arriving impulses may modulate a background spontaneity, may pattern it into periodic bursts, may show facilitation with successive impulses or its opposite, progressively diminished response, may sensitize or habituate, may change the effectiveness of an electrical transmission by a chemical transmitter, may act presynaptically on the axonal terminals about to end on a postsynaptic cell and may change the tuning, that is, the "best" (most effective) frequency (or spatial pattern of stimuli) in a target pathway. I have elsewhere listed *(18,19)* about 50 such variables. Each can exert either positive or negative influence in greater or lesser degree *(20,21)*, with linear or nonlinear input–output functions. Many new integrative variables have since then appeared in the literature, with no end in sight. Even in principle, to discern and allow for each of these variables at each stage in a pathway would take us well beyond the bare concept of circuits as usually used in physics and engineering. Modeling, which is an essential phase of explanation of neural systems, simply has to oversimplify in the face of so many analog variables, in addition to feedback and feed-forward , spontaneous backgrounds for each of some number of tonic states and the prevalence of central control of sense organ sensitivity or other properties, as well as receptivity at each stage in the afferent pathways, in nearly all sensory modalities. We have to deal with very limited segments of the system in order to include the input–output properties of each junction or neuropile, with parallel pathways, plasticity, and state dependence.

7. RELATIVE ROLES OF SPIKES AND SLOW WAVES

Two contrasting views co-exist on the relative importance of continuously graded basic mechanisms of communication between cells in the nervous system vs impulse-based, all-or-none "spikes" and synaptic junctions. The latter is the dominant view and is well established, with a great preponderance of hard evidence from single unit recording in a myriad of places. It fits with the evidence from afferent and efferent peripheral pathways, as well as with efforts to model the neural code in terms of the frequency and number of impulses. The presumptive code on this view is not digital, lacking a segmentation of the time into previously understood bins, but is an analog code with continuous variation of the intervals between impulses. The contrasting minority view recognizes and accepts the near ubiquity of impulses and synapses but underlines the possibility that some—perhaps much—communication between neurons and other cells may not rest on millisecond events with a sharp threshold, conducting nondecrementally over long distances (centimeters), but may depend on graded amplitudes and durations of slower events, on a wider category of cell-to-cell contacts with decrementing, electrotonic spread, and summation of such activity among neighbors. This is much more difficult to model or to find and recognize but it fits with evidence of spikeless neurons from the vertebrate retina and many invertebrate central neurons *(22)* and a variety of situations that have to represent nonsynaptic or parasynaptic communication *(23,24)*. Our understanding of these nonclassical situations is very imperfect but it is this writer's belief that some basic mechanisms quite distinct from the conventional spike and synapse are at work here and there, possibly quite widely spread and even in the mammalian cortex.

8. UNKNOWNS IN HOW CIRCUITS WORK

Quite often, we have details of some part of a circuit but we don't know just what it does for behavior, how it plays its role, or what function it performs on its input. In fact, it can be frustratingly difficult to close the loop –(i.e., to "explain" how the circuit works, as ref. *25* showed for the central pattern generator in quite thoroughly known 30-celled stomatogastric ganglion of lobsters, where every connection among the relevant cells is probably known in detail *[26]*). The very concept and terminology of circuitry may be inappropriate and limiting our understanding because it principally calls attention to the synaptic, that is, nerve impulse connectivity and whether each connection is excitatory or inhibitory, tending to overlook the crucially important dynamic properties, sometimes called the personality properties of each junction, axonal terminal ramification, and twig of the dendritic arbor. There are scores of dynamic properties, each with a range of alternatives, proclivities, and forms of response. The most important message of this chapter is that our ignorance of the ways that a system of neurons works is profound and basic. I expect some major revolution in our understanding of the first principles of coding, representation, and operation, going well beyond the several revolutions we have already weathered *(27,28)*.The connectivity or circuitry, as a metaphor, is probably limiting our thinking and our search for new principles of operation of the organized assemblies of cellular units, their system and subsystem organization into functional units, and forms of reciprocal influence, communication and representation of information. We need to find the language, imagery, and precedents that allow for the major steps in the evolution of nervous systems from the simplest invertebrates to the most complex mammals—a span of formidable magnitude, unapproached by any other in nature except systems of brains (societies).

9. EVOLUTION OF COMPLEXITY

These steps should be at the core of evolution and worthy of research to identify and uncover the dynamical mechanisms by which grades of complexity differ. I think we can point, as an example, to a saltation leading to synchronization of some aspects of neuronal activity, both all-or-none spike and subthreshold, nonspike activity in populations of cells, possibly involving astrocytic neuroglial as well as neuronal elements somewhere between the insect–crustacean–gastropod grade and the cephalopod–elasmobranch–amphibian grade, although we cannot at present propose what this would contribute to the performance, power, or cognitive capacity of the brain and its owner.

There are still so few such propositions that it is almost embarrassing. We need to compare the brains of relatively simple taxa with those of more complex grades of organization, including all kinds of traits, features, and properties, anatomical, physiological, chemical, susceptibility to disease, and to modulation, including the whole ethogram, which means all aspects of behavior, all discriminations the species can make, social, food-wise, predator-wise, place-wise, and all alternative responses in degree and kind, to find all the consequences of evolution. Most will no doubt be "horizontal"—adaptive radiations among groups at about the same grade of complexity and relatively few will, I suspect, be clearly signs of a higher grade—an advance in complexity. But, whether it is a trend or not, there is clearly, in my view, an enormous span of complexity *(29)* between the nervous systems of corals or jellyfish and polychaete worms, between snails or slugs and octopus or cuttlefish, between collembolans or brine shrimp and bees or lobsters, lampreys and manta rays, frogs and alligators, moles and primates. The differences between the simpler and the more complex brains, whether chemical, physiological, or anatomical must include some prime candidates for explaining the advances in behavior and something about how the brain works.

10. NEW PRINCIPLES ARE PROBABLE

The references on comparative neurology *(30–34)* parade examples of differences between taxa, virtually all in the gross or mesoscopic levels: anatomical, histological, and cytological, for example, the corpus callosum, the "six-layered."cerebral cortex; masses of small, cytoplasm-poor, chromatin-rich "globuli" or granule cells; the great variety of kinds of unipolar neurons, defined by their types of branching. Hardly any distinguishing features come from physiology (but *see* refs. *35,36*) or chemistry *(37)*. It seems most likely to this author, that we have to accept the situation that many degrees of freedom exist in any neural system just above the very simplest and least complex invertebrates. This remains true even after we disregard a number of integrative variables as necessary simplifying assumptions for purposes of modeling; this leaves us with a classically insoluble many-body problem. Only a very limited class of questions can be asked, modeled and solved analytically or by systematic parametric exploration of the space of so many variables. I am not despairing or

throwing hands in the air, with a cry that complexity makes it hopeless to explain or understand how nervous systems of moderate complexity actually work. Far from it. I expect
many years of research will continue to reveal secrets of how real brains work in the ample
space between, for example, recognizing visual shapes and playing the piano. We will not
soon run into a limit of what circuit-unraveling can disclose and what transform of the input
code is accomplished by the cerebellar cortex or the olfactory ganglion. What I am saying is
that despite the great advances in brain physiology represented by the enormous literature,
we actually do not understand what is going on and what the stages in the neural pathways
do to the spatiotemporal distribution of activity in the neuronal assembly. We know a great
deal about the pathways and about what sample neurons do in response to defined stimuli.
We have some quantitative information about changes in neuronal excitability at several
stages such as posttetanic potentiation or depression and a little bit about the local field
potential at the time some cells are initiating spikes or are pausing between impulses but
almost nothing at the system level about how to read the code or what the spikes or waves
mean in terms of representing information. We can make plausible models of the way motion might be represented or even visual shape or motor commands but not how retinal
ganglion cells can learn after a few regular flashes to anticipate the time the next flash is due
or fire only when a novel motion occurs, remembering for a short time when the last one
took place. We are used to finding cells that fire much more when a face is presented than to
any other stimulus, whether it is nearer or farther or tilted within limits or full face or side
view but we have no idea what circuit does this. The same is true of cells that fire best when
a certain reaching motion of the hand is seen, even when the motion is that of a neighbor or
in a mirror. We do not expect to find a reasonable model for such cells very soon, but we
could discover how many kinds of complex stimuli are treated in this way, by specialized
recognition cells, and what are the boundary conditions and the teachability of such properties. It is hardly constructive to say so, but this author suspects that some basic feature of
neuronal "personality," signaling or input–output variable has been missed!

REFERENCES

1. Bullock, T.H. (1993) *How Do Brains Work?* Birkhauser, Boston, New York.
2. Leonard, J. (guest ed.) (2001) Simple systems in comparative and integrative neurobiology: a Festschrift for the 85th birthday of Theodore H. Bullock. *Prog. Neurobiol.* **63**:365–485.
3. Ewert J.-P. (1980) *Neuroethology*. Springer Verlag, Berlin, New York.
4. Camhi J.M. (1984) *Neuroethology: Nerve Cells and the Natural Behavior of Animals.* Sinauer Associates, Sunderland, MA.
5. Huber F. and Markl, H. (1983) *Neuroethology and Behavioral Physiology*. Springer Verlag, Berlin, New York.
6. Aoki, K., Ishii, S. and Morita, H. (eds.) (1984) *Animal Behavior: Neurophysiological and Etho-logical Approaches.* Japan Scientific Societies Press, Tokyo; Springer-Verlag, Berlin, New York.
7. Carew, T.J. (2000) *Behavioral Neurobiology*. Sinauer Associates., Sunderland, MA.
8. Zupanc, G. K.H. (2004) *Behavioral Neurobiology*. Oxford University Press, Oxford.
9. Konishi, M. (1989) Birdsong for neurobiologists. *Neuron.* **3**:541–549.
10. Graf, W. and Brunken, W. J. (1984) Elasmobranch oculomotor organization: anatomical and theoretical aspects of the phylogenetic development of vestibulo-oculomotor connectivity. *J. Comp. Neurol.* **227**:569–581.
11. Eckmiller, R. and Mackeben, M. (1978) Pursuit eye movements and their neural control in the monkey. *Pflügers Arch.* **377**:15–23.
12. Bullock, T.H. and Heiligenberg,W. (eds.) (1986) *Electroreception*. John Wiley & Sons, New York, NY.

13. Bullock, T.H. and Zupanc, G.K.H. (2004) Survey of electroreception. In *Electroreception* (Bullock T.H. and Hopkins, C.D., eds.), Springer Verlag, Berlin, New York. (In press)

14. Grant, K., Bell, C. and Han, V. (1996) Sensory expectations and anti-Hebbian synaptic plasticity in cerebellum-like structures. *J. Physiol. (Paris)*. **90**:233–237.

15. Bodznick, D. and Montgomery, J. (2004) Physiology of low frequency electroreception.In *Electroreception* (Bullock, T.H. and Hopkins, C.D., eds.), Springer Verlag, Berlin, New York. (In Press)

16. Kawasaki, M. (2004) Physiology of high frequency electroreception, In *Electroreception* (Bullock, T.H. and Hopkins, C.D., eds.), Springer Verlag, Berlin, New York. (In Press)

17. Basar, E. and Bullock, T.H. (1992) *Induced Rhythms in the Brain*. Birkhäuser, Boston, MA.

18. Bullock, T. H. (1966) Integrative properties of neural tissue. In *Frontiers in Physiological Psychology* (Russell, R.W., ed.), Academic Press, New York, NY, pp. 5–20.

19. Bullock, T. H. with the collaboration of, Orkand, R. and Grinnell, A.D. (1977) *Introduction to Nervous Systems*. W. H. Freeman, San Francisco, CA.

20. Bullock, T.H. (1979) Communication among neurons includes new permutations of molecular, electrical, and mechanical factors. *Behav. Brain Sci.* **2**:419.

21. Bullock, T.H. (1987) The micro-EEG manifests varied degrees of cooperativity among wideband generators. *II Internat. Conf. on Dynamics of Sensory and Cognitive Processing of the Brain, West Berlin*, pp. 11–12.

22. Bullock, T.H. (1981) Spikeless neurones: where do we go from here? In *Neurones Without Impulses* (Roberts, A. and Bush, B.M.H., eds.), Cambridge University Press, Cambridge, MA, pp. 269–284.

23. Schmitt, F.O. (1984) Molecular regulators of brain functions, a new view. *Neuroscience.* **4**:991–1001.

24. Watanabe, A. and Bullock, T.H. (1960) Modulation of activity of one neuron by subthreshold slow potentials in another in lobster cardiac ganglion. *J. Gen. Physiol.* **43**:1031–1045.

25. Selverston, A.I. (1980) Are central pattern generators understandable? *Behav. Brain Sci.* **3**:535–571.

26. Selverston, A.I., Russell, D.F., and Miller, J.P. (1976). The stomatogastric nervous system: structure and function of a small neural network. *Prog. Neurobiol.* **7**:215–290.

27. Bullock, T.H. (1959) Neuron doctrine and electrophysiology. *Science.* **129**:997–1002.

28. Bullock, T.H. (1984) Comparative neuroscience holds promise for quiet revolutions. *Science.* **225**:473–477.

29. Bullock, T. H. (2002) Grades in neural complexity: how large is the span? *Integ. Comp. Biol.* **42**:317–329.

30. Cajal, S.R.Y. (1909) *Histologie des Systeme Nerveuse, de l'homme et des Vertebres*. Consejo Superior de Investigacions Cientificas, Madrid.

31. Nieuwenhuys, R., ten Donkelaar, H.J., and Nicholson, C. (1998) *The Central Nervous System of Vertebrates*. Spring, Berlin, New York.

32. Butler, A.B. and Hodos, W. (1996) *Comparative Vertebrate Neuroanatomy: Evolution and Adaptation*. Wiley-Liss, New York, NY.

33. Ariëns Kappers, C.U., Huber, G.C. and Crosby, E.C. (1936) *The Comparative Anatomy of the Nervous System of Vertebrates, Including Man*. Macmillan, New York, NY.

34. Bullock, T.H. and Horridge, G.A. (1965) *Structure and Function in the Nervous System of Invertebrates*. W. H. Freeman, San Francisco, CA.

35. Bullock, T.H. (1997) Signals and signs in the nervous system: the dynamic anatomy of electrical activity is probably information-rich. *Proc. Natl. Acad. Sci. USA.* **94**:1–6.

36. Bullock, T.H. (2004) Biology of brain waves: natural history and evolution of an information-rich sign of activity. *Clin. EEG.* (In press).

37. Iversen, S. and Iversen, L.L. (1984) *Behavioral Pharmacology*. Oxford University Press, Oxford.

8

Physiological Determinants of Consciousness

Mircea Steriade

1. WHAT IS CONSCIOUSNESS AND WHERE IS IT GENERATED?

It is taken for granted that everyone has a rough idea of what is meant by consciousness, but we know as long as no one asks us to define it *(1)*. This is why some consider that "it is better to avoid a *precise* definition of consciousness because of the dangers of a premature definition" *(2)*. Still, many theoretical writers and even some basic neuroscientists play with the global notion of consciousness in very different ways. For most authors, the notion covers sensations, mental images, thoughts, and volition. Special emphasis was also placed on emotions and feelings as important components of consciousness toward the emergence of self *(3,4)*. In dictionaries, consciousness is simply defined as the state of being awake or, sometimes more subtly, as an organism's *awareness* of its own self and surroundings. Awareness exists in animals, whereas self-awareness and evolved forms of consciousness that implicate complex plans in anticipated future probably restrict this notion to humans. The view that consciousness depends on awareness and arises from a background of brain arousal led most commentators to conclude that consciousness is what abandons us every evening and reappears the next morning when we wake up. However, peculiar types of consciousness, such as mentation with illogical thought and bizarre feelings, occur during dreaming in rapid-eye-movement (REM) sleep, whereas during deep stages of slow-wave sleep (SWS) dream mentation is much closer to real life *(5–7)*.

The difficulties in defining consciousness and deciding whether it can or cannot be explored with physiological tools become insurmountable when introducing its critical ingredient, subjectivity. An elaborated theory *(8)* considered different elements of (a) primary consciousness that consists of sensations, perceptual categorization, and a system for memory and learning; and (b) higher order consciousness that emerges from the above, develops with a concept of self, and culminates with a linguistic system dependent on social interactions. Even basic elements of the first-order consciousness, perceptual experiences, imply subjective states, but it is common agreement that the physiological mechanisms behind the emergence of subjectivity are hidden *(9)* because conscious experience is accessible only to the subject having the experience, not to an external observer *(10)*.

Other opinions state that qualia, which denote subjective experience of any type generated by the nervous system *(11)*, "arise from, fundamentally, properties of single cells, amplified by the organization of circuits specialized in sensory functions" *(12)*. The second part of the sentence, which calls on complex, yet unspecified, neuronal loops, softens the

From: *Integrative Physiology in the Proteomics and Post-Genomics Age*
Edited by: W. Walz © Humana Press Inc., Totowa, NJ

extremely reductionist definition of qualia emerging from single neurons. It is the opinion of this author that intrinsic subjective experiences are not accessible without the benefit of language. Without this human-specific property, neuronal studies of consciousness, viewed as a *global* notion (but *see* Section 3 for some elements lying at the basis of consciousness), become impossible because cellular recordings cannot be systematically made in humans from the multiple brain structures that contribute to the buildup of consciousness. On the other hand, animals can perform behavioral tasks and display motor behaviors, but do not possess the virtue of clearly expressing their subjective states. This is the major obstacle towards the understanding of how action potentials in the geniculo-cortical pathways would give rise to a subjective visual experience, which is just a first step in the complexity of consciousness. Moreover, if verbal reports were accessible, they may fall short under some circumstances. The conclusion is that "exactly how something as ineffable as subjective consciousness can arise from macromolecules, synapses and action potentials may well remain a conundrum" *(13)*.

Only a few neuroscientists would question the fact that consciousness is generated within the brain. A notable exception was Eccles *(14,15)*, a pioneering investigator of synaptic transmission. While Shakespeare oscillated in *The Merchant of Venice* between Aristotle's idea that the acropolis is the heart and the rather vague localization of imagination in the head, subsequent philosophers and physiologists have firmly thought that the seat of different components of consciousness is in the brain. In the 17th century, Gassendi and Willis localized mental functions, among them memory, in cortical circonvolutions *(16)* . These attempts to lower the spirit from immortal levels to senses and their central projections led in the 19th century to Carpenter's idea that consciousness is generated in the cerebral cortex and is projected downward to the thalamus and rostral parts of the mesencephalon *(17)* . This preceded Penfield's concept of a "centrencephalic" system, which was implicated in the unification of physiological and mental functions *(18)* (*see* more recent views on the "centrencephalic" system and its relation with unconsciousness during absence epilepsy in Section 5.).

The evolution of the nervous system provided structural advantages that lie at the basis of consciousness. Thus, the initial biological forms in which motor cells also respond to sensory signals have developed into a stage with specialized cells for input signaling or output reactions, finally reaching the mature state of nervous systems characterized by the appearance of interneurons within virtually all brain structures. Even an elementary sensory system, such as olfaction, which is similar across animal phyla *(19,20)*, is modulated to higher nervous activities. At least in mammals, smelling is dependent on sniffing, which is thought to be an attentional mechanism in olfaction, priming the piriform cortex for the arrival of odor information *(21,22)*. In some degenerative conditions, patients display an olfactory deficit that is partly ascribed to an inability to sniff, not an inability to smell *(23)*. As well, the most stereotyped motor patterns, elaborated at the lowest levels of the central nervous system, are subject to decisive influences arising in higher levels. Earlier views considered that "command neurons" or neuronal networks exerted their functions independently of afferent inputs and superimposed systems. In contrast to these ideas, the command apparatus of locomotion, which can be studied in simple vertebrate systems, such as the lamprey *(24,25)*, includes not only sensory inputs and the spinal cord, with its complex interactions between different types of interneurons and motoneurons, but also a series of supraspinal structures, up to the forebrain *(26)*.

I do not think that the seat of consciousness, as a global entity, can be more precisely specified than considering together reciprocal corticothalamic systems, neuronal circuits in the hippocampus and related structures, and generalized modulatory systems that include, to put it simply, the upper brainstem core, posterior hypothalamus, and basal forebrain. This panoply is quite large and reflects my reluctance to pinpoint more circumscribed brain structures. By contrast, other authors feel comfortable to consider that "consciousness depends crucially on thalamic connections with the cortex" *(2)* or, even more precisely, on reverberations in neuronal loops formed by specific and intralaminar thalamic nuclei with neocortical areas *(27,28)*. It was also suggested that various neocortical areas may play preferential roles in generating conscious phenomena, and the hypothesis was formulated that the primary visual cortex should be excluded from the neuronal assemblies that generate visual awareness because it does not project to the frontal cortex of macaques and, thus, does not have access to fields involved in planning and execution of voluntary motor control, as hypothetically required by visual awareness *(29)*. However, some expressed the view that there is no *a priori* necessity for neurons in the perceptual space to communicate directly with those in decision space *(30)* and others *(31)* pointed to two distinct aspects of consciousness, "phenomenal" and "access," the former being just experience, the latter poised for direct control of action.

The issue is the very numerous brain structures implicated in cognitive and conscious phenomena, such as perirhinal areas and hippocampus in memory, amygdala in the affectivity that colors memory, frontal cortex in decision-making processes, at least four or five ascending activating brainstem and forebrain systems that provide the necessary background of brain activation without which consciousness in wakefulness and dreaming in sleep are not possible, and I enumerated just some of the required structures *(32)*. Do neuronal processes underlying conscious experience extend to most of the brain, as envisaged by James *(1)*? The "dynamic core hypothesis" *(33)* proposed that participating large clusters of neuronal groups are much more strongly interactive among themselves than with the rest of the brain. The search for a neuronal substrate of consciousness led to such an enlargement of brain structures that the "cortical system" was recently considered to include not only the thalamus, but also "the basal ganglia, the cerebellum and many widespread brainstem projection systems" *(34)*. This may not be far from the reality (if one also adds many other structures), but such diversity may prevent experimenters to reveal the *specific* neuronal types that have been hypothesized to generate consciousness (*see* Section 2.).

To summarize, such a multitude of brain areas acting in concert during complex conscious events prevents us to circumscribe the seat of consciousness within thalamocortical or other brain systems.

2. IS IT POSSIBLE TO ASCRIBE CONSCIOUSNESS TO SPECIFIC NEURONAL TYPES?

It is obvious that, if consciousness cannot be localized within circumscribed brain systems (*see* previous discussion), it becomes even more difficult to specify some neuronal types that would be so privileged that their activity gives rise to subjective experience.

Some authors were tempted to specify the neuronal progenitors of visual consciousness. They focused on bursting neurons located in the deep cortical layer V, and thought that "it is only a question of time before specific molecular markers are found" for these neurons *(35)*. Bursting neurons located in layer V have initially been described in cortical in vitro slices and characterized by their intrinsic property to fire high-frequency spike-bursts in response to depolarizing current steps *(36,37)*. Later on, however, such intrinsically bursting (IB) neurons have also been found in more superficial layers, both in vivo *(38)* and in vitro *(39,40)*. Then, IB neurons are not confined to deep cortical layers. More importantly, the firing pattern of IB neurons is transformed into that of regular-spiking (nonbursting) neurons by slight changes in the level of membrane depolarization, as is the case during shifts from SWS to brain-activated states of waking and REM sleep *(41)*. Furthermore, the emphasis placed on IB neurons as generators of awareness is challenged by the fact that, although such neurons are found in a great proportion in cortical slices in vitro or in cortical slabs in vivo *(42,43)*, they represent a negligible proportion (5%) during waking *(41)*, the behavioral state in which awareness arises.

Similar transformations of firing patterns (which are produced by intrinsic cellular properties) by synaptic activities are seen in another cortical cell-type, fast-rhythmic-bursting (FRB) neurons. These neurons fire high-frequency (300-500 Hz) spike-bursts, recurring at fast rates (20–40 Hz), during silent periods of cortical networks, but this pattern is transformed into that of a fast-spiking neuron, with tonic firing and absence of frequency accommodation, during epochs with increased synaptic activity, arising in the thalamus *(44)*. Moreover, far from being confined to superficial cortical layers and limited to pyramidal-shaped neurons as initially assumed *(45)*, FRB neurons are also pyramidal neurons in the deep layers (from where they project to the thalamus) as well as, some of them, local-circuit inhibitory neurons *(44)*.

The above features indicate that the electrophysiological characteristics of the two types of cortical bursting neurons (IB and FRB) are much less inflexible than envisaged by some investigators. In extremely simplified preparations, such as slices maintained in vitro, firing patterns are virtually fixed because there is absence of, or negligible, background synaptic activity. In the intact brain, the intrinsic properties of neurons in neocortex and other central structures are overwhelmed by intense synaptic activity *(32,41)*. As to the bursting neurons in deep layers, a neuronal class that was hypothesized to play a crucial role in awareness *(35)*, their very low proportion during waking casts much doubt on this idea. No particular neuronal type is supposed to possess the privileged role of generating consciousness because this emergent function of the brain required cooperation among multiple cellular types and distributed neuronal systems in numerous structures.

3. SOME ELEMENTS OF CONSCIOUSNESS CAN BE INVESTIGATED AT THE NEURONAL LEVEL

Let us take a more optimistic stance and mention that certain brain functions that lie at the basis of consciousness can be investigated using neuronal recordings in the intact brain. Yet, these studies cannot localize at the neuronal level highly differentiated processes, such as self-awareness, not to mention subjective experiences.

The neuronal systems that maintain an activated state of the cerebrum, which is necessary for most conscious events, have been investigated and the main circuits that are responsible for the waking and dreaming states have been identified. To begin with, none of these four or five systems is necessary and sufficient. One of the most important awakening systems lies in the brainstem reticular (cholinergic, glutamatergic, and some monoamine-containing) cellular aggregates that project to the thalamus, and from the thalamus to the cerebral cortex.

This system was described long ago *(46)* and the brainstem reticular core was viewed as a monolith. Subsequently, the reticular formation was dissected into multiple chemically coded compartments and analyzed at the cellular level *(47)*. Another important system, which releases acetylcholine in the cerebral cortex, arises in the basal forebrain *(48)*. Activation of brain electrical rhythms, which mimics behavioral awakening from sleep, can be obtained by setting into action the ascending brainstem reticular system after either thalamectomy or extensive basal forebrain lesion *(49)*. Thus, none of these two major targets of brainstem-activating projections is necessary and, indeed, the loss of one structure may be compensated by the remaining, intact system. It was long known that rostral midbrain transection produces an acute comatose state, associated with electroencephalogram (EEG) activity and ocular signs (extreme myosis) characteristic for sleep *(50)*. If the animal is maintained an extended length of time (10–20 d), periods of electrical signs typical for waking appear and the myosis becomes less marked *(51,52)*. This evolution is the result of the presence of other (not necessarily redundant) activating systems, in front of the mesencephalic transection, such as dorsal amygdala whose stimulation activates the forebrain even in the absence of the brainstem *(53)* or other activating structures, such as posterior hypothalamus and basal forebrain, which have direct projections to the cerebral cortex. Nonetheless, the mesopontine reticular formation and its projections to thalamocortical systems (notably to thalamic intralaminar nuclei) play a cardinal role in forebrain activation processes, as shown by several lines of evidence: (a) prolonged somnolence after bilateral lesions of thalamic intralaminar nuclei in humans *(54–56)*; (b) a crucial role played by the brainstem-thalamic (intralaminar) system in attention-demanding tasks, beyond diffuse arousal, demonstrated in humans *(57)*; (c) increased activities in mesopontine reticular cholinergic neurons in advance of the most precocious signs indicating transition from SWS to waking or REM sleep *(58)*; and (d) initiation of ponto-geniculo-occipital waves ("the stuff that dreaming is made of") by thalamically projecting mesopontine tegmental neurons *(59)*.

Besides the disclosure of forebrain activation processes during waking and dreaming sleep at the neuronal level, other brain functions that are components of consciousness have also been explored using neuronal recordings. This is the case of the role played by the ventral visual cortical pathway *(60)* and thalamic pulvinar neurons *(61)* in selective attention of non-human primates, different forms of memory related to neuronal activities in the frontal lobe of macaques *(62,63)*, and neuronal basis of fear and emotional processes in the nuclear complex of amygdala *(64)*.

More global methods have been used in humans. The perceptual interpretation of competitive signals from multimodal sensory modalities was studied using event-related functional magnetic resonance imaging (fMRI) *(65)*. The use of fMRI also helped to demonstrate that different visual cortical areas of humans have response properties *(66)* similar to those obtained by single-unit recordings in monkeys *(67)*, namely, nonselective responses in the primary visual cortex, selective responses to meaningful stimuli (such as faces) in later visual areas of the temporal cortex, and sustained activity during memory delays in temporal and prefrontal cortices. Combined fMRI and magnetoencephalography revealed dissociations between the medial and lateral orbitofrontal/prefrontal cortical areas during negative and positive emotional processing *(68)* and disclosed a new positive event during novelty detection in the middle part of the superior temporal gyrus *(69)*. Finally, human scalp EEG provided direct evidence for interactions between the increased neocortical and thalamic fast rhythms during semantic memory recall in humans *(70)*.

Needless to say, human studies cannot identify the specific neuronal circuits and cellular types that have been implicated in consciousness, and experimental studies may reveal neuronal activities in some brain areas, but each of the above mentioned components of consciousness involves so many networks distributed across the cortex and subcortical structures that render systematic simultaneous recordings unfeasible.

4. CONSCIOUSNESS DURING TWO STAGES OF SLEEP

There is general consensus that REM sleep is accompanied by mental activity, that is, dreams with vivid perceptions generated internally and illogical thought. The other state of sleep, with widely synchronized, low-frequency rhythms (SWS), was long regarded as associated with global inhibition of the cerebral cortex and subcortical structures *(71)*, which would underlie an "abject annihilation of consciousness" *(72)*. In contrast to these assumptions that considered that the cerebral cortex is completely quiescent during SWS, even if "occasionally, at places in it, lighted points flash or move but soon subside" *(73,* p.183), studies using intracellular recordings of all four electrophysiologically characterized cortical cell types in naturally sleeping animals showed unexpectedly high levels of spontaneous neuronal activity *throughout* SWS *(41)* (Fig. 1). Although the thalamic gates are closed for signals from the outside world during SWS, because of inhibition of synaptic transmission in the thalamus *(74)*, the responsiveness of cortical neurons and the intracortical dialogue are maintained and even increased during SWS, thus suggesting that this sleep stage may serve important cerebral functions during this sleep stage. The first clear idea relating SWS with plastic activity in the cerebrum postulated that sleep does not concern the fast recovery processes in routine synapses underlying stereotyped activities, but the slow recovery of learned synapses *(75)*. The hypothesis postulating consolidation, during SWS, of memory traces acquired during wakefulness in thalamocortical networks and hippocampus *(76–78)* has recently been tested and corroborated by studies in humans and experimental data in animals.

Human studies have shown that the overnight improvement of discrimination tasks requires several steps, some of them depending on the early SWS stages *(79,80)*. The improvement of visual discrimination skills by early stages of SWS led to the conclusion that procedural memory formation may be associated with oscillations during these stages *(81)*. These authors have also shown that, after training on a declarative learning task, the density of human sleep spindles is significantly higher, compared to the nonlearning control task *(82)*. Dreaming mentation is not confined to REM sleep, but also appears closer to real-life events during SWS sleep *(83)* and the recall rate of dreaming mentation in SWS is quite high *(84)*.

The neuronal operations that may account for the above relations between slow-wave and some forms of mentation as well as improved discrimination after learning tasks should be searched in the rhythmic spike-bursts and spike-trains fired by thalamic and cortical neurons during oscillations that characterize this sleep state, which are behind neuronal plasticity and consolidation of memory traces *(see* below). There are three major types of SWS oscillations: spindles (7–15 Hz), δ waves (1–3 Hz), and slow oscillation (0.5–1 Hz). Spindles arise within the thalamus and the thalamic reticular GABAergic neurons have a pacemaker role in their generation *(85,86)*. There are two components of δ waves, one of them being generated in the thalamus and the other in the cortex *(87)*. The slow oscillation is generated within the neocortex as it survives thalamectomy and high brainstem transection *(88)* and can be recorded even in isolated neocortical slabs in vivo *(43)* as well as in cortical slices in vitro

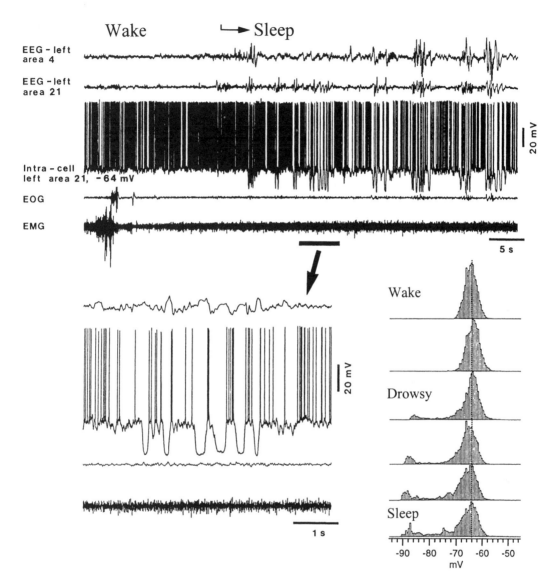

Fig. 1. Natural slow-wave sleep (SWS) is characterized by prolonged hyperpolarizations in neo-cortical neurons, but rich spontaneous firing during the depolarizing phases of the slow oscillation. Chronically implanted cat. Five traces in top panel depict electroencephlogram (EEG) from the depth of left cortical areas 4 (motor) and 21 (visual association), intracellular recording from area 21 regu-lar-spiking neuron (resting membrane potential is indicated), electrooculogram (EOG) and elec-tromyogram (EMG). Transition from waking to SWS is indicated by arrow. Part marked by horizontal bar is expanded below left (arrow). Note relation between the hyperpolarizations and depth-positive EEG field potentials. Below right, histograms of membrane potential (10-s epochs) during the period of transition from waking to SWS depicted above. Note membrane potential around –64 mV during the 20 s of waking and progressively increased tail of hyperpolarizations, from drowsiness to full-blown SWS, up to –90 mV. (Data from experiments by M. Steriade, I. Timofeev and F. Grenier; see details in ref. *41*.)

(89). Despite this variety of rhythms, in the intact brain the cortically generated slow oscillation has the virtue of grouping not only the other SWS oscillations (Fig. 2) but also faster (β and γ) rhythms *(90)*, which are conventionally associated with only waking and REM sleep. This concept of coalescence of brain rhythms, as a corticothalamic effect of the slow oscillation, is now supported by studies on humans *(91)*. Moreover, neocortical sleep oscillations, both spindles and slower rhythms, are related to hippocampal rhythms, and this coactivation may translate short-term hippocampal memory to longer term neocortical stores *(92,93)*.

The rhythmic and synchronized spike-bursts that occur during sleep spindles in thalamocortical neurons depolarize the dendrites of neocortical neurons. The depolarization is associated with Ca^{2+} entry into the dendrites *(94)* and the hypothesis was advanced *(95)* that Ca^{2+} entry may provide an effective signal to efficiently activate Ca^{2+} calmodulin-dependent protein kinase II, which is implicated in synaptic plasticity of excitatory synapses in cortex and other sites in the nervous system *(96)*. Similar phenomena, with Ca^{2+} entry in cortical neurons, may occur in SWS during the rhythmic spike-trains associated with oscillations in the slow (0.5–1 Hz) or δ (1–4 Hz) frequency bands, and could provide the long-time scales needed to mobilize the machinery that was hypothesized as responsible for the consolidation of memory traces acquired during the state of wakefulness *(76)*.

The hypothesis that sleep oscillations lead to neuronal plasticity and consolidation of memory traces has been tested using the experimental model of sleep spindles, namely, augmenting responses *(97)*, which are, like spindles, waxing and waning cortical potentials evoked by thalamic stimuli within the frequency range of spindles, approx 10 Hz. Augmenting responses can be obtained in the thalamus, even in decorticated animals *(98)*, in intact thalamocortical neuronal networks *(99)*, and in the isolated cortex both in vitro *(100)* and in vivo *(101)*. In the intact brain, augmenting responses are modulated by the behavioral state of vigilance, displaying the maximal amplitudes during SWS and being partially or totally obliterated during waking and REM sleep *(102)*.

The neurons that preferentially display augmenting responses in vivo are FRB cells (*see* Section 2.), which fire fast (30–40 Hz) high-frequency (200–500 Hz) spike-bursts and are recorded from all, superficial and deep cortical layers from II to VI. The crucial role played by these neurons in the widespread synchronization of augmenting responses results from their thalamic projections and feedback projections to cortex of the resulting activity *(44)*, even toward areas that are remote from the site where the primary corticothalamic drive originates *(103,104)*. In vitro studies have shown that layer V IB neurons have a major role in the generation of augmenting responses *(100)*. The difference between the results from in vitro and in vivo investigations (namely, preferential role attributed to IB and FRB neurons, respectively) may be ascribed to changing incidences of IB neurons in various experimental conditions, namely, IB neurons may reach very high proportions (up to 40–50%) in slices maintained in vitro *(42)* or in cortical slabs prepared in vivo *(43)*, but the incidence of this neuronal type is lower in the intact cortex and, in naturally alert animals, IB neurons represent less than 5% of sampled neurons *(41)*. The variable incidence of different cell classes under different experimental conditions and the overwhelming effect of synaptic activity on intrinsic neuronal properties have been discussed elsewhere *(32,105)*.

Prolonged rhythmic stimuli, used to elicit cortical augmenting responses, lead to progressively reduced inhibition, depolarization and enhanced neuronal responsiveness (Fig. 3A). The cellular mechanism of these responses may depend on the activation of high-threshold Ca^{2+} channels *(106,107)*. The self-sustained oscillations that are identical to the frequency

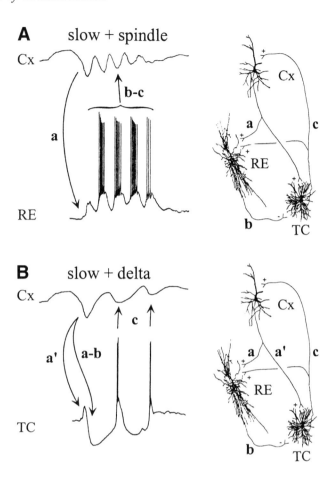

Fig. 2. Coalescence of the cortical slow oscillation with other slow-wave sleep (SWS) rhythms generated in the thalamus. In the left column, two traces represent field potential from the depth of association cortical area V and intracellular recording from thalamic reticular neuron (top and bottom traces, respectively); below, traces represent field potential from the depth of cortical area V and intracellular recording from thalamocortical ventrolateral neuron. In the right column, circuits involved in the generation of the respective SWS pattern. The synaptic projections are indicated with small letters, corresponding to the arrows at left, which indicate the time sequence of the events. (A) Combination of slow oscillation with a spindle sequence. The depolarizing phase of the field slow oscillation (depth-negative, downward deflection, also called K-complex) in the cortex (*Cx*) travels through the corticothalamic pathway (*a*) and triggers in the thalamic reticular nucleus (*RE*) a spindle sequence that is transferred to thalamocortical cells (*ThCx*) of the dorsal thalamus (*b*) and thereafter back to the cortex (*c*) where it shapes the tail of the slow oscillatory cycle. (B) Modulation of slow oscillation by a sequence of clock-like delta waves originating in the thalamus by interplay between two inward currents (I_H and I_T) of thalamocortical neurons. The synchronous activity of cortical neurons during the slow oscillation (depth-negative peak of cortical field potential) travels along the corticothalamic pathway (*a'*) eliciting an excitatory postsynaptic potential, curtailed by an inhibitory postsynaptic potential produced along the cortico-RE (*a*) and RE-ThCx (*b*) projections. The hyperpolarization of the thalamocortical cell generates a sequence of low-threshold potentials crowned by high-frequency spike-bursts at δ frequency that may reach the cortex through the thalamocortical link (*c*). Intracellular staining of three neuronal types (reconstructed) modified from Steriade et al. *(77)*, diagrams modified from Amzica and Steriade *(130)*, and intracellular recordings by Steriade et al. *(76)* and Contreras and Steriade *(131)*.

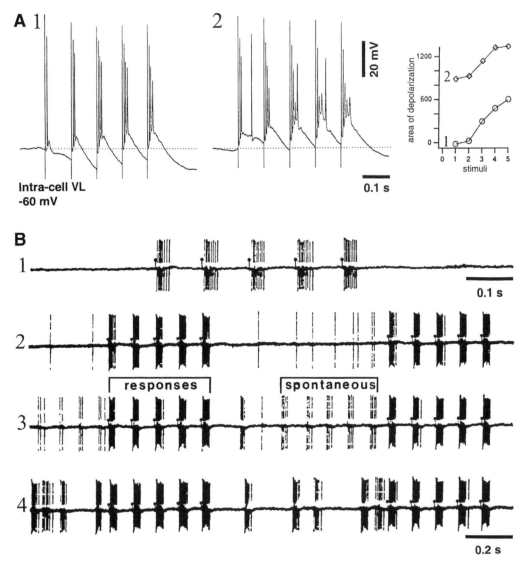

Fig. 3. Short-term plasticity during and after intrathalamic and corticothalamic augmenting responses. (**A**) Decorticated cat under anesthesia. Intracellular recording from thalamocortical vent-rolateral (VL) neuron. Pulse-trains consisting of five stimuli at 10 Hz were applied to the thalamic VL nucleus every 2 s. The type of augmenting illustrated here is high-threshold, occurring at a depo-larized level. Responses to four pulse-trains (*1–4*) are illustrated (*1* and *2* were separated by 2 s; *3* and *4* were spearated by 2 s and followed 14 s after *2*). With repetition of pulse-trains, inhibitory postsyn-aptic potentials (IPSPs) elicited by preceding stimuli in the train were progressively reduced until their complete obliteration and spike-bursts contained more action potentials with spike inactivation. The graph depicts the increased area of depolarization from the first to the fifth responses in each pulse-train as well as from pulse-train *1* to pulse-trains *3* and *4*. (**B**) Brainstem-transected cat. Corti-cally evoked spike-bursts in thalamic VL neuron (*1*). Motor cortex stimulation was applied with pulse-trains at 10 Hz delivered every 1.3 s. In *1*, the pattern of cortically evoked responses at the onset of rhythmic pulse-trains (faster speed than in *2* to *4*). *2–4*, responses at later stages of stimula-tion. Stimuli are marked by dots. In *2* to *4*, stimuli and evoked spike-bursts are aligned. Note progres-sive appearance of spontaneous spike-bursts resembling the evoked ones, as a form of "memory" in the corticothalamic circuit. (Modified from refs. *98* and *102*.)

of responses elicited during the prior period of stimulation (Fig. 3B) are regarded as the "memory" of thalamocortical and corticothalamic neuronal circuits *(108)*. In the intact brain, cortical stimuli applied during the depolarizing envelope of spindle sequences accompanied by firing elicit an enhancement of the control response, which may last from tens of seconds to several minutes *(101)*. One mechanism that may explain this increased responsiveness is the high-frequency firing in response to rhythmic, repeated pulse-trains, which would result in activation of high-threshold Ca^{2+} currents and enhanced $[Ca^{2+}]_I$ that, in association with synaptic volleys reaching the neuron, may activate protein kinase A *(109)* and/or *Ras*/mitogen-activated protein kinase *(110)*, which are involved in memory consolidation.

The above experimental data and human observations show that, far from being a behavioral state during which all mental activity is suspended, the neuronal circuitry in corticothalamic systems is endowed during slow-wave sleep with the capacity to store information acquired during the waking state. Beyond a certain limit, however, normal aspects of neuronal plasticity, which are beneficial for memory consolidation, develop into paroxysmal states that are electrographically similar to those of some epileptic fits.

5. NEURONAL BASIS OF UNCONSCIOUSNESS DURING SOME EPILEPTIC SEIZURES

The dramatic increase in excitability of cortical FRB neurons (*see* Sections 3. and 4.), which occurs during rhythmic repetition of thalamic or cortical volleys in the frequency range of sleep spindles, mimics epileptiform responses *(108,111)*. This raises the intriguing question of the transformation, during SWS, from plasticity underlying normal mnemonic functions into abnormal forms that underlie the pathological synchronization of brain electrical activity, as occurring in absence seizures and Lennox-Gastaut syndrome *(112)*.

The term *absence* (unconsciousness) is valid only for those seizures that occur during wakefulness because during SWS, when there is an increased incidence of such paroxysms, or in experimental animals under anesthesia, the subjects prone to these seizures are already quite absent. At variance with the classical spike-wave complexes at approx 3 Hz that characterize absence seizures, the spike-wave complexes in the Lennox-Gastaut syndrome are relatively slower (1.5–2.5 Hz) and they are often associated with fast runs at 10–20 Hz, but up to 30 Hz *(113)*. As is also the case with the sleep-related absence seizures, the close relation between Lennox-Gastaut seizures and the behavioral state of SWS was repeatedly mentioned in human and animal studies *(113–115)*. Most investigators pointed to intracortical neuronal networks as generators of these seizures because of focal disturbances of cortical metabolism and favorable outcome for overall seizure frequency after callosotomy or focal corticectomy *(116)*. Experimental studies also demonstrated the cortical origin of spike-wave seizures because their cortical occurrence was not prevented by thalamectomy *(117)*.

Thus, these more recent data do not support the concept of a "centrencephalic system" *(118)* that would have explained the conventional definition of absence seizures as suddenly generalized and bilaterally synchronous. Nor do recent research support the claim that so-called "absence" seizures may be induced in cortex by electrical stimulation of midline thalamic nuclei at 3 Hz, as proposed in an earlier study conducted at a critical level of barbiturate anesthesia *(119)*. In that study, only spike-wave-like *responses* were evoked in cortex, but no self-sustained activity. As to the presumed "centrencephalic" system that would produce bilaterally synchronous spike-wave complexes, it should be mentioned that there are no bilaterally projecting thalamic neurons. Brainstem core neurons with generalized projec-

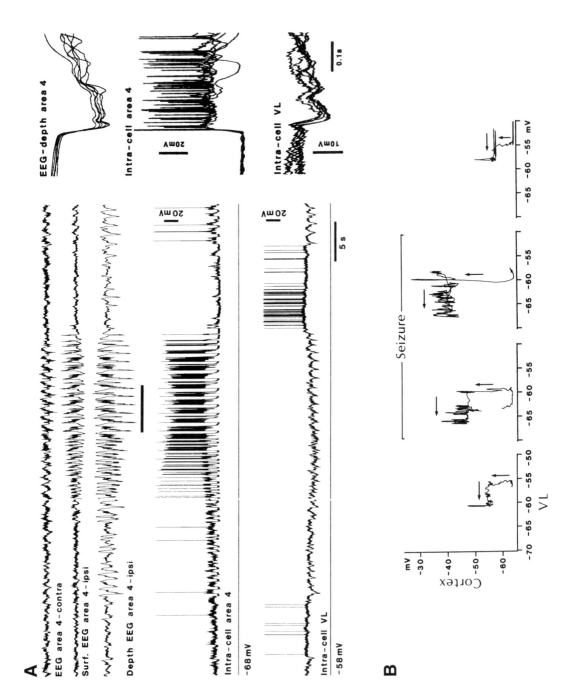

tions disrupt, rather than produce, spike-wave seizures *(120)*. Experimental studies point to the progressive build-up of spike-wave/polyspike-wave seizures at approx 3 Hz that obey the rule of synaptic circuits, sequentially distributed through short- and long-range circuits in corticocortical and corticothalamic synaptic networks *(121)*. Similarly, EEG studies and toposcopic analyses in humans and animals have also indicated that some spike-wave seizures are locally generated and result from multiple, independent cortical foci *(122,123)* and topographical analyses of spike-wave complexes in humans showed that the EEG "spike" component propagates from one hemisphere to another with time-lags as short as 15 ms *(124,125)*, which cannot be estimated by visual inspection. This explains why absence seizures are less detrimental than *grand-mal* epilepsy that implicates more widespread neuronal manifestations. Indeed, one of the characteristics of spike-wave seizures is that there is little or no disruption of cognitive abilities after an ictal event.

The unconsciousness during absence seizures is likely due to steady inhibition of thalamocortical neurons during cortically generated spike-wave seizures *(126)*. This finding, first reported with dual intracellular recordings from neocortical and thalamic relay neurons in vivo (Fig. 4) was corroborated by data from genetically determined absence seizures in rats *(127,128)*. The prolonged inhibition of thalamocortical neurons during spike-wave seizures is due to the faithful following of paroxysmal depolarizations generated in cortex by tha-

Fig. 4. *(continued from facng page)* Inhibition of thalamocortical (TC) neurons during cortically generated spike-wave seizures, Dual intracellular recordings from neocortical (area 4) and TC (ventrolateral [VL]) neurons demonstrating hyperpolarization of VL neuron during spike-wave seizure depolarization and spike-bursts in area 4 neuron. Cat under ketamine-xylazine anesthesia. (A) Five traces depict simultaneous recordings of electroencephalogram (EEG) from the skull over the right cortical area 4, surface- and depth-EEGs from the left area 4, as well as intracellular activities of left area 4 cortical neuron and thalamic VL neuron (below each intracellular trace, current monitor). The seizure was initiated by a series of EEG waves at 0.9 Hz in the depth of left area 4, continued with spike-wave discharges at 2 Hz, and ended with high-amplitude, periodic EEG sequences consisting of wavelets at 14 Hz. All these periods were faithfully reflected in the intracellular activity of the cortical neuron, whereas the thalamic VL neuron displayed a tonic hyperpolarization throughout the seizure, with phasic sequences of IPSPs related to the large cortical paroxysmal depolarizations and spike-bursts occurring at the end of the seizure. Note disinhibition of the VL neuron after cessation of cortical seizure. The part indicated by horizontal bar (below the depth-EEG trace) is expanded at right (superimposition of six successive traces). Note spiky depth-negative EEG deflections associated with depolarization of cortical neuron and rhythmic IPSPs of the thalamic VL neuron. (B) Phase relations between simultaneously intracellularly recorded cortical neuron (area 4) and thalamic (VL) neuron are preserved during the development from sleep to seizure activity. The four parts represent: one sleep period prior to seizure, two periods during the early and late parts of the seizure, and one period after the paroxysmal activity. Phase plots of averaged membrane voltage of area 4 cortical neuron (ordinate) against that of the VL neuron (abscissa). The development from sleep to seizure did not change the phase relations between neurons, but accentuated the amplitude of the elements constituting the normal (sleep) oscillatory behavior preceding the epileptic seizure. Cortical depolarization (upward arrows) preceded the hyperpolarization of the VL neuron (left-directed arrows) in the four periods, although the amplitude of membrane excursions were considerably enhanced during the seizure. (Modified from ref. *126* and unpublished data by M. Steriade and D. Contreras.)

lamic reticular GABAergic neurons *(126,129)* and is operational in blocking sensory signals from the outside world in their route to cortex *(130)*.

6. CONCLUSIONS

Some elements of consciousness (such as memory, selective attention, decision making, and forebrain generalized activation that permits conscious thinking during the state of wakefulness) can be investigated at the neuronal level, mainly in animal experiments. However, the global entity of consciousness, which includes, of necessity, subjective states that are expressed through language, cannot be reduced, as some have postulated, to precise neuronal types, with specific electrophysiological features, located in different cortical layers. The fine properties of the hypothesized "awareness-generating" neurons require intracellular recordings, but systematic neuronal recordings cannot be done in humans. The large panoply of cerebral structures that is required for the generation of conscious phenomena transcends the usually considered thalamocortical systems to obligatorily include a series of many other brain structures, among them are perirhinal cortices and hippocampus for memory, amygdala nuclear complex for coloring memory with affectivity, and four or five generalized activating systems located in the brainstem core and forebrain. Simultaneous neuronal recordings from all these sites, during states of consciousness, are not feasible. States of vigilance that are conventionally regarded as associated with complete annihilation of consciousness, such as SWS, are accompanied by a series of brain oscillations that, because of rhythmic spike-trains and spike-bursts in cortical and thalamic networks, develop neuronal plasticity that is operational in consolidating memory traces acquired during the state of wakefulness.

REFERENCES

1. James, W. (1890) *The Principles of Psychology.* Henry Holt, New York, NY.
2. Crick, F. (1994) *The Astonishing Hypothesis.* Charles Scribner's Sons, New York, NY.
3. Damasio, A.R. (1999) *The Feelings of What Happens: Body and Emotion in the Making of Consciousness.* Harcourt, Brace & Co, New York, NY.
4. Damasio, A.R. (2003) *Looking for Spinoza: Joy, Sorrow, and the Feeling Brain.* Harcourt/ Heinemann, New York, NY.
5. Hobson, J.A. and Steriade, M. (1986) Neuronal basis of behavioral state control. In *Handbook of Physiology* (Vol. IV) (Mountcastle, V.B. and Bloom, F.E., eds.), American Physiological Society, Bethesda, MD, pp. 701–823.
6. Kahn, D., Pace-Schott, E.F. and Hobson, J.A. (1997) Consciousness in waking and dreaming: the roles of neuronal oscillations and neuromodulation in determining similarities and differences. *Neuroscience.* **78**:13–38.
7. Hobson, J.A., Pace-Schott, E. and Stickgold, R. (2000) Dreaming and the brain: toward a cognitive neuroscience of conscious states. *Brain Behav. Sci.* **23**:793–842.
8. Edelman, G.M. (1989) *The Remembered Present: A Biological Theory of Consciousness.* Basic Books, New York, NY.
9. Damasio, A.R. and Damasio, H. (1996) Making images and creating subjectivity. In *The Mind-Brain Continuum* (Llinás, R. and Churchland, P.S., eds.), The MIT Press, Cambridge, MA, pp. 19–27.
10. Libet, B. (1998) Do the models offer testable proposals of brain functions for conscious experience? In *Consciousness: At the Frontiers of Neuroscience* (*Advances in Neurology*, vol. 77) (Jasper, H.H., Descarries, L., Castelucci, V.F. and Rossignol, S., eds.), Lippincott-Raven, Philadelphia, PA, pp. 213–217.

11. Smart, J.J.C. (1959) Sensations and brain processes. *Philos. Rev.* **68**:141–156.
12. Llinás, R.R. (2001) *I of the Vortex.* The MIT Press, Cambridge, MA.
13. Nichols, M.J. and Newsome, W.T. (1999) The neurobiology of cognition. *Nature.* **402**(Suppl.): C35–C38.
14. Popper, K.R. and Eccles, J.C. (1977) *The Self and Its Brain.* Springer, Berlin.
15. Eccles, J.C. (1994) *How the Self ControlsI Its Brain.* Springer, Berlin.
16. Soury, J. (1899) *Le Système Nerveux Central. Histoire Critique des Théories et des Doctrines* (vol. 1). Carré et Naud, Paris.
17. Walshe, F.M.R. (1957) The brain-stem conceived as the "highest level" of function in the nervous system: with particular reference to the "automatic apparatus" of Carpenter (1850) and to the "centrencephalic integrating system" of Penfield. *Brain.* **80**:510–539.
18. Penfield, W. and Rasmussen, T. (1950) *The Cerebral Cortex of Man. A Clinical Study of Localization of Function.* Macmillan, New York, NY.
19. Zattore, R.J., Jones-Gotman, M., Evans, A.C. and Meyer, E. (1992) Functional localization and lateralization of human olfactory cortex. *Nature.* **360**:339–341.
20. Laurent, G. (1996) Dynamical representation of odors by oscillating and evolving neural assemblies. *Trends Neurosci.* **19**:489–496.
21. Laing, D.G. (1983) Natural sniffing gives optimum odor perception for humans. *Perception.* **12**:99–117.
22. Sobel, N., Prabhakaran, V., Desmond, J.E., et al. (1998) Sniffing and smelling: separate subsystems in the human olfactory cortex. *Nature.* **392**:282–286.
23. Doty, R.L., Deems, D.A. and Stellar, S. (1988) Olfactory dysfunction in Parkinsonism: a general deficit unrelated to neurologic signs, disease stage, or disease duration. *Neurology.* **38**:1237–1244.
24. Grillner, S. and Matsushima, T. (1991) The neural network underlying locomotion in lamprey-synaptic and cellular mechanisms. *Neuron.* **7**:1–5.
25. Viana di Prisco, G., Pearlstein, E., Robitaille, R. and Dubuc, R. (1997) Role of sensory-evoked NMDA plateau potentials in the initiation of locomotion. *Science.* **278**:1122–1125.
26. Grillner, S. (1997) Ion channels and locomotion. *Science.* **278**:1087–1088.
27. Llinás, R.R. and Paré, D. (1991) Of dreaming and wakefulness. *Neuroscience.* **44**:521–535.
28. Llinás, R., Ribary, U., Joliot, M. and Wang, X.J. (1993) Content and context in temporal thalamocortical binding. In *Temporal Coding in the Brain* (Buzsáki, G., Llinás, R., Singer, W., Berthoz A. and Christen, Y., eds.), Springer, Berlin, pp. 251–272.
29. Crick, F. and Koch, C. (1995) Are we aware of neural activity in primary visual cortex? *Nature.* **375**:121–123.
30. Pollen, D.A. (1995) Cortical areas in visual awareness. *Nature.* **377**:293–294.
31. Block, N. (1996) How can we find the neuronal correlates of consciousness? *Trends Neurosci.* **19**:456–459.
32. Steriade, M. (2001) *The Intact and Sliced Brain.* MIT Press, Cambridge, MA.
33. Tononi, G. and Edelman, G.M. (1998) Consciousness and complexity. *Science.* **282**:1846–1851.
34. Crick, F. and Koch, C. (2003) A framework for consciousness. *Nat. Neurosci.* **6**:119–126.
35. Koch, C. (1998) The neuroanatomy of visual consciousness, in *Consciousness: At the Frontiers of Neuroscience* (*Advances in Neurology*, vol. 77) (Jasper, H.H., Descarries, L., Castelucci, V.F. and Rossignol, S., eds.), Lippincott-Raven, Philadelphia, PA, pp. 229–241.
36. Connors, B.W., Gutnick, M.J. and Prince, D.A. (1982) Electrophysiological properties of neocortical neurons in vitro. *J. Neurophysiol.* **48**:1302–1320.
37. Connors, B.W. and Amitai, Y. (1995) Functions of local circuits in neocortex: synchrony and laminae. In *The Cortical Neuron* (Gutnick, M.J. and Mody, I., eds), Oxford University Press, New York, Oxford, pp. 123–140.
38. Steriade, M., Nuñez, A. and Amzica, F. (1993) A novel slow (<1 Hz) oscillation of neocortical neurons in vivo: depolarizing and hyperpolarizing components. *J. Neurosci.* **13**:3252–3265.

39. Chen, W., Zhang, J.J., Hu, G.Y. and Wu, C.P. (1996) Electrophysiological and morphological properties of pyramidal and non-pyramidal neurons in the cat motor cortex in vitro. *Neuroscience.* **73**:39–55.

40. Nishimura, Y., Kitagawa, H., Saitoh, K., et al. (1996) The burst firing in the layer III and V pyramidal neurons of the cat sensorimotor cortex in vitro. *Brain Res.* **727**:212–216.

41. Steriade, M., Timofeev, I. and Grenier, F. (2001) Natural waking and sleep states: a view from inside neocortical neurons. *J. Neurophysiol.* **85**:1969–1985.

42. Yang, C.R., Seamans, J.K. and Gorelova, N. (1996) Electrophysiological and morphological properties of layers V-VI principal pyramidal cells in rat prefrontal cortex in vitro. *J. Neurosci.* **16**:1904–1921.

43. Timofeev, I., Grenier, F., Bazhenov, M., Sejnowski, T.J. and Steriade, M. (2000) Origin of slow oscillations in deafferented cortical slabs. *Cereb. Cortex* **10**:1185–1199.

44. Steriade, M., Timofeev, I., Dürmüller, N. and Grenier, F. (1998) Dynamic properties of corticothalamic neurons and local cortical interneurons generating fast rhythmic (30–40 Hz) spike bursts. *J. Neurophysiol.* **79**:483–490.

45. Gray, C.M. and McCormick, D.A. (1996) Chattering cells: superficial pyramidal neurons contributing to the generation of synchronous oscillations in the visual cortex. *Science.* **274**:109–113.

46. Moruzzi, G. and Magoun, H.W. (1949) Brain stem reticular formation and activation of the EEG. *Electroencephalogr. Clin. Neurophysiol.* **1**:455–473.

47. Steriade, M. and McCarley, R.W. (1990) *Brainstem Control of Wakefulness and Sleep.* Plenum, New York, NY.

48. Mesulam, M.M., Mufson, E.J., Levey, A.I. and Wainer, B.H. (1983) Cholinergic innervation of cortex by the basal forebrain: cytochemistry and cortical connections of the septal area, diagonal band nuclei, nucleus basalis (substantia innominata), and hypothalamus in the rhesus monkey. *J. Comp. Neurol.* **214**:170–197.

49. Steriade, M., Amzica, F. and Nuñez, A. (1993) Cholinergic and noradrenergic modulation of the slow (~0.3 Hz) oscillation in neocortical cells. *J. Neurophysiol.* **70**:1384–1400.

50. Bremer, F. (1935) Cerveau "isolé" et physiologie du sommeil. *C. R. Soc. Biol. (Paris).* **118**:1235–1241.

51. Batsel, H.L. (1964) Spontaneous desynchronization in the chronic cat *"cerveau isolé".* Arch. Ital. Biol. *102:547–566.*

52. Villablanca, J. (1965) The electrocorticogram in the chronic cerveau isolé cat. *Electroencephalogr. Clin. Neurophysiol.* **19**:576–586.

53. Kreindler, A. and Steriade, M. (1964) EEG patterns of arousal and sleep induced by stimulating various amygdaloid levels in the cat. *Arch. Ital. Biol.* **102**:576–586.

54. Façon, E., Steriade, M. and Wertheimer, N. (1958) Hypersomnie prolongée engendrée par des lésions bilatérales du système activateur médial: le syndrome thrombotique de la bifurcation du tronc basilaire. *Rev. Neurol. (Paris).* **98**:117–133.

55. Castaigne, P., Buge, A., Escourolle, R. and Mason, M. (1962) Ramollissement pédonculaire médian, tegmentothalamique avec ophtalmoplégie et hypersomnie. *Rev. Neurol. (Paris).* **106**:357–367.

56. Plum, F. (1991) Coma and related global disturbances of the human conscious state. In *Cerebral Cortex* (vol. 9, *Normal and Altered States of Function*) (Peters, A. and Jones, E.G., eds.) Plenum, New York, NY, pp. 359–425.

57. Kinomura, S., Larsson, J., Gulyás, B. and Roland, P. (1996) Activation by attention of the human reticular formation and thalamic intralaminar nuclei. *Science.* **271**:512–515.

58. Steriade, M., Datta, S., Paré, D., Oakson, G. and Curró Dossi, R. (1990) Neuronal activities in brainstem cholinergic nuclei related to tonic activation processes in thalamocortical systems. *J. Neurosci.* **10**:2541–2559.

59. Steriade, M., Paré, D., Datta, S., Oakson, G. and Curró Dossi, R. (1990c) Different cellular types in mesopontine cholinergic nuclei related to ponto-geniculo-occipital waves. *J. Neurosci.* **10**:2560–2579.

60. Desimone, R. and Duncan, J. (1995) Neural mechanisms of selective attention. *Ann. Rev. Neurosci.* **18**:193–222.
61. Robinson, D.L. and Cowie, R.J. (1997) The primate pulvinar: structural, functional, and behavioral components of visual salience. In *Thalamus* (vol. 2, *Experimental and Clinical Aspects*) (Steriade, M., Jones, E.G. and McCormick, D.A., eds.), Elsevier, Oxford, pp. 53–92.
62. Fuster, J.M. (1996) Network memory. *Trends Neurosci.* **20**:451–459.
63. Goldman-Rakic, P.S. (1996) Regional and cellular fractionation of working memory. *Proc. Natl. Acad. Sci. USA.* **93**:13473–13480.
64. LeDoux, J.E. (1996) *The Emotional Brain.* Simon and Schuster, New York, NY.
65. Bushara, K.O., Hanakawa, T., Immisch, I., Toma, K., Kansaku, K. and Haller, M. (2003) Neural correlates of cross-modal binding. *Nat. Neurosci.* **6**:190–195.
66. Ungerleider, L.G., Courtney, S.M. and Haxby, J.V. (1998) A neural system for human visual working memory. *Proc. Natl. Acad. Sci. USA* **95**:883–890.
67. Wilson, F.A., Scalaidhe, S.P. and Goldman-Rakic, P.S. (1993) Dissociation of object and spatial processing domains in primate prefrontal cortex. *Science.* **260**:1955–1958.
68. Northoff, G., Richter, A., Gessner, M., et al. (2000) Functional dissociation between medial and lateral prefrontal cortical spatiotemporal activation in negative and positive emotions: a combined fMRI/MEG study. *Cereb. Cortex* **10**:93–107.
69. Opitz, B., Mecklinger, A., Friederici, A.D. and von Cramon, D.Y. (1999) The functional neuroanatomy of novelty processing: integrating ERP and fMRI results. *Cereb. Cortex* **9**:379–391.
70. Slotnick, S.D., Moo, L.R., Kraut, M.A., Lesser, R.P. and Hart J Jr. (2002) Interactions between thalamic and cortical rhythms during semantic memory recall in human. *Proc. Natl. Acad. Sci. USA.* **99**:6440–6443.
71. Pavlov, I.P. (1923) "Innere Hemmung" der bedingten Reflexe und der Schlaf - ein und derselbe Prozess. *Skand. Arch. Physiol.* **44**:42–58.
72. Eccles, J.C. (1961) Chairman's opening remarks. In *The Nature of Sleep* (Wolstenholme, G.E.W. and O'Connor, M., eds.), Churchill, London, pp. 1–3.
73. Sherrington, C.S. (1955) *Man on his Nature.* Doubleday, New York, NY.
74. Steriade, M., Jones, E.G. and Llinás, R.R. (1990) *Thalamic Oscillations and Signaling.* Wiley-Interscience, New York, NY.
75. Moruzzi, G. (1966) The functional significance of sleep with particular regard to the brain mechanisms underlying consciousness. In *Brain and Conscious Experience* (J.C. Eccles, ed.), Springer, New York, NY, pp. 345–379.
76. Steriade, M., Contreras, D., Curró Dossi, R. and Nuñez, A. (1993) The slow (<1 Hz) oscillation in reticular thalamic and thalamocortical neurons: scenario of sleep rhythm generation in interacting thalamic and neocortical networks. *J. Neurosci.* **13**:3284–3299.
77. Steriade, M., McCormick, D.A. and Sejnowski, T.J. (1993) Thalamocortical oscillation in the sleeping and aroused brain. *Science.* **262**:679–685.
78. Buzsáki, G. (1996) The hippocampo-neocortical dialogue. *Cereb. Cortex* **6**:81–92.
79. Stickgold, R., James, L. and Hobson, J.A. (2000) Visual discrimination learning requires sleep after training. *Nat. Neurosci.* **3**:1237–1238.
80. Stickgold, R., Whitbee, D., Schirmer, B., Patel, V. and Hobson, J.A. (2000) Visual discrimination improvement. A multi-step process occurring during sleep. *J. Cogn. Neurosci.* **12**:246–254.
81. Gais, S., Plihal, W., Wagner, U. and Born, J. (2000) Early sleep triggers memory for early visual discrimination skills. *Nat. Neurosci.* **3**:1335–1339.
82. Gais, S., Mölle, M., Helms, K. and Born, J. (2002) Learning-dependent increases in sleep density. *J. Neurosci.* **22**:6830–6834.
83. Hobson, J.A., Pace-Schott, E. and Stickgold, R. (2000) Dreaming and the brain: toward a cognitive neuroscience of conscious states. *Brain Behav. Sci.* **23**:793–842.
84. Nielsen, T. (2000) Cognition in REM and NREM sleep. *Brain Behav. Sci.* **23**:851–866.
85. Steriade, M., Deschênes, M., Domich, L. and Mulle, C. (1985) Abolition of spindle oscillations in thalamic neurons disconnected from nucleus reticularis thalami. *J. Neurophysiol.* **54**:1473–1497.

86. Steriade, M., Domich, L., Oakson, G. and Deschênes, M. (1987) The deafferented reticularis thalami nucleus generates spindle rhythmicity. *J. Neurophysiol.* **57**:260–273.

87. Steriade, M. (1999) Cellular substrates of brain rhythms. In *Electroencephalography: Basic Principles, Clinical Applications, and Related Fields* (4th ed.) (Niedermeyer, E. and Lopes Da Silva, F., eds.), Williams & Wilkins, Baltimore, MD, pp. 28–75.

88. Steriade, M., Nuñez, A. and Amzica, F. (1993) Intracellular analysis of relations between the slow (<1 Hz) neocortical oscillation and other sleep rhythms. *J. Neurosci.* **13**:3266–3283.

89. Sanchez-Vives, M.V. and McCormick, D.A. (2000) Cellular and network mechanisms of rhythmic recurrent activity in neocortex. *Nat. Neurosci.* **3**:1027–1034.

90. Steriade, M., Amzica, F. and Contreras, D. (1996) Synchronization of fast (30–40 Hz) spontaneous cortical rhythms during brain activation. *J. Neurosci.* **16**:392–417.

91. Mölle, M., Marshall, L., Gais, S. and Born, J. (2002) Grouping of spindle activity during slow oscillations in human non-REM sleep. *J. Neurosci.* **22**:10941–10947.

92. Siapas, A.G. and Wilson, M.A. (1998) Coordinated interactions between hippocampal ripples and cortical spindles during slow-wave sleep. *Neuron.* **21**:1123–1128.

93. Sirota, A., Csicsvari, J., Buhl, D. and Buzsáki, G. (2003) Communication between neocortex and hippocampus during sleep in rodents. *Proc. Natl. Acad. Sci. USA.* **100**:2065–2069.

94. Yuste, R. and Tank, D.W. (1996) Dendritic integration in mammalian neurons, a century after Cajal. *Neuron.* **16**:701–716.

95. Destexhe, A. and Sejnowski, T.J. (2001) *Thalamocortical Assembly.* Oxford University Press, Oxford.

96. Soderling, T.R. and Derkach, V.A. (2000) Postsynaptic protein phosphorylation and LTP. *Trends Neurosci.* **23**:75–80.

97. Morison, R.S. and Dempsey, E.W. (1942) Mechanism of thalamocortical augmentation and repetition. *Am. J. Physiol.* **138**:297–308.

98. Steriade, M. and Timofeev, I. (1997) Short-term plasticity during intrathalamic augmenting responses in decorticated cats. *J. Neurosci.* **17**:3778–3795.

99. Steriade, M., Timofeev, I., Grenier, F. and Dürmüller, N. (1998) Role of thalamic and cortical neurons in augmenting responses: dual intracellular recordings *in vivo. J. Neurosci.* **18**:6425–6443.

100. Castro-Alamancos, M.A. and Connors, B.W. (1996) Cellular mechanisms of the augmenting response: short-term plasticity in a thalamocortical pathway. *J. Neurosci.* **16**:7742–7756.

101. Timofeev, I., Grenier, F., Bazhenov, M., Houweling, A., Sejnowski, T.J. and Steriade, M. (2002) Short- and medium-term plasticity associated with augmenting responses in cortical slabs and spindles in intact cortex of cats in vivo. *J. Physiol. (Lond.)* **542**:583–598 .

102. Steriade, M. (1991) Alertness, quiet sleep, dreaming. In *Cerebral Cortex* (vol. 9, *Normal and Altered States of Function*) (Peters, A. and Jones, E.G., eds.), Plenum, New York, NY, pp. 279–357.

103. Bazhenov, M., Timofeev, I., Steriade, M. and Sejnowski, T.J. (1998) Computational models of thalamocortical augmenting responses. *J. Neurosci.* **18**:6444–6465.

104. Kato, N. (1990) Cortico-thalamo-cortical projection between visual cortices. *Brain Res.* **509**:150–152.

105. Steriade, M. (2001) Impact of network activities on neuronal properties in corticothalamic systems. *J. Neurophysiol.* **86**:1–39.

106. Hernández-Cruz, A. and Pape, H.C. (1989) Identification of two calcium currents in acutely dissociated neurons from the rat lateral geniculate nucleus. *J. Neurophysiol.* **61**:1270–1283.

107. Kammermeier, P.J. and Jones, S.W. (1997) High-voltage-activated calcium currents in neurons acutely isolated from the ventrobasal nucleus of the rat thalamus. *J. Neurophysiol.* **77**:465–475.

108. Steriade, M. and Timofeev, I. (2003) Neuronal plasticity in thalamocortical networks during sleep and waking oscillations. *Neuron.* **37**:563–576. .

109. Abel, T., Nguyen, P.V., Barad, M., Deuel, T.A., Kandel, E.R. and Bourtchouladze, R. (1997) Genetic demonstration of a role for PKA in the late phase of LTP and in hippocampus-based long-term memory. *Cell.* **88**:615–626.

110. Dolmetsch, R.E., Pajvani, U., Fife, K., Spotts, J.M. and Greenberg, M.E. (2001) Signaling to the nucleus by an L-type calcium channel-calmodulin complex through the MAP kinase pathway. *Science.* **294**:333–339.

111. Steriade, M. and Timofeev, I. (2003) Neuronal plasticity during sleep oscillations in cortico-thalamic systems. In *Sleep and Brain Plasticity* (Maquet, P., Stickgold, R. and Smith, C.S., eds.), Oxford University Press, Oxford, pp. 271–291.

112. Steriade, M. (2003) *Neuronal Substrates of Sleep and Epilepsy.* Cambridge University Press, Cambridge, UK.

113. Niedermeyer, E. (1999) Abnormal EEG patterns (epileptic and paroxysmal). In *Electroencephalography: Basic Principles, Clinical Applications and Related Fields* (4th ed.) (Niedermeyer, E. and Lopes da Silva, F., eds.), Williams & Wilkins, Baltimore, MD, pp. 235–260.

114. Halasz, P. (1991) Runs of rapid spikes in sleep: a characteristic EEG expression of generalized malignant epileptic encephalopathies. A conceptual review with new pharmacological data. *Epilepsy Res.* **2**(Suppl.):49–71.

115. Steriade, M., Amzica, F., Neckelmann, D. and Timofeev, I. (1998) Spike-wave complexes and fast runs of cortically generated seizures. II. Extra- and intracellular patterns. *J. Neurophysiol.* **80**:1456–1479.

116. Reutens, D.C., Bye, A.M., Hopkins, I.J., et al. (1993) Corpus callosotomy for intractable epilepsy: seizure outcome and prognostic factors. *Epilepsia.* **34**:904–909.

117. Steriade, M. and Contreras, D. (1998) Spike-wave complexes and fast runs of cortically generated seizures. I. Role of neocortex and thalamus. *J. Neurophysiol.* **80**:1439–1455.

118. Penfield, W. and Jasper, H.H. (1954) *Epilepsy and the Functional Anatomy of the Human Brain.* Little, Brown, Boston, MA.

119. Jasper, H.H. and Droogleever-Fortuyn, J. (1949) Experimental studies on the functional anatomy of petit-mal epilepsy. *Res. Publ. Ass. Nerv. Ment. Dis.* **26**:272–298.

120. Danober, L., Depaulis, A., Vergnes, M. and Marescaux, C. (1995) Mesopontine cholinergic control over generalized non-convulsive seizures in a genetic model of absence epilepsy in the rat. *Neuroscience.* **69**:1183–1193.

121. Steriade, M. and Amzica, F. (1994) Dynamic coupling among neocortical neurons during evoked and spontaneous spike-wave seizure activity. *J. Neurophysiol.* **72**:2051–2069.

122. Jasper, H.H. and Hawkes, W.A. (1938) Electroencephalography. IV. Localization of seizure waves in epilepsy. *Arch. Neurol. (Chic.)* **39**:885–901.

123. Petsche, H. (1962) Pathophysiologie und Klinik des Petit-Mal. *Wiener Zeitschr. Nervenheilkrank.* **19**:345–442.

124. Lemieux, J.F. and Blume, W.T. (1986) Topographical evolution of spike-wave complexes. *Brain Res.* **373**:275–287.

125. Kobayashi, K., Nishibayashi, N., Ohtsuka, Y., Oka, E. and Ohtahara, S. (1994) Epilepsy with electrical status epilepticus during slow sleep and secondary bilateral synchrony. *Epilepsia.* **35**:1097–1103.

126. Steriade, M. and Contreras, D. (1995) Relations between cortical and thalamic cellular events during transition from sleep pattern to paroxysmal activity. *J. Neurosci.* **15**:623–642.

127. Pinault, D., Leresche, N., Charpier, S., Deniau, J.M., Marescaux, C., Vergnes, M. and Crunelli, V. (1998) Intracellular recordings in thalamic neurones during spontaneous spike and wave discharges in rats with absence epilepsy. *J. Physiol. (Lond.).* **509**:449–456.

128. Crunelli, V. and Leresche, N. (2002) Childhood absence epilepsy: genes, channels, neurons and networks. *Nat. Rev. Neurosci.* **3**:371–382.

129. Timofeev, I., Grenier, F. and Steriade, M. (1998) Spike-wave complexes and fast runs of cortically generated seizures. IV. Paroxysmal fast runs in cortical and thalamic neurons. *J. Neurophysiol.* **80**:1495–1513.
130. Amzica, F. and Steriade, M. (2002) The functional significance of K-complexes. *Sleep Med. Rev.* **6**:139–149.
131. Contreras, D. and Steriade, M. (1996) Spindle oscillation: the role of corticothalamic feedback in a thalamically generated rhythm. *J. Physiol. (Lond.).* **490**:159–179.

Frank W. Booth and P. Darrell Neufer

1. INTRODUCTION

The successful coordination of hundreds of biological processes that allow the unanesthe-tized human to undergo maximal aerobic work is one of the elegant marvels in modern biology. It is difficult to identify other examples of normal function that are more complex in integrative or systems biology. Aerobic physical activity has long been recognized as pro-ducing a high metabolic state that can exceed approx 25 times resting metabolism in world-class athletes. It is no wonder, then, that the coordination of multiple processes is required. Even the average individual increases metabolic rate four- to eightfold in a slow run. Great strides have been made in deciphering the intercommunications among systems, organs/ tissues, cells, organelles, genes, and molecules to support the many-fold increase in meta-bolic demand imposed by physical activity. Numerous factors, such as autonomic nervous system activity, the rate of blood flow, the delivery of a complex mix of substrates and hormones in the plasma, and elimination of metabolic byproducts such as heat and carbon dioxide, contribute to the complexity of the required integration that enables moderate-in-tensity exercise to occur without harm in an individual accustomed to physical activity. Attempting to understand how all these factors converge, regulate, and then minimize the disruption in homeostasis is a prime example of integrative physiology. The curious biolo-gist can only be awed by the physiological complexity of how the human body is able to support physical activity.

The value of the science of physical activity extends far beyond its historical application to Olympic athletes or to a complex integration of normal function as a part of the discipline of exercise science. As a consequence of engineering physical activity out of daily life, there has been a dramatic increase in 35 chronic health disorders (1), the most alarming of which is the epidemic rise in the prevalence of obesity and type 2 diabetes in developed countries. As in any other area of medicine where deciphering the molecular basis of disease is obliga-tory to improving prevention and cure, physical inactivity as a direct or indirect cause of 35 chronic health conditions must be simultaneously understood at the molecular and integra-tive levels. The World Health Organization has concluded that 1.9 million premature deaths each year result from physical inactivity.

This chapter is divided into five sections. The presentation begins with classical examples of integration during aerobic types of physical activity. We then discuss some of the major accomplishments in exercise research; and focus on selected examples of how changes in

From: *Integrative Physiology in the Proteomics and Post-Genomics Age*
Edited by: W. Walz © Humana Press Inc., Totowa, NJ

mRNA and protein expression during physical activity may contribute to new concepts about inter-organ communication during physical activity. We also consider what should be considered as normal physiological levels for the human genome; and finally, we present concepts on the dysfunctions occurring in normal physiological cycles as a result of a reduction in the daily level of physical activity by industrialization and technology.

2. CLASSICAL INTEGRATION IN PHYSICAL ACTIVITY

There is perhaps no other nonpathological or nontraumatic event that represents such an abrupt and dramatic challenge to physiological homeostasis as that evoked by a simple bout of exercise. Suddenly, skeletal muscle, which at rest serves primarily as a supportive tissue to other organs, becomes the center of attention—the tissue for which all other organs now must provide metabolic support. Although skeletal muscle can meet on its own the initial metabolic demand imposed by exercise, the ability to sustain physical activity beyond even just 1 to 3 min requires an adequate delivery of O_2 and metabolic substrates to the working muscle, as well as the efficient removal of metabolic by products (i.e., CO_2 and heat). However, the support provided to skeletal muscle by other organ systems is not without its limits. A prime example is the thermoregulatory system, which will begin to override the demands of working muscle for blood flow when exercise is performed in a hot environment, diverting blood flow away from working skeletal muscle to the skin to facilitate heat loss, thereby creating a competition for blood supply that ultimately hastens the development of fatigue. Other safeguards are also in place to ensure, for example, that the metabolic needs of critical organs are not sacrificed when exercise continues for long periods of time. Thus, it is evident that the physiological responses to exercise, as well as the internal protective mechanisms to diminish the adverse homeostatic disruptions in other organs created by exercise, represent the integration of multiple organ systems and regulatory mechanisms *(2)*.

2.1. Metabolic Responses During Endurance Types of Exercise

One of the major challenges to maintain homeostasis during endurance types of exercise (defined as rhythmic movements of limbs for 20 or more continuous minutes) is to ensure that muscle pH and free adenosine triphosphate (ATP) are kept near to resting levels. In order to accomplish this, multiple integration of processes must occur *(2)*. Amazingly respiratory minute volume increases sufficiently to maintain arterial PO_2, cardiac output increases such that perfusion pressure (mean arterial blood pressure minus mixed venous pressure) actually increases, and blood flow is redistributed to working skeletal muscle. As a result of these physiological responses, muscle PO_2 and pH are maintained near resting levels during exercise up to 70% of maximal work capacity. When work loads increase above 70% of maximum, muscle PO_2 still remains fairly stable while pH will begin to decline, although pH does not drop below 6.4 (from its resting level of 6.9). The decrease in pH arises primarily from an increase in muscle lactate due to the mass action effect of glycolysis increasing pyruvate's concentration and conversion to lactate. Although the increase in muscle blood flow is sufficient to remove the metabolic CO_2 produced, muscle temperature does rise owing to incomplete heat transfer from muscle to the blood.

The aforementioned respiratory and cardiovascular responses also contribute to the provision of substrates required to meet the energy needs of the working muscle. Creatine phosphate and existing ATP represent readily available forms of energy in skeletal muscle, but their supply is limited and becomes depleted within the first 3 to 5 min of exercise. Intramuscular stores of glycogen constitute the predominant energy source utilized during heavy muscular work as well as during the initial stages of prolonged exercise. Muscle glycogenolysis is stimulated directly by contraction-induced increases in Ca^{2+} and by β-adrenergic receptor-mediated activation of protein kinase A, reflecting sympathetic stimulated release of epinephrine from the adrenal gland and norepinephrine to the working muscle. As the rate of glycogen utilization decreases beyond the initial stages of exercise, blood glucose becomes an increasingly important substrate for muscle metabolism. Interestingly, contractile activity is able to stimulate glucose uptake in muscle by a mechanism that, although not completely understood, is independent of insulin, the other primary means of activating muscle glucose uptake (discussed in more detail later in the chapter). Reflecting the integrated nature of the metabolic response to exercise, epinephrine also stimulates glucose production in the liver, activating both hepatic glycogenolysis and gluconeogensis (from lactate), and the release of free fatty acids (FFA) from adipose tissue by activating lipolysis. Intramuscular triglyceride represents another potential source of energy for working muscle, although its overall contribution to metabolism during moderate intensity exercise of up to 2 h duration remains controversial. Other extramuscular sources of energy include circulating triglycerides and amino acids, although the contribution of each is thought to be very minor. Thus, the integration of metabolism during exercise incorporates sympathetic and endocrine responses with circulatory changes.

The precise nature of the metabolic control exerted during acute exercise is illustrated by maintenance of blood glucose concentration. Owing to the precise control exerted over hepatic glucose production (via glycogenolysis and gluconeogenesis), blood glucose concentration remains remarkably stable over a wide range of exercise intensities and glucose utilization rates. However, as exercise duration extends beyond 90 to 120 min (i.e., as during marathon running, triathelon, etc.), liver glycogen content becomes depleted. Because gluconeogenesis cannot compensate for the decline in glycogenolysis, hepatic glucose output no longer matches peripheral glucose utilization and arterial glucose concentration begins to fall. At this point, intermediary metabolism reaches a critical juncture because the brain almost exclusively is dependent on blood glucose for its energy needs. One solution, of course, would be for a mechanism to be in place to ensure that even a small drop in blood glucose concentration cause exercise fatigue. Although such a mechanism would appear to be compatible with immediate survival (and perhaps relief) of the individual, an inability to sustain moderate physical activity beyond approx 2 h would not be favored through evolution. For example, in hunter–gatherer societies, fatigue would severely limit the ability to obtain food. Moreover, liver glycogen levels also become depleted after just 24 h of fasting, implying that hunting activities would also not be possible during a famine that extended beyond 1 d, clearly incompatible with survival of the species. Thus, additional integrative metabolic responses must exist.

Recent research has suggested that a different strategy has evolved to conserve glucose utilization, allowing physical activity to continue despite a decline in whole-body carbohydrate content. When exercise duration continues beyond 60 to 90 min, the contribution of glucose to total oxidative metabolism in exercising muscle steadily declines whereas the contribution from FFA steadily increases. As a result, blood glucose levels remain fairly steady and hypoglycemia is avoided for a longer period of exercise. Recent research suggests that the mechanism responsible for this shift in substrate utilization may involve the induction of a newly discovered protein kinase gene known as pyruvate dehydrogenase kinase 4 (PDK4).

PDK4 is one of four isozymes identified in mammalian tissues that catalyze the phosphorylation of the mitochondrial pyruvate dehydrogenase (PDH) enzyme *(3)*. Phosphorylation inactivates the PDH enzyme complex, inhibiting the conversion of pyruvate to acetyl coenzyme A (CoA) and thus preventing products of glycolysis (originating from glucose or glycogen) from entering the mitochondria for oxidation. It is important to realize that the PDH reaction is the committed step to the complete oxidation and, therefore, irreversible loss of three-carbon compounds from the body. Inhibiting this pivotal step allows conservation of the three-carbon source by promoting the conversion of muscle pyruvate to lactate, a major precursor for hepatic gluconeogenesis. Although PDH is subject to allosteric regulation, inhibition via PDK-mediated phosphorylation also represents an important means of control. Interestingly, transcription of the PDK4 gene, the predominant isoform expressed in skeletal muscle, is nearly undetectable under resting conditions, but increases by as much as 200-fold during prolonged low-intensity exercise *(4)*. Given the large mass of skeletal muscle that is involved in whole-body exercise and the high degree of glucose oxidation associated with moderate-intensity exercise, it is intriguing to consider the possibility that induction of the PDK4 gene in working muscle during the latter stages of prolonged exercise may represent a natural mechanism for restricting glucose oxidation, thereby defending blood glucose and preserving the remaining whole-body carbohydrate reserves. Such a mechanism would permit hepatic glucose production via gluconeogenesis to more aptly keep pace with the remaining glucose utilization, ensuring that the brain continues to be supplied with sufficient glucose. As mentioned, the reliance on FFA oxidation progressively increases during exercise to compensate for the decreased carbon supply coming from carbohydrates. Whether PDK4-mediated inhibition of PDH also directly facilitates the transition to FFA oxidation remains to be determined. Nonetheless, carbohydrate and lipid metabolism are uniquely integrated during aerobic exercise.

2.2. Central Nervous System and Exercise

Higher brain centers initiate skeletal muscle contraction through efferent motor nerves. To support the metabolic requirements of the working skeletal muscles, the heart must be signaled to increase cardiac output. Two major control mechanisms (central and peripheral) integrate the response of the cardiovascular system to exercise *(5)*. First, neural activity that is responsible for recruiting motor units in the cerebral cortex activates, in parallel, cardiovascular control centers in the ventrolateral medulla (a feedforward mechanism) to cause immediate changes in the efferent activity of the sympathetic and parasympathetic nervous systems to the heart and blood vessels as well as motor outflow to the respiratory muscles. Increased sympathetic outflow to the heart increases both heart rate and stroke volume. As a result, cardiac output may increase from a resting level of approx 5 L/min to as much as 30 L/min in a well trained athlete. Direct sympathetic stimulation also induces vasoconstriction in select regions of the vasculature. For example, bood flow to the splanchnic bed decreases from approx 1 L/min at rest to only approx 0.2 L/min during maximal exercise. Thus, sympathetic-mediated vasoconstriction, which does not occur in arteries feeding skeletal muscle, redistributes blood flow away from inactive organ beds to the working skeletal muscle and/or skin. In fact, blood propelled to the working skeletal muscle increases from about 1 L/min at rest to as much as 26 L/min during maximal exercise in an aerobically trained person (maximal cardiac output of 30 L/min).

The second control mechanism regulating the cardiovascular response to exercise is a neural feedback system. Contracting skeletal muscles feedback the intensity of the contractile effort to the same control centers in the medulla that received the feed forward signal.

Contracting skeletal muscle initiates depolarization of group III (sensing the amount of muscle mass and force developed) and group IV (sensing some metabolite produced by the contraction) afferents. Thus, a peripheral feed back network is transmitted from skeletal muscle via afferent nerves in the spinal cord to cardiovascular integrative centers in the brain stem, modifying sympathetic output to the cardiovascular system. The central and peripheral drives appear to be redundant and it remains to be determined how the two integrate to make the appropriate increase in the cardiopulmonary systems during exercise.

Remarkably, when physically active individuals become sedentary for a few weeks and perform an exercise task at the same absolute submaximal intensity of exercise (e.g., same running speed), heart rate becomes higher and stroke volume lower while cardiac output is unchanged. This observation implies a change in integration to match cardiac output to work intensity. Delineation of the molecular integration in central neural regions is only beginning to be studied.

3. SELECTED MAJOR ACCOMPLISHMENTS IN EXERCISE PHYSIOLOGY RESEARCH

3.1. Historical Contribution

In 1967, John Holloszy published what has become a classic study in exercise physiology *(6)*. This paper found that muscle metabolic protein activities and concentration were not fixed, but could be increased by daily treadmill running in research animals. The genesis for this research came from comparative studies between wild and domesticated animals and from the adaptive changes in muscle mitochondria observed in response to thyroid hormone. Dr. Holloszy put forward the hypothesis that the differences between respiratory enzyme levels in muscles from active and inactive animals might be the result of, not only genetic differences, but also an adaptive process sensing muscle usage and adjusting metabolic pathways to allow work to be performed longer before fatigue. Holloszy observed in skeletal muscle of rats that experienced daily treadmill running increases in the capacity to oxidize pyruvate, the maximal velocity of oxidative enzymes, and the concentration of an electron transport chain protein.

In addition, the integrative interpretation of the data in this paper contributed to the scope of the finding. Holloszy suggested that the reduction in blood lactate at any given submaximal work load after endurance training was a consequence of the adaptive increase in skeletal muscle mitochondria, diminishing the increase in glycolysis and pyruvate levels during aerobic exercise. Thus, the concept proposed by Holloszy integrated biochemical adaptations from a specific tissue to a whole-body explanation of why rises in blood lactate are attenuated in the endurance-trained individuals, permitting humans to perform physical activity longer at the same submaximal work rate before exhaustion.

3.2. Exercise Activates Gene Expression Transiently in Skeletal Muscle

The necessity to better understand the mechanisms of how skeletal muscle adapts to increased physical activity has opened up a whole new field of research. Studies in both rodents and humans have shown that skeletal muscle possesses a remarkable capacity to adapt to the increases in metabolic and physical demand imposed by endurance training. These adaptive responses are reflected by easily measurable changes in the expression of specific proteins, the magnitude of which depends on such factors as the intensity, duration, and frequency of exercise during the training program. Although rapid changes (days) in the expression of mitochondrial proteins can be generated experimentally in animals by continuous stimulation of the motor nerve, an endurance training regimen in humans requires several weeks of daily training to achieve a measurable change in mitochondrial protein concentration. Thus, the temporal nature of intermittent exercise bouts (performed for only 2–10% of a day) as a stimulus experienced by myofibers makes elucidation of the underlying cellular and molecular mechanisms a relatively unique and challenging problem.

Exercise activates the transcription of select genes in skeletal muscle. Although some metabolic genes are activated during the exercise bout itself (depending on factors such as duration, energy state, etc.), transcriptional activation of several other metabolic genes peaks during the initial few hours after exercise and returns to baseline within less than 24 h *(7,8)*. The magnitude of activation varies considerably, ranging from approximately two- to threefold for genes encoding for mitochondrial enzymes to more than100-fold for several immediate early and metabolic-stress genes. Although not extensively studied, it appears that transcriptional activation is followed by corresponding transient changes in both mRNA and protein. The length of time that the expression of a given gene product remains elevated after exercise likely depends on the half-life of that particular mRNA and/or protein. In other words, the increase in expression of gene products with short half-lives (i.e., hours) may be dramatic but temporary, returning to baseline within 1 d after exercise. In fact, it appears that changes in mRNAs for a subset of regulatory proteins may be maximal after a single bout of exercise. In contrast, the exercise-induced increase in expression of other gene products with longer half-lives (i.e., >24 h) may be less dramatic but will be evident for much longer periods of time after the exercise bout. The underlying basis for many of the latter cellular adaptations to endurance training in skeletal muscle arise from the cumulative effects of transient changes in gene expression that occur in response to each individual exercise bout. These studies establish the temporal framework for future research to determine the cellular and molecular mechanisms by which myofibers sense, respond and ultimately integrate adaptations to changes in daily physical activity.

3.3. Ca²⁺ Mediated Regulation of Gene Expression in Skeletal Muscle

Although calcium has long been recognized as a second messenger that modulates the activity of various signaling kinases for cross-bridge cycling, it has only recently been delineated how transient spikes in intracellular free Ca^{2+} associated with contractile activity in skeletal muscle could be linked to the regulation of specialized cellular functions such as the regulation of gene expression. The conceptual breakthrough came from work being conducted on receptor signaling in T cells and the calcium-dependent protein phosphatase calcineurin, also known as protein phosphate 2B. Activation of calcineurin in T cells occurs in response to sustained, low amplitude increases in intracellular calcium while brief, high amplitude spikes in calcium fail to activate the enzyme. Once activated, calcineurin dephosphorylates a protein known as nuclear factor of activated T cells (NFAT) that, in turn, translocates to the nucleus to, in combination with other transcription factors, activate the transcription of target genes *(9)*. The fact that contracting slow-twitch myofibers maintained relatively high oscillatory concentrations of calcium, whereas fast-twitch myofibers are characterized by sharp transient increases in calcium, raised the possibility that differential calcineurin signaling could be involved in regulating muscle fiber phenotype. Indeed, transgenic mice expressing activated calcineurin specifically in skeletal muscle display an approximate twofold increase in the number of slow-twitch fibers in the predominantly fast-twitch gastrocnemius muscle, whereas mice lacking calcineurin have a reduced number of slow-twitch fibers *(10)*. Although impressive, the number of fibers with the altered phenotype in these genetic models of altered calcineurin activity represents only a small percentage of the total number of fibers in the muscles analyzed. Fiber-type transformations, strictly defined by shifts in fibers immunostaining for myosin type I vs type II isoform expression, rarely occur in humans and are possible in animals only when a predominantly fast-twitch muscle is subjected to continuous motor nerve stimulation for several months. Thus, calcineurin does not increase the percentage of type I fibers in humans during aerobic training.

Far more physiologically relevant than fiber-type transitions is the remarkable capacity of all myofibers to adapt metabolically to changes in contractile activity. For example, endurance-exercise training promotes mitochondrial biogenesis in skeletal muscle, increasing both muscle and whole-body oxidative capacity and resistance to fatigue. Ca^{2+}-mediated signaling events have also been implicated in the regulation of mitochondrial biogenesis (Fig. 1). Mice engineered to express a constitutively active form of calcium/calmodulin-dependent protein kinase IV (CaMK IV) in skeletal muscle are characterized by elevated mitochondrial DNA content, higher levels of mitochondrial enzymes required for oxidative metabolism, and reduced susceptibility to fatigue *(11)*.

CaMK IV also activates expression of the peroxisome proliferator-activated receptor α coactivator 1α (PGC-1 α), a transcriptional coactivator protein that is thought to coordinate the activation of nuclear genes that encode for mitochondrial enzymes . In humans, exercise increases both CaMK II (but not CaMK IV) activity *(12)* and PGC-1 αexpression in muscle *(13,14)*. Constitutive overexpression of PGC-1 αin skeletal muscle of transgenic mice causes the muscles to take on a much redder appearance *(15)*, presumably resulting from corresponding increases in the expression of heme containing proteins (myoglobin and cytochrome proteins) and in overall mitochondrial content. Interestingly, PGC-1 is thought to be a master regulator, not only of mitochondrial biogenesis in skeletal muscle, but also of other aspects of energy metabolism including gluconeogenesis in liver and adaptive thermogenesis in brown fat *(16)*. Thus, PGC-1 may play an important role in multiple tissues integrating and coordinating the metabolic adaptive responses to exercise.

3.4. AMPK As an Energy Sensor

AMP-activated protein kinase (AMPK) is a highly conserved energy sensing kinase found in all eukaryotic cells *(17)*. AMPK is activated allosterically by increases in intracellular AMP concentration. Although the resting concentration of AMP in skeletal muscle is typically 100-fold lower than ATP (a competitive inhibitor of AMPK), AMP concentration increases rapidly under conditions of accelerated ATP utilization during contraction, due in part to the adenylate kinase reaction (ADP + ADP → ATP + AMP). AMPK is also activated by phosphorylation of a regulatory site in the catalytic domain, a reaction catalyzed by an upstream kinase known as AMPK kinase, which is also activated by AMP. Thus, AMPK represents an acute sensor of the AMP/ATP ratio in cells that, once activated, initiates a series of signaling events to promote both the acute and long-term restoration and stabilization of energy levels in the cell (Fig. 2). Acute activation of AMPK in skeletal muscle is believed to stimulate fatty acid oxidation by (a)catalyzing the phosphorylation and inactivation of acetyl-CoA carboxylase (ACC), the enzyme catalyzing the formation of malonyl-CoA, and (b) the phosphorylation and activation of malonyl-CoA decarboxylase, a major

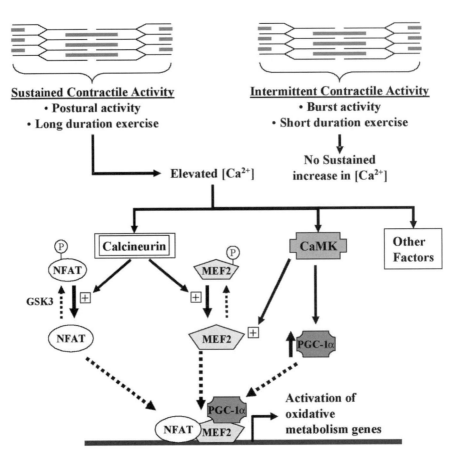

Fig. 1. Calcium-mediated signaling mechanisms implicated in the adaptive response to endurance exercise training. Sustained increases in intracellular calcium concentration induced by endurance activity activate calcineurin, a Ca^{2+}-sensitive serine-theonine phosphatase, and Ca^{2+}-calmodulin dependent protein kinase (CaMK). Calcineurin catalyzes the dephosphorylation and activation of nuclear factor of activated T cells (NFAT) and myocyte-enhancing factor (MEF2), leading to the activation of genes that encode for proteins involved in oxidative metabolism. CaMK also promotes gene activation through MEF2 and via upregulation of the peroxisome proliferator-activated receptor gamma coactivator 1 (PGC-1α), a putative master regulator of mitochondrial biogenesis. See text for detailed description. (Depiction based on refs. *10* and *11*.)

enzyme responsible for malonyl CoA degradation. The net effect is a reduction in the concentration of cytosolic malonyl-CoA, which relieves malonyl-CoA-mediated inhibition of carnitine palmitoyl-CoA transferase I (CPTI), the enzyme responsible for initiating the transport of fatty acids into the mitochondria. AMPK has also been implicated in the acute activation of contraction-stimulated glucose transport in skeletal muscle *(18)*, although recent evidence indicates that AMPK is, at most, only partially involved in this process (discussed later). AMPK does, however, phosphorylate glycogen synthase in skeletal muscle, a response that is speculated to be important for the inactivation of glycogen synthase during exercise in an intensity-dependent manner *(19)*.

AMPK signaling may extend well beyond control of acute substrate utilization to chronic adaptive changes. Indeed, the AMPK catalytic subunit found in skeletal muscle preferen-

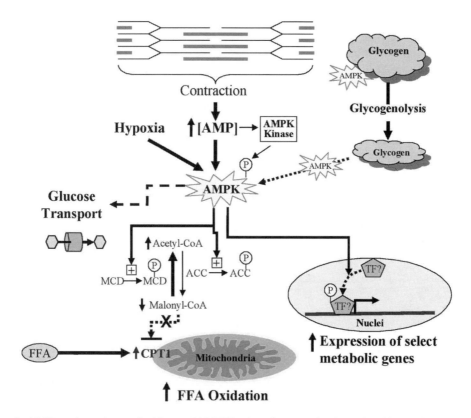

Fig. 2. AMP-activated protein kinase (AMPK) signaling mechanisms implicated in the adaptive response to endurance exercise and exercise training. Increases in (AMP) induced by contractile activity or hypoxia activate AMPK and the upstream AMPK kinase. Activation of AMPK by hypoxia (and the AMP analogue AICAR) stimulates GLUT4-mediated glucose transport, activation by contractile activity appears to only partially account for the contraction-induced activation of glucose transport. AMPK also promotes fatty acid oxidation by catalyzing the phosphorylation of both acetyl-CoA carboxylase (ACC) and malonyl-CoA dehydrogenase (MCD), decreasing the concentration of malonyl-CoA, an allosteric inhibitor of the mitochondrial fatty acid transport enzyme carnitine palmitoyltransferase I (CPT1). Activation of AMPK also leads to the transcriptional activation of genes encoding for proteins involved in both glucose and oxidative metabolism. See text for detailed description. (Depiction based on refs. *17* and *19–21*.)

tially localizes to the nucleus when expressed in cell culture, and has been shown to directly phosphorylate the nuclear protein p300, a transcriptional coactivator that interacts with a variety of transcription factors. In fully conscious and mobile rats, acute activation of AMPK via a 1 h infusion of the adenosine analogue AICAR increases transcription of the GLUT4 and hexokinase II genes *(20)*, whereas chronic activation of AMPK elicited by daily injections of AICAR leads to an overall increase in expression of GLUT4, hexokinase II and several mitochondrial enzymes in skeletal muscle *(21)*. Thus, AMPK signaling has been implicated in integrating the regulation of both substrate uptake and utilization by working muscle as well as the expression of those genes that encode for the proteins mediating these processes.

3.5. Insulin-Independent Translocation of GLUT4

Physicians have known for more than 100 yr that exercise is beneficial for diabetic patients because exercise lowers blood glucose and stimulates the uptake of glucose by muscle independent of insulin. Normally after a meal, insulin is released into the circulation and binds to the insulin receptor located on the plasma membrane of skeletal muscle and adipose cells. Insulin binding activates a tyrosine kinase intrinsic to the insulin receptor, initiating a signaling cascade that results in phosphorylation of insulin receptor substrate-1 (IRS1), followed by activation of phosphatidylinositol 3-kinase (PI3-kinase). Through mechanisms yet to be defined, activation of PI3-kinase ultimately leads to the translocation of the glucose transporter protein GLUT4 to the plasma membrane, thereby permitting glucose entry into the cell. Contractile activity in skeletal muscle also stimulates the translocation of GLUT4 to the plasma membrane by a process that does not involve PI3-kinase signaling *(22)*. Other mechanisms that stimulate GLUT4 translocation and glucose transport in skeletal muscle include hypoxia and AMPK activation induced by AICAR (Fig. 2). Activation of glucose transport by hypoxia and AICAR also does not require PI3-kinase activation. In addition, all three mechanisms are additive to the effects of insulin. These findings led to the suggestion that contraction and hypoxia stimulate glucose transport by a signaling pathway distinct from that activated by insulin, and that AMPK may be the protein that activates this alternative pathway. However, recent data obtained from transgenic mice engineered to express a dominant negative form of AMPK, and in mice with selective elimination of the β1 or β2 catalytic subunits of AMPK, reveal that although AMPK is required for 100% of the hypoxia- and AICAR-stimulated glucose transport, it is only partially responsible for the exercise-induced stimulation of glucose transport in skeletal muscle *(19,23)*. Thus, at least two signaling pathways (AMPK and an unidentified) integrate contractile-induced stimulation of glucose uptake. The potential evolutionary significance of separate insulin and exercise signaling pathways for GLUT4 translocation is discussed later.

4. INTEGRATION AT THE GENOMIC AND PROTEOMIC LEVELS IN PHYSICAL ACTIVITY

4.1. Skeletal Muscle-Brain

The potential importance of exercise on neuroplasticity in the hippocampus region of the brain has recently been explored *(24)*. The hippocampus serves an integrative role in the central nervous system (CNS) by receiving sensory information and projecting that information throughout the brain. Although most neurons are terminally differentiated and not replaced when they die, the dentate gyrus of the hippocampus undergoes neurogenesis, a process that is influenced by a variety of environmental factors. Among these factors, voluntary wheel running in rodents has been shown to be positively correlated with increased hippocampal neurogenesis, learning, and memory (reflected by a maze performance test). At the cellular level, voluntary running increases hippocampal expression of brain-derived neurotrophic factor (BDNF), a member of the nerve growth factor family of neurotrophins. Interesting, increases in hippocampal BDNF expression are directly related to the distance run, are significantly higher in mice selectively bred 25 generations for high voluntary wheel running activity, and are associated with increased activation of synapsin I and cyclic AMP response element-binding protein (CREB), two downstream effectors of BDNF *(25,26)*. Voluntary running also prevents the decline in BDNF expression associated with high fat feeding *(26)*. Thus, whole-body physical activity clearly impacts portions of the CNS, raising the intriguing possibility that those that choose to be physically active enhance the potential for hippocampal neuroplasticity, which may contribute to a variety of hippocampus-related functions, including learning and memory.

Another example of the integration among organs comes from studies of mice engineered to overexpress the GLUT4 glucose transporter protein specifically in skeletal muscle. Amazingly, overexpression of this single protein only in skeletal muscle, not in the brain, alters the will of the animal to voluntarily run on a wheel. Male mice overexpressing GLUT4 and having free access to an exercising wheel voluntarily ran fourfold longer distances each day than did wild-type mice *(27)*. Thus, an unknown direct or indirect link must exist from skeletal muscle to central behavioral neurons controlling the desire to run in voluntary wheels. Remarkably, the desire to run in the wheel by GLUT4-overexpressing mice did not extend to cage activity (movement in the cage outside the running wheel). In addition, GLUT4-overexpressing mice ate 45% more, yet their body weights were moderately decreased when compared with the wild-type. The latter observation implies that physically active mice have a different set point for body weight than mice prohibited from access to running wheels. The delineation of the integration between exercise, appetite and the body-weight set point is poorly understood.

4.2. IL-6 Release From Muscle

For years, researchers have wondered whether there may be a factor(s) released by contracting muscle during exercise that would help to coordinate the activities of other organs to support the metabolic needs of skeletal muscle. In the early 1960s, it was observed that blood and lymphatic fluid transferred from exercising dogs, both normal and diabetic, into resting dogs induced hypoglycemia in the resting animals, suggesting that exercise caused the liberation of a hypoglycemic factor that activated the glucose transport system *(28)*. In spinal cord-injured individuals, it has been shown that electrical stimulation of paralyzed muscles evokes similar physiological responses in other organs despite the complete lack of afferent or efferent nerve impulses *(29)*. Recent work suggests that interleukin (IL)-6 may represent one such "exercise factor" *(30)* from skeletal muscle integrating metabolism. IL-6 is released from exercising muscle at a rate that increases exponentially during endurance exercise, generating increases in circulating IL-6 concentrations of more than 100-fold. The release of IL-6 is not due to muscle damage and only a minor portion can be attributed to sympathetic stimulation. Production of IL-6 arises from transcriptional activation of the IL-6 gene and appears to originate from contracting myofibers rather than from other cell types present in muscle. The remarkably high rates of IL-6 release from working muscle, particularly during the latter stages of long duration exercise (3–6 h), raised the hypothesis that regulation of IL-6 production may be sensitive to muscle glycogen content. Indeed, lowering muscle glycogen content in one leg on the days preceding exercise markedly accelerates the transcriptional activation and production of IL-6 during exercise relative to the leg with normal glycogen levels. Although somewhat controversial, carbohydrate ingestion during exercise does not appear to attenuate the production of IL-6. Because the rate of muscle glycogenolysis was unaffected by carbohydrate ingestion, these findings collectively suggest that regulation of IL-6 expression and release from working muscle may represent an important mechanism for signaling the level of muscle glycogen reserves to other organs.

What is the function of IL-6 release from exercising muscle? An initial and attractive hypothesis was that IL-6 release from exercising muscle signals the need for glucose production by the liver. Although there is some indirect evidence to suggest that IL-6 stimulates glucose production in hepatocytes, infusion of IL-6 in resting healthy humans at a rate sufficient to increase plasma IL-6 to levels corresponding to those obtained by prolonged exercise failed to have any affect on hepatic glucose output. Although other factors may be required to affect hepatic glucose production, more recent evidence suggests that IL-6's main effect may be to stimulate lipolysis in adipose tissue, releasing FFAs into the circulation to provide an alternative fuel source to carbohydrate for exercising muscle *(30)*. Thus, skeletal muscle may communicate its glycogen level during exercise to adipocytes.

Another integrative event whose mechanism of communication from muscle to the liver is poorly described is the preferential refilling of glycogen stores in skeletal muscle before the liver when both have been depleted by long-term exercise. At the whole-body level, priority has been placed on muscle over liver in the need for glycogen fuel depots. The next section describes the current understanding of the cellular integration by muscle glycogen levels.

4.3. Muscle Glycogen and Control of Gene Expression

In skeletal muscle, it is now well established that glycogen content significantly influences the regulation of insulin-stimulated glucose transport and insulin signaling as well as contraction-induced glucose transport activity. As alluded to previously, recent work in humans provides evidence that the transcriptional activation of several metabolic genes in muscle in response to exercise is inversely related to muscle glycogen content. In other words, it is as if glycogen content serves as a fuel gauge in skeletal muscle and that signaling and gene regulatory mechanisms are activated when glycogen content is low and persist until glycogen content is restored.

Glycogen metabolism is regulated by the activity of the glycogen synthase and glycogen phosphorylase enzymes. Both enzymes are tightly controlled by allosteric effectors and covalent modification via phosphorylation. Activation of the classic protein kinase A pathway leads to the phosphorylation of both enzymes and the overall stimulation of glycogenolysis. Conversely, dephosphorylation of sites within the glycogen synthase and glycogen phosphorylase enzymes activates glycogen synthesis. In skeletal muscle, the dephosphorylation of glycogen synthesis is catalyzed by a glycogen bound form of protein phosphatase-1 (PP1G) that is composed of a catalytic subunit (PP1c) complexed with a glycogen-targeting subunit. From a regulatory standpoint, it is interesting to note that a number of enzymes regulating glycogen metabolism, including PP1 and glycogen synthase kinase 3, interact not only with proteins that direct activity toward glycogen, but also targeting proteins specific for the sacroplasmic reticulum, myofibrils, and the nucleus. This implies that the intracellular location of regulatory enzymes such as PP1 and GSK3, as determined by reversible binding with noncatalytic targeting proteins, plays an important role in determining substrate specificity and enzymatic activity. Along these same lines, it has recently been demonstrated that AMPK also binds to glycogen *(31)*. Given that high glycogen content represses the activation of AMPK by AICAR or exercise *(19)*, it is tempting to speculate that as glycogen content declines during prolonged exercise, enzymes typically bound to the glycogen scaffold (e.g., AMPK, PP1c, GSK3) may be released and free to associate with different targeting proteins that direct enzyme activity to other parts of the cell, particularly the nucleus to regulate gene expression (Fig. 2). Indeed, deletion of the glycogen-binding domain in the yeast Snf1 kinase (the yeast homologue of mammalian AMPK) completely abolishes glycogen binding and activates a transcriptional program (glycogen synthases, etc.) typically induced by low glycogen content *(32)*. Establishing the molecular link between changes in glycogen concentration and the control of other cellular processes such as glucose uptake and gene expression in skeletal muscle represents an exciting area for future research.

5. WHAT IS A NORMAL PHYSIOLOGICAL VALUE? WHAT THEN IS THE CORRECT CONTROL GROUP?

Physiology is defined as the branch of biology dealing with the functions and activities of living organisms and their parts, including all physical and chemical processes. According to most physiology textbooks, many of the chemical variables in the healthy human are actively maintained within narrow physiological limits. Thus, the "narrow limits" define "normal" physiological range. According to the text "Physiology" by Berne and Levy *(32a)*, the list of physiological values with narrow ranges is long; including body temperature, blood pressure, the ionic composition of blood plasma, blood glucose levels, the oxygen and carbon dioxide content of blood, and many others. Although some normal values are unaffected by the habitual level of physical activity, others differ markedly between sedentary and physically active states. Thus, a question can be raised as to what is the correct value for a measurement when physical activity places the measured item outside of the narrow physiological range for sedentary living? For example, the normal range for resting heart rate is considered for nonphysically active individuals to be approx 72 beats per minute (bpm)while US President George Bush, who is physically active, has a resting heart rate of 49 bpm. The question then is whether 49 or 72 bpm should be considered a normal resting heart rate?

To answer the aforementioned question, the question is asked does majority rule? In other words, is the normal physiological value set by what the majority of individuals' measured, or by the value present in the healthiest of people? One case arguing against majority determination is that the majority of Americans are overweight (60% of adults have a body mass index [BMI] over 25), yet the normal values for BMI are 18.5 to 24.9. Consequently precedence exists for a normal value to occur in less than 50% of the population.

The question of what is a "normal" physiological value has great impact on interpretations of genomic and proteomic responses. For example, the question is posed here as to what is normal heart size in wild-type mice? The functional interpretation of cardiac mass and function in genetically manipulated mice will depend on what is considered the "normal" value for cardiac mass in wild-type mice. Cardiac mass differs depending on whether the mice are housed with access to running wheels or not. When voluntary running wheels are placed in cages, mice run an average distance of 7 km each night. At the fourth week of voluntary running, heart mass is 13% smaller in mice who are denied the opportunity to run voluntarily. Hearts that are smaller in size must have some genomic and proteomic differences or else their heart sizes would be the same. Mice permitted access to running wheels over their entire life-spans (weaning to 33 mo of age) have 50% fewer changes in mRNA levels (137 vs 62) in their hearts as compared to the sedentary population without access to running wheels. Of the 42 changes in gene expression that were common to the voluntary running and sedentary groups, 72% displayed smaller fold changes as a result of the lifetime of exercise. Bronikowski et al. *(33)* suggest that adaptive physiological mechanisms that are induced by exercise retard many effects of aging on heart muscle at the transcriptional level.

Our interpretation is that the physical inactivity due to a lack of voluntary running enhances changes in gene expression with aging. Thus, the question is what is the normal value for cardiac function? Is the heart mass of a mouse who wishes to run, but is not allowed, normal? Or is the heart mass of mice who are allowed voluntary running the normal group, much as their ancestors performed in the wild for tens of thousands years?

Another example of applying genomics and proteomics to better understand physiological function concerns the mechanisms underlying the increased myocardial contractility of hearts from physically active vs sedentary animals. After 11 wk of treadmill training on motor-driven wheels, atrial myosin light-chain (aMLC) mRNA was found to be 2.7-fold greater in ventricles of rats *(34)*. The observation was next confirmed by real-time polymerase chain reaction, and then at the protein level with two-dimensional gel electrophoresis; $13.4 \pm 2.1\%$ of the total ventricular myosin light chain-1 was aMLC-1 protein in the ventricles of physically trained mice as compared to no detectable level of aMLC-1 protein in ventricles of caged rats. Other studies have found that higher percentages of aMLC-1 protein are associated with increased Ca^{2+} sensitivities for tension and maximal shortening velocities. Because exercise training also increases cardiac contractility, the hypothesis has been put forward that the training-induced increase in aMLC-1 protein expression may be an underlying molecular mechanism for the improved myocardial contractile function associated with aerobic exercise training.

Thus, cardiac size, contractility, and aMLC-1 protein are greater in the physically active mice than in their sedentary cage mates. The initial question posed was which is more representative of normal physiology? The answer favored here is that the normal genomic and proteomic value should be what was representative of the environment during which genes were selected to support the function of life. Mice who are allowed voluntary physical activity in running wheels are more representative of the environment experienced in the wild by mice, which have been physically active for their entire existence. Thus, the normal values for wild-type mice/rats in genomic and proteomic studies should be heart sizes that are 13% larger than caged mice/rats, myocardial contractility that is 20% greater, and 13% of total MLC protein in the ventricle that is aMLC *(34,35)*. Other large differences in the levels of certain mRNAs and proteins exist between animals housed with or without access to voluntary running activity. For example, mice with access to voluntary running wheels had 80% higher ANF mRNA and 150% more calcineurin phosphatase activity in cardiac ventricles than sedentary mice *(35,35a)*.

Because voluntary running in wheels while caged more closely approaches the natural level of physical activity experienced by mice for their entire existence on Earth, and caging of mice has occurred only in the last 100 yr, the representative biological state for genomics and proteomics is more closely approximated by mouse that can voluntary run, not by caging.

6. Pathophysiological Integration When Physical Activity is Absent

The examples presented in the previous section raised questions regarding the proper controls for genomic and proteomic experiments in the field of human integrative physiology. Physical activity has markedly decreased in the last century for those humans who have been fortunate enough not to have to hunt or gather their own food and, as automation and computerization has progressively eliminated manual labor activities. Estimates are of an 80% reduction in that portion of the caloric expenditure used for physical activity. Today's sedentary office worker expends 4.4 kcal/kg/d in physical activity, whereas hunter–gatherer males were estimated to use 19.6 to 24.7 kcal/kg/d to complete daily tasks. Presented in a slightly different context, the male office worker expends approx 300 kcal for total daily physical activity as compared to approx 900 to 1800 kcal expended for daily work, according to estimates from two hunter–gatherer measurements. Amazingly, these estimates in the reduction in physical activity are the caloric equivalent to 60 to 150 lb of fat in 1 yr. Obviously, the decrease in physical activity by humans has occurred in a period (the past century) when gene sequences have not changed. Therefore, it is probable that representative normal genomic and proteomic values for *Homo sapiens* are those from physically active, not sedentary, individuals. The aforementioned information suggests that genes were selected to support physical work when exercise was required to search for food and thus survival, and thus a representative gene expression level would be when physical activity is 19.6 to 24.7 kcal/kg/d.

An outcome of the estimated 80% reduction in daily physical activity experienced by humans in sedentary societies as compared to their ancestors of 100 yr ago is a lower cardiorespiratory/aerobic fitness. Recent male hunter–gatherers were found to have 40% greater maximal oxygen consumptions rates (an index of cardiorespiratory fitness) than sedentary populations (52.5 mL/kg/min vs 37.5 mL/kg/min *[36]*). Women showed the same difference, but with lower absolute values than males. The clinical significance of low cardiorespiratory fitness is that all-cause mortality is higher (i.e., the poorer the aerobic fitness, the higher is the probability of death).

Cardiovascular fitness is determined by the capacity of multiple organ systems to integrate the transfer of oxygen from the atmosphere to the working skeletal muscles (includes neural, pulmonary, cardiac maximal pumping capacity, capillarization, enzymatic systems in skeletal muscle to oxidize glucose and fatty acids, and skeletal muscle mass together). Although there is an inheritance factor, cardiovascular fitness is not fixed, but malleable by physical activity, increasing in a period of weeks with increases in daily physical activity, and decreasing after a few weeks of lessened physical activity. Illustrative of the speed and magnitude of changes in cardiovascular fitness is the extreme case of physical inactivity, continuous bed rest. Twenty days of bed rest dropped cardiovascular fitness 28% in healthy young men. In the same subjects, resting heart volume decreased 11%, maximal cardiac output fell 26%, and maximal stoke volume declined by 29% *(37)*. The 30% lesser cardiovascular fitness (52.5 mL/kg/min to 37.5 mL/kg/min) in the current sedentary person as compared to the hunter–gatherer is likely a consequence of the 80% decline in daily caloric expenditure for physical activity.

Likely, a higher level of cardiovascular fitness is more representative for the baseline integration of biological systems at rest and during exercise. Other authors indicate that physical inactivity has altered physiological baselines. Bougneres *(38)* wrote that modern sedentary life and caloric abundance have created new physiological conditions capable of changing the level of expression of a number of genes involved in fuel metabolism and bodyweight regulation.

The question then is raised as to whether there are biological links in today's sedentary individual from decreased caloric expenditure to an increased risk of chronic health conditions (such as atherosclerosis, cancer, obesity, type 2 diabetes). The answer to this question can be formulated from the hypotheses that many of the metabolic dysfunctions occurring in sedentary societies are related to alterations in the expression of metabolic genes as a result of (a) severe reductions in caloric expenditure, (b) the lack of cycling of metabolic proteins that normally occurs during feast–famine, and/or (c) the periodicity of metabolic events related to the reduction and consequent replenishment of fuel stores during and after physical activity, respectively. These hypotheses are based on the assumptions that (a) caloric expenditure was higher in humans 10,000 yr ago as compared to the sedentary individual of today (discussed previously), (b) biochemical processes that cycled over a period of days have been evolutionarily selected in humans to support survival during a period of famine, and (c) a rise and fall in fuel stores (glycogen and triglyceride) have cycled with physical activity and recovery from exercise for most of human existence. The assumption, until recently, has been that the biological consequences of a positive caloric balance and the lack of physical activity are relatively benign. However, the recent onsets of the obesity and type 2 diabetes epidemics, and the association of obesity and physical inactivity with multiple deleterious health conditions, now require a biological explanation at the level of organ/ tissue integration and in molecular regulation to provide a scientific basis for prevention and treatment of these epidemics.

How may the lack of cycling associated with feast–famine and physical activity–recovery lead to the development of chronic health disorders? The combined absences of starvation and physical activity have removed two major impetuses for the turnover of carbohydrate and triglyceride stores *(39)*. Instead, stores of skeletal muscle glycogen and adipocyte triglyceride now stall at high levels in the physically inactive (i.e., they do not turnover sufficiently to evoke the ancient cycling of proteins that regulate the processes of fuel anabolism and catabolism). For example, in physically active individuals, the highest insulin sensitivity occurs during exercise and falls off when muscle glycogen stores are restored following exercise. A lack of periodic physical activity removes exercise-induced cycling of insulin sensitivity (i.e., a stall in insulin resistance occurs without a cycling to insulin sensitivity). Some key endocrine functions stall above their normal values. For example, a stall in hyperglycemia and hyperinsulinemia occur in pre-diabetic states. Although meals and sleep would cycle blood glucose and insulin in hyperglycemic and hyperinsulinemic states, postmeal levels of blood glucose and insulin would not fall into normal range unless either starvation or exercise occur. Thresholds for the altered expression (thus allowing cycling) of specific regulatory proteins are hypothesized to be reached only in the normal plasma range of glucose and insulin. Thus, the biochemical and molecular regulatory events related to the cycling of glycogen and fat stores are largely blunted in the fed, sedentary individual. An important future question is to explain how normal physiological control becomes so dysfunctional as to lead to overt clinical chronic health conditions and premature death. Molecular medicine is based on the premise that knowledge of molecular causes improves prevention and treatment.

Physical inactivity triggers prediabetes. Physically active individuals who stop exercise for 10 d exhibit a doubling in the area under their insulin curves during an oral glucose tolerance test (a sugar meal causes blood insulin to spike, but the rise in plasma insulin is twice as great after 10 d of inactivity). Similar results have been produced in healthy young subjects who undergo 3 to 4 d of continuous bedrest; they exhibit increased plasma levels of glucose and insulin after a standardized glucose meal. The data that physical inactivity triggers prediabetes confirm the report in which insulin-stimulated glucose uptake was correlated with maximal cardiovascular fitness. In Native Americans and Euro-Americans, physical inactivity was responsible for approx 25% of the variance in insulin-stimulated glucose uptake (obesity and genes accounted for the remaining approx 25% and approx 50%, respectively) *(40)*. Others suggest that insulin resistance was once of evolutionary use to our species and is now responsible for the worldwide epidemic of type 2 diabetes. In fact, it is possible that insulin resistance may have conveyed a selective advantage for populations under conditions when whole-body glucose availability was scarce by redirecting glucose away from muscles, ensuring that sufficient glucose was present to support the needs of the brain, fetus, and mammary gland.

Other events also minimize the uptake of blood glucose into working muscle, and thus conserve the limited stores of body carbohydrate (body glycogen stores can supply only about 25% of a single day's calories for resting metabolism). Physically trained humans and wild animals have higher mitochondrial concentrations, myoglobin, and capillaries than the sedentary individual and caged animal, respectively. The biological significance of a greater density of mitochondria in skeletal muscle is that a greater percentage of ATP is obtained from FFAs during exercise at a given workload, which lessens the depletion of the limited stores of glycogen, saving whole body glucose. Higher mitochondrial activities and densities have been associated with lower prevalences of insulin resistance, type 2 diabetes, and obesity. In summary, biochemical adaptations in skeletal muscle to aerobic training play roles in lessening triglyceride storage in adipose tissue, inhibiting pancreatic β-cell production of insulin, increasing insulin sensitivity, improving postprandial glucose and triglyceride removal from the blood, and more closely approximate the environment under which gene expression evolved.

7. SUMMARY

Selected examples have been given for complex integrations that are occurring at the cellular, tissue, system, and whole-body level in order to support physical activity. The dramatic decline in daily physical activity that has occurred in the past 100 yr is unprecedented in human history and has clearly contributed to the proliferation of many chronic diseases. As more molecular information in multiple organs comparing physically active and inactive animals is gathered at the cellular level, it is becoming apparent enormous differences in gene expression exist. Future research now needs to employ genomics, proteomics, bioinformatics, and systems biology to explain the molecular mechanisms by which habitual physical activity plays a protective role in human health.

ACKNOWLEDGMENTS

This work was supported by Public Health Service grants AR19393 (FB) and AR45372 (PDN).

REFERENCES

1. Chakravarthy, M.V. and Booth, F.W. (2003) *Exercise*. Philadelphia: Elsevier.
2. Rowell, L.B. and Shepherd, J.T. (1996) *The Handbook of Physiology.* Bethesda, MD: American Physiological Society.
3. Sugden, M.C. and Holness, M.J. (2003) Recent advances in mechanisms regulating glucose oxidation at the level of the pyruvate dehydrogenase complex by PDKs. *Am. J. Physiol. Endocrinol. Metab.* **284**: E855–E862.
4. Pilegaard, H. and Neufer, P.D. (2004) Transcriptional regulation of PDK4 in skeletal muscle during and after exercise. Proc. Nutr. Soc. **63**, 211–226.
5. Waldrop, T.G., Eldridge, F.L., Iwanmoto, G.A. and Mitchell, J.H. (1996) Central neural control of respiration and circulation during exercise. In *The Handbook of Physiology. Exercise: Regulation and Integration of Multiple Systems* (Rowell, L.B. and Shepherd, J.T., ed.), American Physiological Society, Bethesda, MD, pp. 333–380.

6. Holloszy, J.O. (1967) Biochemical adaptations in muscle. Effects of exercise on mitochondrial oxygen uptake and respiratory enzyme activity in skeletal muscle. *J. Biol. Chem.* **242**, 2278–2282.

7. Hildebrandt, A.L., Pilegaard, H. and Neufer, P.D. (2003) Differential transcriptional activation of select metabolic genes in response to variations in exercise intensity and duration. *Am. J. Physiol. Endocrinol. Metab.* **285**, E1021–E1027.

8. Pilegaard, H., Ordway, G.A., Saltin, B. and Neufer, P.D. (2000) Transcriptional regulation of gene expression in human skeletal muscle during recovery from exercise. *Am. J. Physiol. Endocrinol. Metab.* **279**, E806–E814.

9. Crabtree, G.R. and Olson, E.N. (2002) NFAT signaling: choreographing the social lives of cells. *Cell.* **109,** Suppl: S67–79.

10. Bassel-Duby, R. and Olson, E.N. (2003) Role of calcineurin in striated muscle: development, adaptation, and disease. *Biochem. Biophys. Res. Commun.* **311,** 1133–1141.

11. Wu, H., Kanatous, S.B., Thurmond, F.A., et al. (2002) Regulation of mitochondrial biogenesis in skeletal muscle by CaMK. *Science.* **296**, 349–352.

12. Rose, A.J. and Hargreaves, M. (2003) Exercise increases Ca2+-calmodulin-dependent protein kinase II activity in human skeletal muscle. *J. Physiol.* **553**, 303–309.

13. Norrbom, J., Sundberg, C.J., Ameln, H., Kraus, W.E., Jansson, E. and Gustafsson, T. (2004) PGC-1α mRNA expression is influenced by metabolic perturbation in exercising human skeletal muscle. *J. Appl. Physiol.* **96**, 189–194.

14. Pilegaard, H., Saltin, B. and Neufer, P.D. (2003) Exercise induces transient transcriptional activation of the PGC-1α gene in human skeletal muscle. *J. Physiol.* **546**, 851–858.

15. Lin, J., Wu, H., Tarr, P.T., et al. (2002) Transcriptional co-activator PGC-1α drives the formation of slow-twitch muscle fibres. *Nature.* **418**, 797–801.

16. Puigserver, P. and Spiegelman, B.M. (2003) Peroxisome proliferator-activated receptor-γ coactivator 1 α (PGC-1 α): transcriptional coactivator and metabolic regulator. *Endocr. Rev.* **24**, 78–90.

17. Hardie, D.G., Scott, J.W., Pan, D.A. and Hudson, E.R. (2003) Management of cellular energy by the AMP-activated protein kinase system. *FEBS. Lett.* **546**, 113–120.

18. Merrill, G.F., Kurth, E.J., Hardie, D.G. and Winder, W.W. (1997) AICA riboside increases AMP-activated protein kinase, fatty acid oxidation, and glucose uptake in rat muscle. *Am. J. Physiol.* **273,** E1107–E1112.

19. Wojtaszewski, J.F., Nielsen, J.N., Jorgensen, S.B., Frosig, C., Birk, J.B. and Richter, E.A. (2003) Transgenic models - a scientific tool to understand exercise-induced metabolism: the regulatory role of AMPK (5'-AMP-activated protein kinase) in glucose transport and glycogen synthase activity in skeletal muscle. *Biochem. Soc. Trans.* **31**: 1290–1294.

20. Stoppani, J., Hildebrandt, A.L., Sakamoto, K., Cameron-Smith, D., Goodyear, L.J. and Neufer, P.D. (2002) AMP-activated protein kinase activates transcription of the UCP3 and HKII genes in rat skeletal muscle. *Am. J. Physiol. Endocrinol. Metab.* **283**: E1239–E1248.

21. Winder ,W.W. (2001) Energy-sensing and signaling by AMP-activated protein kinase in skeletal muscle. *J. Appl. Physiol.* **91**, 1017–1028.

22. Sakamoto, K. and Goodyear, L.J. (2002) Invited review: intracellular signaling in contracting skeletal muscle. *J. Appl. Physiol.* **93**, 369–383.

23. Mu, J., Brozinick, J.T, Jr., Valladares, O., Bucan, M. and Birnbaum, M.J. (2001) A role for AMP-activated protein kinase in contraction- and hypoxia- regulated glucose transport in skeletal muscle. *Mol. Cell.* **7**, 1085–1094.

24. Cotman, C.W. and Berchtold, N.C. (2002) Exercise: a behavioral intervention to enhance brain health and plasticity. *Trends Neurosci.* **25**, 295–301.

25. Johnson, R.A., Rhodes, J.S., Jeffrey, S.L., Garland, T., Jr. and Mitchell, G.S. (2003) Hippocampal brain-derived neurotrophic factor but not neurotrophin-3 increases more in mice selected for increased voluntary wheel running. *Neuroscience.* **121**, 1–7.

26. Molteni, R., Wu, A., Vaynman, S., Ying, Z., Barnard, R.J and Gomez-Pinilla, F. (2004) Exercise reverses the harmful effects of consumption of a high-fat diet on synaptic and behavioral plasticity associated to the action of brain-derived neurotrophic factor. *Neuroscience.* **123**, 429–440.

27. Tsao, T.S., Li, J., Chang, K.S., et al. (2001) Metabolic adaptations in skeletal muscle overexpressing GLUT4: effects on muscle and physical activity. *FASEB. J* **15**: 958–969.

28. Goldstein, M.S. (1961) Humoral nature of the hypoglycemic factor of muscular work. *Diabetes.* **10**, 232–234.

29. Kjaer, M., Secher, N.H., Bangsbo, J., Perko, G., Horn, A., Mohr, T. and Galbo, H. (1996) Hormonal and metabolic responses to electrically induced cycling during epidural anesthesia in humans. *J. Appl. Physiol.* **80**, 2156–2162.

30. Pedersen, B.K., Steensberg, A., Fischer, C., et al. (2003) Searching for the exercise factor: is IL-6 a candidate? *J. Muscle Res. Cell Motil.* **24**, 113–119.

31. Polekhina, G., Gupta, A., Michell, B.J., et al. (2003) AMPK beta Subunit Targets Metabolic Stress Sensing to Glycogen. *Curr. Biol.* **13**, 867–871.

32. Wiatrowski, H.A., Van Denderen, B.J., Berkey, C.D., Kemp, B.E., Stapleton, D. and Carlson, M. (2004) Mutations in the gal83 glycogen-binding domain activate the snf1/gal83 kinase pathway by a glycogen-independent mechanism. *Mol. Cell. Biol.* **24**: 352–361.

32a. Berne, R.M. and Levy, M.N. (1998) Physiology, 4th Ed. St. Louis:Mosby, p. xi.

33. Bronikowski, A.M., Carter, P.A., Morgan, T.J., et al. (2003) Lifelong voluntary exercise in the mouse prevents age-related alterations in gene expression in the heart. *Physiol. Genomics.* **12**, 129–138.

34. Diffee, G.M., Seversen, E.A., Stein, T.D. and Johnson, J.A. (2003) Microarray expression analysis of effects of exercise training: increase in atrial MLC-1 in rat ventricles. *Am. J. Physiol. Heart Circ. Physiol.* **284**, H830–H837.

35. Allen, D.L., Harrison, B.C., Maass, A., Bell, M.L., Byrnes, W.C. and Leinwand, L.A. (2001) Cardiac and skeletal muscle adaptations to voluntary wheel running in the mouse. *J. Appl. Physiol.* **90**, 1900–1908.

35a. Eto, Y., Yonekura, K., Sonoda, M., et al. (2000) Calcineurin is activated in rat hearts with physiological left ventricular hypertrophy induced by voluntary exercise training. *Circulation* **101**, 2134–2137.

36. Cordain, L., Gotshall, R.W. and Eaton, S.B. (1997) Evolutionary aspects of exercise. *World Rev. Nutr. Diet.* **81**, 49–60.

37. Saltin, B., Blomqvist, G., Mitchell, J.H., Johnson, R.L., Jr., Wildenthal, K. and Chapman, C.B. (1968) Response to exercise after bed rest and after training. *Circulation.* **38**: VII1–78.

38. Bougneres, P. (2002) Genetics of obesity and type 2 diabetes: tracking pathogenic traits during the predisease period. *Diabetes.* **51,** Suppl 3: S295–303.

39. Chakravarthy, M.V. and Booth, F.W. (2004) Eating, exercise, and "thrifty" genotypes: connecting the dots toward an evolutionary understanding of modern chronic diseases. *J. Appl. Physiol.* **96**, 3–10.

40. Bogardus, C., Lillioja, S., Mott, D.M., Hollenbeck, C. and Reaven, G. (1985) Relationship between degree of obesity and in vivo insulin action in man. *Am. J. Physiol.* **248**, E286–E291.

10

Biochemical Adaptation to Extreme Environments

Kenneth B. Storey and Janet M. Storey

1. INTRODUCTION

Physiology can be viewed as the collection of mechanisms and processes that allows organisms to deal with challenges from both internal (e.g., exercise, growth, reproduction) and external (e.g., variations in temperature, oxygen and water availability, salinity, pressure, radiation, heavy metals, etc.) sources. In this chapter we focus on solutions to some of the external challenges to life in extreme environments. This subject is a huge one because life on Earth has radiated into every conceivable environment, from the frigid Antarctic to boiling hot springs, from the ocean depths to the tops of mountains, from hypersaline lakes to the driest deserts, and many more. We mainly consider biochemical and molecular solutions by vertebrate animals to environmental challenges of low oxygen and low temperature because these hold lessons that can be applied to the human condition and medical concerns. However, the reader should be aware that the extremes of vertebrate life are bested on every front by the capabilities of invertebrates, plants, bacteria, and archaea and many excellent resources explore life at the extremes from different perspectives; selected texts include those by Hochachka and Somero *(1)*, Schmid-Neilsen *(2)*, Ashcroft *(3)*, Margesin and Schinner *(4)*, Willmer et al. *(5)*, Lutz et al. *(6)*, and Gerday and Glansdorff *(7)*.

In general, adaptive responses to environmental stresses are needed for two main reasons. First, all biological molecules and all biochemical reactions are directly susceptible to perturbation by multiple environmental parameters including temperature, pressure, pH, ionic strength, solute concentrations, water availability, radiation, and attack by free radicals. Second, to sustain life, all cells must maintain adequate energy turnover by maintaining sufficient energy currencies, primarily adenosine triphosphate (ATP), that is used to drive thermodynamically unfavorable reactions, and reduced nicotinamide adenine dinucleotide phosphate (NADPH), that is used for reductive biosynthesis *(1)*. Depending on the circumstances, adaptation can be undertaken at behavioral, physiological, and/or biochemical levels and can address one (or more) of three goals: compensation, conservation, and protection.

Compensatory responses often deal with short-term or relatively mild stresses under which the organism aims to maintain normal functions. An example of a compensatory response is the rapid increase in ventilation rate and the release of stored erythrocytes that occurs in response to hypoxia (low oxygen) challenge. In man, this is well-known as the first response

From: *Integrative Physiology in the Proteomics and Post-Genomics Age*
Edited by: W. Walz © Humana Press Inc., Totowa, NJ

to movement from low to high altitude. Compensation can also be employed for long-term adaptation. For example, homeoviscous adaptation is a common response to temperature change in ectotherms that alters the proportions of saturated, mono- and poly-unsaturated fatty acids in cellular membranes to compensate for temperature effects on membrane fluidity. In fish, a switch from warm to cold water triggers a rapid upregulation of desaturase enzymes that raise monounsaturated levels in membranes at the expense of saturated fatty acids *(8)*.

Conservation responses are typically enacted when organisms face a stress that is too strong or too prolonged and, therefore, incompatible with maintaining normal life. Conservation often includes behavioural responses that move the organism to a sheltered site (e.g., desert toads dig underground during the dry season to limit their exposure to desiccating air; many turtles spend the winter under water to avoid freezing temperatures on land) as well as strong metabolic rate depression (MRD). MRD frequently combines adaptations at both physiological (e.g., reduced rates of heart beat and breathing) and biochemical (e.g., selective inhibition of nonessential metabolic functions) levels to reduce basal metabolic rate to as little as 1 to 20 % of the previous resting rate *(9,10)*. MRD is often brought into play when the stress compromises an organism's access to exogenous fuels (either foodstuffs or oxygen) and thereby greatly extends the time that the organism can survive using only its fixed internal fuel reserves.

Protective responses are typically enacted against stresses that challenge the physical integrity of living organisms. For example, desiccation or high salinity dehydrate cells causing metabolic damage from high ionic strength and compression stress on membranes owing to cell-volume collapse. Exposure to subzero temperatures brings the risk of freezing and the massive physical and metabolic destruction that can arise from the propagation of ice crystals through a body. Protective responses often involve the proliferation of low-molecular-weight metabolites (e.g., high plasma urea limit body water loss in desert amphibians; high glycerol acts as an antifreeze for cold-hardy insects) or the synthesis of new proteins (e.g., heat shock or cold shock proteins that stabilize protein conformations, antifreeze proteins that inhibit ice crystal formation) *(4,11,12)*. These address both cell-volume concerns and the physical stability and conformation of macromolecules. Protective responses often go hand- in- hand with conservation responses.

At the biochemical level, virtually all aspects of cell function and metabolism can be the target of adaptive change in response to environmental stress *(13)* and can include elements such as those that follow:

1. Increasing or decreasing the expression of selected genes to produce corresponding changes in the levels of various proteins.
2. Elaborating novel genes and proteins that address stress-specific concerns.
3. Altering the kinetic and regulatory properties of enzymes and functional proteins.
4. Changing enzyme susceptibility to posttranslational modification, in particular the effects of reversible phosphorylation by protein kinases and protein phosphatases.
5. Modifying sensing and signaling mechanisms by changes to cell surface receptors, signal transduction cascades, cross-talk between signaling pathways, and the targets of signals including proteins, transcription factors, and genes.
6. Changing protein–protein binding interactions that alter the composition of enzyme/protein complexes or localize enzyme/protein function via associations with specific binding proteins or subcellular structures.
7. Changing membrane composition to compensate for changes in environmental factors including temperature, pH, and ionic composition.

8. Producing protective molecules that can defend cell volume and/or stabilize protein or membrane structure/function.

We discuss examples of many of these strategies of biochemical adaptation throughout the remainder of this chapter. In doing so, we draw most of our examples from discussions of three animal strategies for dealing with extreme environments: anoxia tolerance, hibernation, and freeze tolerance. Our treatment of each of these topics is in no way comprehensive, for each is a huge field of its own. We begin with a brief overview of each of these strategies, but then focus on two areas of major new research: the molecular mechanisms of MRD, and the use of new technologies in genomics to provide a comprehensive assessment of the full range of metabolic adaptations that underlies animal survival in extreme environments.

2. ANOXIA TOLERANCE

2.1. Low Oxygen Injuries and Strategies for Survival in Oxygen-Sensitive Organisms

Humans lead a highly oxygen-dependent existence—an interruption of oxygen supply to the most oxygen-sensitive human organ, the brain, for more than about 4 to 5 min can cause irreparable damage. Many other organisms similarly rely on the high rates of ATP generation possible from oxidative phosphorylation to fuel energy-expensive lifestyles such as the homeothermy of mammals and birds and the muscle power requirements flying insects or jetting squid. For such organisms, oxygen limitation can have grave consequences. Their response to hypoxia (low oxygen) is an immediate implementation of compensatory mechanisms that both increase oxygen delivery to tissues (e.g., increased ventilation rate, release of erythrocytes from spleen) and elevate ATP output from anaerobic sources (e.g., activate glycolysis, creatine phosphate hydrolysis). Gene expression is also activated under the regulation of the hypoxia-inducible factor 1 (HIF-1) to provide more long-lasting compensatory responses. Genes activated by HIF-1 include those for vascular endothelial growth factor (to stimulate capillary growth), erythropoietin (to stimulate red blood cell synthesis), several glycolytic enzymes (to increase glycolytic capacity), and glucose transporters (to enhance substrate uptake) *(14)*.

These compensatory responses work well for dealing with mild hypoxia or ischemia (reduced blood flow) but fail as responses to severe hypoxia, anoxia (no oxygen) or severe ischemia (blood flow stopped). This is because an elevated glycolytic rate is rarely sufficient to sustain cellular ATP demands for very long. For example, ischemic mouse brain shows an almost instantaneous increase in glycolytic rate of four- to sevenfold but this only partially compensates for the much lower ATP yield from glycolysis compared with the full oxidation of glucose by brain mitochondria (a net of 2 ATP is produced per 1 glucose catabolized to 2 lactate vs 36 ATP if glucose is catabolized to CO_2 and H_2O). As a result, within 5 min of oxygen deprivation, as much as 90% of the ATP is depleted in mammalian brain because the rate of ATP output from glycolysis cannot keep pace with the unaltered rate of ATP consumption by energy-consuming cell functions.

Indeed, this imbalance between ATP-producing and ATP-consuming processes is the root of low oxygen injuries in all oxygen-sensitive systems as well as the primary reason for the implementation of conservation responses by anoxia-tolerant species. Chief among the causes of metabolic failure is the inability of oxygen-limited cells to meet the ATP demands of the ion pumps that are involved in maintaining membrane potential difference and the many sensing and signaling functions that rely on transmembrane ion gradients *(15,16)*. For example, the sodium-potassium ATPase alone uses 5 to 40% of cellular ATP turnover in mammals, depending on cell type *(17)*. Membrane potential difference is maintained by a balance between the actions of ATP-dependent ion pumps that move ions against their concentration gradients and facilitative ion channels that allow ions to move down their concentration gradients. If ion pumps fail because of ATP limitation, then membrane depolarization quickly occurs. Depolarization results in a rapid uptake of Na^+ and water (and a loss of K^+) from cells and is followed by an influx of Ca^{2+} through voltage-gated Ca^{2+} channels. Transient elevations of cytosolic Ca^{2+} are critical signaling mechanisms for many cell functions but sustained high Ca^{2+} triggers a range of pathological changes including the activation of phospholipases and proteases that lead to damage and death of cells *(18,19)*.

Not only is oxygen deprivation damaging, but another set of injuries arise when ischemic tissues are reperfused with oxygenated blood. Reperfusion injury, such as occurs after heart attack or stroke, is caused by a burst of reactive oxygen species (ROS) generation, chiefly superoxide radicals, from a highly reduced electron transport chain when oxygen is restored. ROS production can overwhelm existing antioxidant defenses of cells and cause oxidative damage to macromolecules including DNA, proteins, and membrane lipids *(20)*. Indirect damage can also arise such as from an inability of sarco- or endoplasmic reticulum membranes that are damaged by peroxidation to properly re-sequester Ca^{2+}, thereby exacerbating the Ca^{2+}-mediated damage that occurred under anoxia/ischemia.

2.2. Facultative Anaerobiosis

Unlike the oxygen-sensitive species discussed earlier, many organisms are well equipped to survive the environmental extreme of anoxia. Indeed, some organisms are obligate anaerobes, whereas others can survive equally well in the presence or absence of oxygen. Among vertebrates, the premier facultative anaerobes are various species of freshwater turtles of the *Chrysemys* and *Trachemys* genera. These hibernate underwater to escape freezing temperatures but in doing so cannot breathe with lungs. However, turtles can survive in cold, deoxygenated water for as long as three months using anaerobic glycolysis as their only source of ATP generation *(21)*. Carp and goldfish also use well-developed anoxia tolerance to support winter survival in small ice-locked ponds where oxygen in the water is depleted by the respiration of all organisms present. Many types of invertebrates are also excellent facultative anaerobes, the best-studied of these being molluscs and annelids of the marine intertidal zone; these gill-breathing animals have full access to oxygen when under water but switch to anaerobic metabolism each time the tide recedes *(22)*.

Given the multiple forms of metabolic injury that can arise because of oxygen limitation in oxygen-sensitive organisms, it is clear that facultative anaerobes must address a variety of issues in order to survive periods of oxygen deprivation. The overriding strategy for anaerobiosis is not compensation but conservation—organisms use strategies to minimize their ATP use, optimize the time that fixed internal fuel reserves can fuel metabolism, and limit the disruption of cellular homeostasis. The following five main categories of biochemical adaptation have been identified *(9,10,13,16)*:

1. *Fuel supply*: Facultative anaerobes maintain large reserves of glycogen in their tissues and marine invertebrates also maintain substantial pools of fermentable amino acids (e.g., aspartate, glutamate).

2. *Enhance ATP yield*: Anaerobic ATP production by the basic glycolytic pathway can be supplemented with other reactions that increase the ATP yield per glucose catabolized. For example, many marine molluscs catabolize glucose to succinate and propionate, with additional substrate-level phosphorylation reactions increasing the yield to 4 or 6 ATP per glucose, compared with 2 ATP per glucose converted to lactate (Fig. 1).

3. *Minimize cytotoxicity*: Cells generating ATP from anaerobic glycolysis ending in lactate production soon undergo significant acidification as well as major end-product accumulation. Solutions include enhanced buffering capacity (e.g., turtles release calcium carbonate from their shell and bone to buffer acid build-up and also move huge amounts of lactate into their shells for storage), making less acidic end-products (e.g., synthesis of succinate or propionate generates a much lower proton load than does lactate output), or making products that can be excreted easily (e.g., carp and goldfish catabolize lactate to ethanol + CO_2 in their skeletal muscles and then excrete both across the gills).

4. *MRD*: A coordinated and strong reduction in the rates of ATP consumption by multiple cell functions reduces ATP demand into line with ATP output from fermentative pathways and greatly extends the time that fixed internal reserves of fermentable fuels can sustain anaerobic survival.

5. *Antioxidant defense*: Well-developed enzymatic and metabolite antioxidant defenses minimize oxidative stress during the transition from anaerobiosis back to aerobic life. For example, anoxia-tolerant freshwater turtles show the highest constitutive antioxidant defenses among ectotherms (closely comparable to mammalian levels of defense) and anoxic or ischemic stresses frequently induce the synthesis of antioxidants in species that encounter low oxygen stress less frequently *(23,24)*.

3. HIBERNATION

A distinguishing characteristic of mammals and birds is endothermy, heating the body from internal biochemical reactions to permit homeothermy, the maintenance of a high and near constant core body temperature (Tb). Endothermy is very costly, the metabolic rate of mammals being four to seven times higher than that of comparably sized reptiles. This must be supported by equally higher rates of fuel consumption, supplied by foraging or, if food supply is limiting, by food caches or body fuel reserves (chiefly adipose). When environmental temperature falls in the autumn and winter, so to does the metabolic rate, Tb and food needs of an ectothermic (cold-blooded) organism. However, the opposite is true of mammals—they lose body heat faster at colder temperatures, thereby necessitating a higher metabolic rate and greater fuel consumption to compensate. Some mammals can meet this challenge and minimize the extent of metabolic compensation needed by strategies including enhanced body insulation, huddling in groups, counter-current heat exchangers in extremities, and so forth *(5)*. However, for others, the combination of cold temperatures and lack of food availability makes winter survival as a homeotherm impossible. The problem is particularly acute for species such as insectivorous bats or grazing herbivores (e.g., ground squirrels, marmots) that have little or no access to edible food in the winter.

For many small mammals, the solution to life in extremely cold environments is hibernation. By abandoning homeothermy and allowing Tb to fall, tracking environmental temperature, they gain tremendous energy savings, sufficient to sustain life until spring. For example, ground squirrels save as much as 88% of the energy that would otherwise be needed to maintain a Tb of 37°C over the winter *(25)*. During hibernation, all aspects of the animal's physiology slow dramatically. Metabolic rate can be as low as 1 to 5% (at Tb = 0–5°C) of the normal resting rate at 37°C. Heart beat in ground squirrels can drop from 200–300 to just 5–

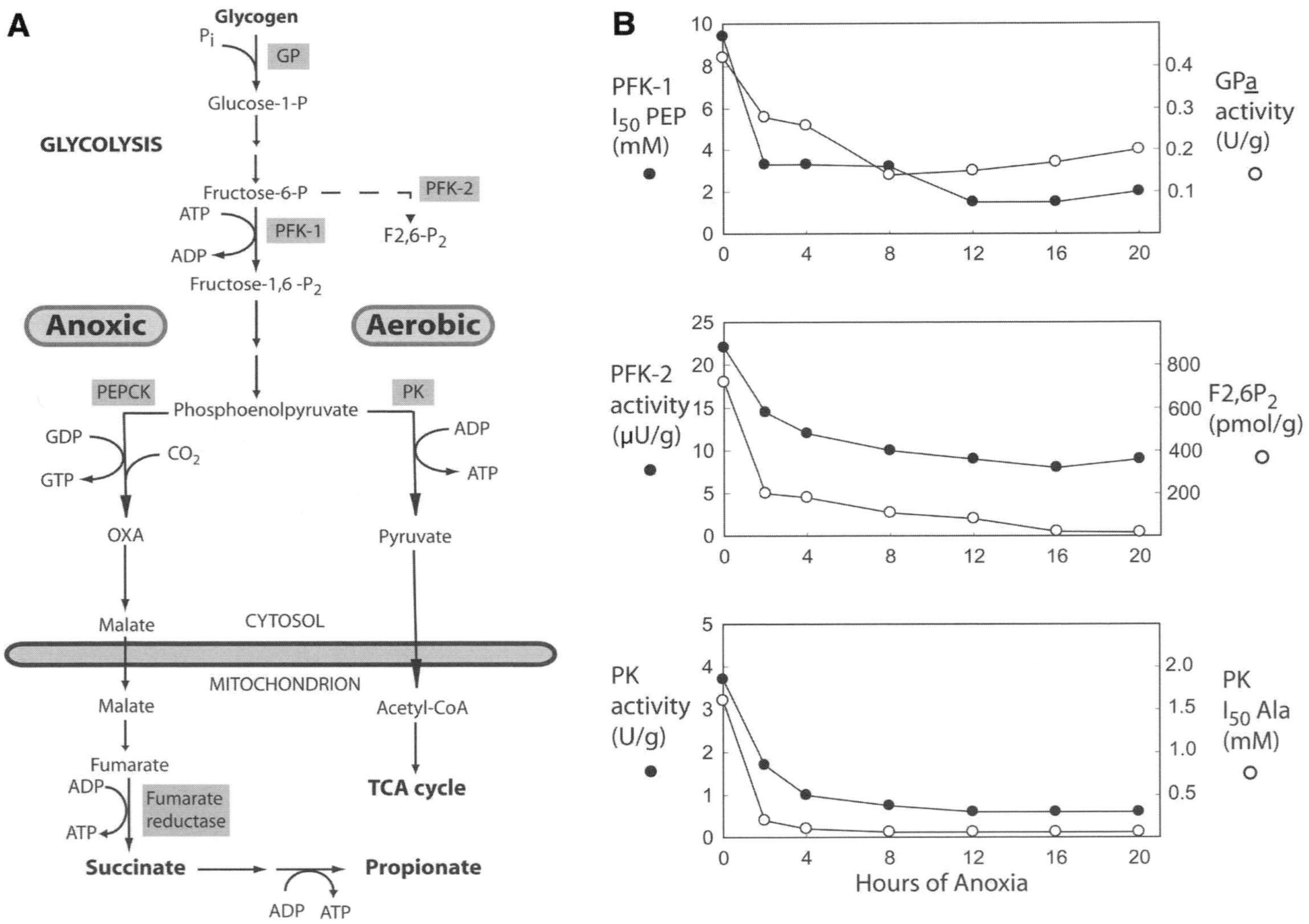

A

GLYCOLYSIS

Glycogen → GP → Glucose-1-P → Fructose-6-P → PFK-1 → Fructose-1,6-P_2 → → → Phosphoenolpyruvate

P_i

PFK-2

F2,6-P_2

Aerobic

PK: Phosphoenolpyruvate → Pyruvate (ADP → ATP)

Pyruvate → Acetyl-CoA → **TCA cycle**

Anoxic

PEPCK: Phosphoenolpyruvate + CO_2 → OXA (GDP → GTP)

OXA → Malate

CYTOSOL | MITOCHONDRION

Malate → Fumarate → Fumarate reductase → **Succinate** → **Propionate** (ADP → ATP)

174

B

PFK-1 I_{50} PEP (mM) ●

GP$\underline{a}$ activity (U/g) ○

PFK-2 activity (μU/g) ●

F2,6P_2 (pmol/g) ○

PK activity (U/g) ●

PK I_{50} Ala (mM) ○

Hours of Anoxia

10 beats per minute. Breathing rate similarly declines and sometimes includes long periods of apnea (breath-hold). Hibernation is not continuous but consists of a series of torpor bouts that in midwinter typically stretch to 1 to 3 wk and are interspersed with brief periods of arousal, generally lasting 6 to 24 h, when the animal uses nonshivering thermogenesis by brown adipose tissue to return its Tb to 37°C. Arousals are by far the greatest energy expenditures of the winter season although the signals that trigger them and their purpose are still not well understood.

3.1. Hypothermic and Ischemic Injury

From our point of view as nonhibernating mammals, there are multiple risks involved with the hibernation strategy. Hypothermia is a serious problem for most mammals; for example, humans undergo severe, often lethal, metabolic injuries if our core Tb drops below about 25°C. Hypothermic injury arises from two main factors. The first is the differential effects of temperature change on cellular reaction rates that culminate in a mismatch between the net rates of ATP-producing and ATP-utilizing reactions. The result is that energy currencies are depleted and the major manifestation of this energy crisis is membrane depolarization, which sets off a chain of catastrophic events that are much the same as those described earlier for anoxia-induced energy failure *(15)*. The second main effect of hypothermia is a decrease in lipid fluidity in both membranes and adipose depots as temperature declines. Membrane lipid fluidity is crucial for allowing protein movements within membranes and the protein conformational changes that are associated with receptor and transporter functions, whereas adipose depots must remain fluid in order for triglycerides to be mobilized as fuels. Normally, the composition of mammalian lipids is optimized for 37°C function and they solidify at about room temperature. Other problems associated with chilling of nonhibernators include the ischemia that develops at extremely low blood-flow rates and a greatly increased risk of blood clotting at low flow rates.

3.2. The Solutions for Hibernators

Summer-active individuals of hibernating species are just as susceptible to hypothermia-induced membrane depolarization as are nonhibernating species. Hence, the preparations for

---

Fig. 1. *(continued from facing page)* Control of glycolysis in anoxia tolerant marine molluscs. (A) The glycolytic pathway showing aerobic and anoxic routes of carbohydrate catabolism determined by the fate of phosphoenolpyruvate (PEP) at the PEP branchpoint. The aerobic route feeds PEP via an active pyruvate kinase (PK) into the tricarboxylic acid (TCA) cycle. The anaerobic route, facilitated by phosphorylation-mediated inhibition of PK, feeds PEP via PEP carboxykinase (PEPCK) into reactions of succinate and propionate synthesis that are linked with substrate-level phosphorylation of ADP to increase the ATP yield of anaerobic metabolism. (B) Overall glycolytic rate depression under anoxia is mediated by reversible phosphorylation control at multiple enzyme loci. Data from gill of the whelk, *Busycon canaliculatum*, over the course of 20 hours of anoxia exposure show coordinated reduction in (1) the activity of the active phosphorylated a form of glycogen phosphorylase (GPa), (2) the I_{50} value for PEP of 6-phosphofructo-1-kinase (PFK-1)(one of the enzyme kinetic parameters modified by anoxia-induced phosphorylation of PFK-1), (3) the activity of 6-phosphofructo-2-kinase (PFK-2), (4) levels of the PFK-2 product, fructose-2,6-bisphosphate (F2,6P$_2$), (5) the activity of PK and (6) the I_{50} value for L-alanine of PK. Enzyme activities are in units (or micro-units) per gram wet mass and concentrations are in millimolar. Enzyme data modified from Storey *(45)*.

hibernation must include adaptations that address the variety of potential problems noted previously as well as ensure that adequate fuel reserves, in the form of adipose depots, are accumulated. The main mechanisms known to be involved in mammalian hibernation are discussed here.

3.2.1. MRD

This is the primary conservation strategy of hibernation. Strong MRD causes the fall in Tb which is unopposed because of an accompanying reduction of the hypothalmic set point for Tb (i.e., the equivalent of lowering a thermostat). Coordination is key to reestablishing balanced rates of ATP production vs ATP use during torpor and selectivity is applied to reorder cellular priorities and shut down various functions that are not needed in the torpid state.

3.2.2. Fuel Accumulation

Although some hibernating species cache food in their burrows and can eat between torpor bouts, most do not. In late summer, animals enter a phase of hyperphagia and lay down huge reserves of lipids in white adipose depots, increasing body mass by 50% or more. Sufficient fuel must be laid down to support winter torpor, periodic arousals, and considerable activity in the spring before eating resumes. Normal hormonal controls on satiety and lipid storage by adipose tissue are overridden during this period to alter the body mass set point; for example, the production of leptin is reduced despite a metabolic situation (rising adiposity) that should elevate levels of this hormone *(26)*. Diet selection is also employed to ensure that lipid depots acquire elevated levels of polyunsaturated fatty acids (PUFAs) *(27)*. A high PUFA content (particularly linoleic 18:2 and α-linolenic 18:3 fatty acids) is needed to maintain the fluidity of lipid depots down to at least 5°C. However, although the composition of depot lipids is modified before hibernation, homeoviscous adaptation of membranes does not occur and it is not yet clear how functionality of hibernator membranes is maintained in the cold. Indeed, it is possible that impaired membrane function at low Tb values contributes to MRD.

3.2.3. Fuel Metabolism

Both seasonal and hibernation-induced adjustments are made to switch organs over to the use of lipids as the primary fuel supply during the winter with a strong accompanying suppression of carbohydrate use *(28)*. For example, entry into a torpor bout triggers the upregulation of fatty acid binding proteins (that provide intracellular transport of fatty acids) and of pyruvate dehydrogenase (PDH) kinase, the enzyme that phosphorylates and inhibits PDH thereby suppressing carbohydrate use during torpor (Fig. 2) *(29,30)*. Ketogenesis by liver is also enhanced to supplement fuel supply for the brain and minimize the need for muscle proteolysis to supply amino acids for gluconeogenesis.

3.2.4. Thermogenesis

Arousal from bouts of torpor is dependent on high rates of nonshivering thermogenesis by brown adipose tissue (BAT) that is found in large masses in the interscapular region, the perirenal area, and surrounds the aorta and heart of the hibernator. BAT proliferation and differentiation is responsive to multiple signals including insulin and insulin-dependent growth factor (IGF-I) that are particularly involved in longer term seasonal responses (and are mediated by protein kinase B) and noradrenaline that is responsible for acute activation of nonshivering thermogenesis *(31)*. Noradrenaline acts via β3-adrenergic receptors on the BAT plasma membrane to activate protein kinase A which, in turn, triggers lipolysis (activation of hormone-sensitive lipase) and the upregulation of gene expression, particularly the expression of uncoupling protein 1 (UCP1) that is the key to thermogenesis. The net action of UCP1 is as a protonophore that allows proton re-entry into the mitochondrial matrix without driving ATP synthesis by the F_1F_0-ATP synthase *(32)*. Hence, the energy that would normally be trapped in ATP is released as heat. However, recent work ahs shown that UCP1 does not actually carry protons itself. Its physiological substrates are free fatty acid (FFA) anions that it transports out of the mitochondrial matrix. FFAs are protonated in the acidic intermembrane space and then neutral FFA-H diffuse back and dissociate in the more basic pH environment of the matrix. The net effect is that protons re-enter the matrix without driving ATP synthesis.

3.2.5. Differential Temperature Controls on Metabolism

The Tb of hibernators can vary from 37°C in euthermia to near 0°C in torpor but hibernators do not have the option of major metabolic restructuring (e.g., homeoviscous adaptation of membranes) to resculpt metabolism for low temperature function because they must always be prepared for a rapid arousal back to 37°C. However, the effects of temperature change on different enzymes and proteins can be employed to achieve different metabolic outcomes. Several proteins that are key to the hibernation phenotype show temperature-insensitive properties that allow them to function well over the full range of possible Tb values. For example, UCP1 shows temperature-independent properties with regard to both the maximal binding capacity and the dissociation constant (Kd) for guanosine diphosphate, its major allosteric regulator. Hibernator fatty acid binding protein (FABP), that plays a key role in energy metabolism by transporting fatty acid substrates through the cytoplasm to the mitochondria for oxidation, also shows temperature insensitive dissociation constants for both natural and artificial substrates, whereas rat FABP has reduced substrate binding abilities at low temperature *(33)*. Various other proteins and enzymes show temperature sensitive properties that can enhance, suppress or radically alter function at low temperature, in some cases contributing to MRD and in others supporting an altered function for the enzyme in the torpid state *(28)*. For example, ground squirrel liver glutamate dehydrogenase not only undergoes a stable modification between euthermic and hibernating states but the properties of the hibernating form (particularly sensitivities to ADP and guanosine triphosphate as allosteric effectors) strongly poise the enzyme for a glutamate-utilizing function at low Tb that would aid gluconeogenesis during torpor *(34)*.

4. FREEZE TOLERANCE

Hibernating mammals will let their Tb fall to near 0°C, but if ambient temperature in their hibernaculum falls below 0°C they activate a low level of thermogenesis to keep their bodies from freezing. However, ectothermic animals have no such recourse when exposed to sub-zero temperatures. They have only two choices for enduring this temperature extreme: (a) freeze avoidance—antifreeze mechanisms are used to allow body fluids to remain liquid (supercooled), sometimes to temperatures as low as –40°C, or (b) freeze tolerance—controlled growth of ice in extracellular spaces is allowed coupled with antifreeze protection of the cytosol. Freeze avoidance is used by hundreds of species of terrestrial arthropods (e.g., insects, spiders, ticks, etc.) and other invertebrates and also by coldwater marine teleosts that live in seawater that is colder than the freezing point (FP) of fish blood *(11,35)*. Freeze avoidance relies on three main strategies: (a) the proliferation of antifreeze proteins that bind

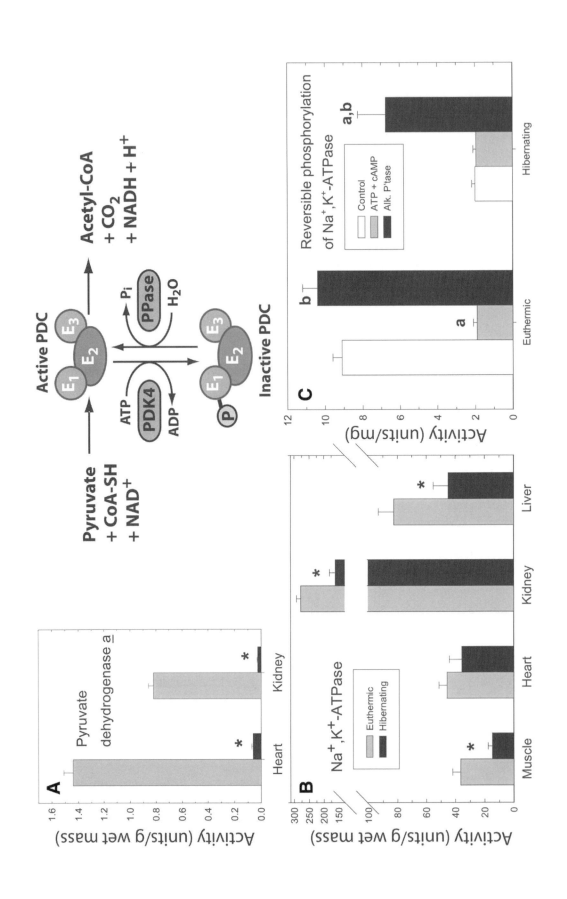

to microscopic ice crystals and keep them from growing, (b) the accumulation of high concentrations (often 2 molar or more) of polyhydric alcohols or sugars that provide colligative suppression of the FP and supercooling point (SCP) of body fluids, and (c) where possible, reducing the probability of ice nucleation by reducing body water content (dehydration), shielding with a cuticle, epiphragm or cocoon that prevents ice contact with body tissues, or clearing potential nucleators from the body (e.g., voiding the gut to get rid of bacteria) *(11)*. Other ectotherms have developed freeze tolerance and in doing so have had to address multiple injurious consequences of ice formation in biological tissues.

4.1. Freezing Damage

For most organisms on Earth, internal ice formation is highly damaging and frequently lethal. Damage can arise in several ways *(36–38)*. Ice crystals can cause direct physical injury to cells and tissues by shearing and squeezing stresses as ice grows among cells and by ice expansion that can burst delicate capillaries so that on thawing the organism suffers severe internal bleeding. Ice growth inside of cells destroys subcellular architecture and compartmentation and damage is so severe that even freeze-tolerant organisms do not endure intracellular freezing (the only good documented exception to this rule is the Antarctic nematode, *Panagrolaimus davidi [39]*). Ice growth in extracellular spaces also places volume and osmotic stresses on cells. Growing crystals exclude solutes so that remaining extracellular fluid becomes highly concentrated. This sets up a steep osmotic gradient across the plasma membrane that causes water to flow out and cells shrink. Dehydration, elevated ionic strength, and cell-volume reduction all have negative effects on cell morphology and metabolic functions, and this is compounded during thawing when cells can swell so quickly that they can burst. Shrinkage below the critical minimum cell volume also causes irreparable damage to cell membranes owing to compression stress. Finally, freezing disrupts cellular energetics because the delivery of oxygen and fuel supplies is cut off when ice forms in plasma and extracellular spaces. Hence, freezing is an anoxic and ischemic stress and although the low Tb of a frozen animal means that metabolic rate is also very low, the duration of a freezing episode can stretch to weeks or months and strain anaerobic capacities.

Fig. 2. *(continued from opposite page)* Effect of hibernation on the activities of (**A**) the active form of pyruvate dehydrogenase (PDHa) and (**B**) Na⁺K⁺ATPase in tissues of ground squirrels (*Spermophilus lateralis*). (**C**) Effects of incubation under conditions that stimulate endogenous protein kinase A (10 mM ATP, 10 mM MgCl₂, 0.3 mM cAMP) and subsequent alkaline phosphatase treatment (10 units) on Na⁺K⁺-ATPase activity in skeletal muscle extracts from euthermic and hibernating ground squirrels. Data are means ± SEM, *n* = 4. Also shown is the reaction of the pyruvate dehydrogenase complex (PDC) and its interconversion between active (dephosphorylated) and inactive (phosphorylated) forms by the actions of PDH kinase versus phosphatase. The PDH complex is composed of three enzymes: E₁ is pyruvate dehydrogenase (phosphorylated by pyruvate dehydrogenase kinase [PDK]), E₂ is dihydrolipoyl transacetylase, and E₃ is dihydrolipoyl dehydrogenase. *, Significantly different from the corresponding euthermic value, *p* < 0.05. a, Significantly different from the untreated control sample; b, significantly different from the protein kinase treated sample, *p* < 0.05. Enzyme data compiled from Brooks and Storey *(48)* and MacDonald and Storey *(49)*.

4.2. Natural Freezing Survival

Natural freeze tolerance is well developed in several species of North American frogs that hibernate on land as well as some reptiles, a variety of intertidal invertebrates, and hundreds of insect species *(36,37)*. The main vertebrate model that we use is the wood frog, *Rana sylvatica (38)*. Known principles of freeze tolerance include protective strategies that regulate ice growth and preserve cellular macromolecules as well as conservation strategies that ensure metabolic survival during freezing.

4.2.1. Regulation of Ice Growth

Freeze-tolerant animals typically ensure that freezing is initiated close to the equilibrium FP so that the rate of ice formation is slow and there is plenty of time to initiate metabolic adjustments. Freezing may be triggered by contact with environmental ice across a water-permeable skin or via the action of nucleators, most commonly specific ice-nucleating proteins that are added to the blood of the organism or bacteria in skin or gut that have ice-nucleating abilities *(11,38)*. By triggering ice growth in extracellular and extra-organ spaces (e.g., the abdominal cavity in frogs), the probability of intracellular ice nucleation is drastically reduced. The plasma of many freeze-tolerant animals also contains proteins with antifreeze actions (when assayed in vitro) but their function in vivo appears to be to limit the size of crystal growth and inhibit recrystallization.

4.2.2. Cell-Volume Regulation

Freeze-tolerant animals frequently endure the conversion of up to 65 to 70% of total body water into extracellular ice and cells/organs undergo substantial dehydration and shrinkage. To prevent cell-volume reduction beyond a critical minimum, high concentrations of low-molecular-weight carbohydrates are typically accumulated to provide colligative limitation of cell water loss. Glycerol and other polyhydric alcohols often do this job but wood frogs use glucose and show extreme freeze-induced hyperglycemia with glucose rising to 150 to 300 mM in plasma and organs compared with approx 5 mM in unfrozen controls *(36,38)*. Other protectants (e.g., trehalose, proline) protect/stabilize cell membranes against compression stress during volume reduction.

4.2.3. Energetics

Extracellular freezing imposes an ischemic state on all cells/organs of frozen animals and therefore freeze-tolerant animals also show well-developed ischemia/anoxia resistance including pathways of fermentative ATP generation, regulated MRD, and antioxidant defenses to provide protection against reperfusion damage during thawing *(38)*.

4.2.4. Vital Signs

Freezing halts all vital signs including heart beat, breathing, muscle movement, and nerve transmission. All are sequentially reactivated during/after thawing (heart beat being the first sign of reanimation) but the molecular mechanisms underlying these processes are still largely unknown.

5. METABOLIC RATE DEPRESSION

Although alien to human physiology, the ability to strongly suppress metabolic rate and sink into a hypometabolic state is a life-saving mechanism for many organisms that must endure extreme environmental stress on a periodic or seasonal basis. When faced with environmental extremes that threaten normal life, limit food availability or impose severe challenges to their physiology, this conservation strategy allows organisms to endure until conditions are again conducive for active life. Examples of hypometabolism abound as animal responses to high and low temperature, oxygen deprivation, food restriction and water limitation and, coupled with various protective adaptations that apply to specific cases, MRD is the underlying principle of stress survival in phenomena including anaerobiosis, estivation, diapause, torpor, dormancy, hibernation and anhydrobiosis (also called cryptobiosis). Among animals, MRD can range from the nightly torpor (a 20–30% reduction in metabolic rate for a few hours) that eases the energy budget of small birds and mammals in cold environments to many months of seasonal dormancy in response to winter cold (hibernation) or summer drought (estivation), and even to many years in a virtual ametabolic state for cryptobiotes (e.g., many seeds, spores, cysts, eggs; the most frequently studied example of cryptobiosis is the brine shrimp, *Artemia*). The molecular mechanisms of MRD have been the subject of much recent study in our lab and others. By analyzing the mechanisms of MRD in four situations—anaerobiosis, hibernation, estivation, and freeze tolerance—we have documented the principles of metabolic control that extend across phylogeny to regulate MRD in multiple states *(9,10)*.

MRD has three main principles: (a) both intrinsic mechanisms within cells and extrinsic mechanisms are involved, (b) the rates of energy-producing and energy-consuming cellular processes are suppressed in a coordinated manner so that a new lower net rate of ATP turnover can be sustained over the long term, and (c) cellular priorities are reorganized to give precedence to key functions (e.g., maintenance of membrane potential difference) and more strongly suppress functions that are less essential (e.g., protein synthesis) under energy-restricted conditions. Extrinsic influences on MRD include a general suppression of physiological functions (heart rate, breathing, digestion, muscle movement) in the hypometabolic state as well as the hypercapnia (due to apnoic breathing patterns) and reduced cytosolic pH that typically accompany entry into hypometabolism and that both suppress metabolic functions. Hypothermia can also be a factor; for instance, many ectotherms voluntarily seek cooler temperatures when challenged by hypoxia and thereby use a decrease in body temperature to help reduce tissue demands for oxygen. An ancient hypoxia–hypothermia interaction may also contribute to the mechanism of MRD that allows small mammals to coordinate the suppression of metabolic rate and Tb as they sink into the hibernating state *(40,41)*.

Intrinsic mechanisms of MRD within cells account for at least half of the total MRD. Intrinsic mechanisms lead to both a net decrease in cellular metabolic rate and a reorganization of priorities for ATP use. An example of these principles comes from studies with isolated hepatocytes of anoxia-tolerant turtles *(42)*. Five main ATP-consuming processes were identified in hepatocytes and under normoxic conditions the fractional use of cellular ATP turnover was calculated as 28, 36, 17, 3, and 17% for ion pumping by the Na^+K^+-ATPase, protein synthesis, protein degradation, urea synthesis, and gluconeogenesis, respectively. However, when cells were incubated under anoxic conditions, not only did total ATP turnover decrease by 94% but each of the ATP-consuming processes was differently affected with the result that the fractional use of ATP turnover by the five processes in anoxic cells shifted to 62, 21, 9, 8, and 0 %, respectively. In other words, the sodium/potassium pump became the dominant energy sink in the energy-restricted anoxic state. Priorities are also rearranged between organs in hypometabolic states. For example, when the capacity for protein synthesis was analyzed in different organs of hibernating vs euthermic ground squirrels, the rate of synthesis was unaltered in brown adipose tissue but was reduced by 66% in brain and 85% in kidney (all rates measured in cell extracts in vitro at 37°C) *(43,44)*.

5.1. Regulation by Reversible Protein Phosphorylation

What are the molecular mechanisms of intrinsic MRD? By far the most important mechanism is reversible protein phosphorylation (RPP) carried out by the actions of protein kinases or protein phosphatases on enzymes and functional proteins. Key advantages of RPP as an instrument of MRD include the following:

1. RPP can induce major changes to the activity states of selected regulatory enzymes and functional proteins, often including virtual on–off control.
2. Thousands of proteins in cells are susceptible to RPP so the mechanism provides an excellent way of coordinating the responses by multiple cell functions.
3. Signal transduction cascades involving protein kinases and protein phosphatases are fast and allow a rapid suppression of the activities of multiple ATP-utilizing functions and an equally fast reversal to re-establish normal cell functions during arousal from the hypometabolic state
4. Major changes in the activity states of enzymes and pathways are achieved without the need to change the overall amounts of proteins via synthesis or degradation, another factor that benefits fast recovery from the hypometabolic state.

Recognition of the role of RPP as a mechanism of metabolic suppression arose initially from studies of the control of glycolysis in anoxia-tolerant marine molluscs *(9,22)*. Strong anoxia-induced inhibition of pyruvate kinase (PK) was found to be responsible for rerouting carbohydrate flux at the phosphoenolpyruvate (PEP) branchpoint under anoxia (Fig. 1A). This directs PEP away from the aerobic route of carbohydrate degradation (via PK and into the tricarboxylic acid cycle) and into the PEP carboxykinase reaction and onwards into the reactions of anaerobic succinate synthesis. The suppression of PK activity was traced to anoxia-induced phosphorylation of the enzyme that caused a strong decrease in enzyme activity and major changes in enzyme properties that virtually shut off PK in vivo under anoxic conditions (Fig. 1B) *(45)*. For example, anoxia-induced phosphorylation of whelk muscle PK had the following strong effects: substrate affinity decreased (K_m for PEP rose 12-fold), enzyme sensitivity to the activator, fructose-1,6-bisphosphate decreased (K_a increased 26-fold), and sensitivity to inhibition by L-alanine rose (I_{50} decreased by 490-fold) *(46)*. Inhibitory control is further enhanced by the major accumulation of alanine as an initial product of anaerobic metabolism in marine molluscs.

It was next determined that RPP provided not just PK control in marine molluscs under anoxia but also coordinated an overall suppression of glycolysis (as part of the overall MRD) by targeting enzymes including glycogen phosphorylase, 6-phosphofructo-1-kinase (PFK-1), and 6-phosphofructo-2-kinase (PFK-2; that produces the potent activator of PFK-1, fructose-2,6-bisphosphate) *(45)*. Figure 1B shows these coordinated responses to anoxia by enzymes in gill of the whelk, *Busycon canaliculatum*. The realization that RPP had an even broader role in orchestrating MRD in multiple situations came when the same coordinated phosphorylation of glycolytic enzymes was found to regulate the suppression of carbohydrate catabolism during anoxia exposure in vertebrates (turtles and goldfish) and during estivation, an aerobic dormancy, in land snails and desert toads *(10,12,45,47)*.

Studies with hibernating mammals and estivating snails next confirmed that RPP also controlled enzymes, not just of glycolysis, but also of aerobic substrate catabolism. Strong inhibition of PDH, the enzyme that gates carbohydrate entry into the tricarboxylic acid cycle was documented in both situations *(12,28)*. In hibernators, for example, the amount of PDH present in the dephosphorylated active form dropped from 60 to 80% in euthermia to less than 5% in hibernation (Fig. 2A) *(48)*. This supports carbohydrate sparing during torpor and the shift to a primary reliance on lipid oxidation for energy generation. Furthermore, gene-screening studies found a strong hibernation-induced upregulation of PDH kinase, the enzyme that phosphorylates and turns off PDH *(29)*.

5.2. Ion-Motive ATPases and Channel Arrest

The examples presented here deal with catabolic enzymes involved in ATP production but hypometabolism requires a balanced suppression of the rates of both ATP-producing and ATP-utilizing pathways so that a new lower rate of ATP turnover is established. Studies with hibernators and anoxia-tolerant turtles have led the way in analyzing the controls on ATP-utilizing reactions in hypometabolic states. As noted earlier, ion-motive ATPases are huge consumers of cellular energy and, therefore, key potential targets for achieving metabolic suppression. Indeed, analysis of Na/K-ATPase in ground squirrel tissues showed that activity of the enzyme was strongly suppressed during hibernation. Activities were just 40 to -60% of the corresponding euthermic values when quantified at the same temperature (25°C) *(49)*. Again, the mechanism involved is RPP. Figure 2B shows the results of incubation studies with extracts of ground squirrel tissues. Conditions that promoted the activity of protein kinases suppressed Na/K-ATPase activity in extracts from euthermic tissues, whereas treatment with alkaline phosphatase restored activity and also elevated activity in tissue extracts from hibernating animals. Similar results were found for Ca-ATPase of skeletal muscle sarcoplasmic reticulum (SR); maximal activity was reduced by 50% during hibernation along with comparable reductions in other proteins involved in SR calcium signaling including the SR calcium-release channel (ryanodine receptor) and SR calcium-binding proteins (e.g., sarcalumenin, calsequestrin) *(50)*.

Suppression of the activities of ion-motive ATPases must be balanced by concomitant suppression of the ion movements through oppositely directed ion channels in order to maintain membrane potential difference in a hypometabolic state. Channel arrest has received the most attention in studies of vertebrate brain hypoxia/anoxia tolerance, specifically in studies with anoxia tolerant turtles *(51)*. Oxygen-sensitive potassium, sodium, and calcium channels have been identified in neuronal plasma membranes controlled by mechanisms including RPP, redox regulation, Ca^{2+}-dependent regulation, and regulation by neuromodulators acting via G protein-coupled receptors. Adenosine is a particularly important neuromodulator in this regard. It accumulates quickly in turtle brain during anoxia exposure and acts via adenosine A1 receptors to suppress excitatory neurotransmission that is mediated largely by Ca^{2+} entry through *N*-methyl-D-aspartate (NMDA) receptors *(52)*. Patch-clamp studies using cells from normoxic brain showed that both anoxia exposure and adenosine perfusion reduced the NMDA receptor open probability by 65%, an effect that was antagonized by an A1 receptor blocker. Hence, the activities of both ion channels and ion motive ATPases are suppressed during anoxia in anoxia-tolerant brains, contributing to the overall MRD. Overexpression of adenosine A1 receptors is also known to increase myocardial tolerance of ischemia in transgenic mice *(53)*, suggesting a wide role for adenosine signaling in channel arrest in hypoxia or ischemia. In addition, we have recently documented putative upregulation of the genes for adenosine A1 receptors and 5'-nucleotidase (that synthesizes adenosine from adenosine monophosphate) as a response to freezing in freeze-tolerant wood frogs, which suggests that adenosine-mediated responses may also aid ischemia resistance in frozen tissues *(54)*. Furthermore, a universal role for adenosine signaling in both aerobic and anaerobic forms of MRD might now be suggested based on recent studies with hibernating hamsters. Shiomi and Tamura *(55)* found that intracerebroventricular injections of adenosine or an adenosine A1 receptor agonist produced profound hypothermia in hamsters on a time course coincident with the normal descent of Tb during entry into natural hibernation whereas an adenosine A1 receptor antagonist could elevate Tb and interrupt hibernation.

5.3. Regulation of Protein Biosynthesis in Arrested States

Another major energy expenditure in cells is protein synthesis, requiring approx 5 ATP equivalents per peptide bond formed and consuming as much as 40% of total ATP turnover in some organs such as liver. Suppression of protein biosynthesis has been widely documented in hypometabolic states, occurring in an organ-specific manner in hibernating ground squirrels (discussed earlier) with a major reduction in the fractional use of cellular ATP by protein synthesis in hypometabolic states (as discussed earlier for turtle anoxia). Suppression can occur rapidly and before cells experience energy stress; for example, the rate of ^{3}H-leucine incorporation into protein in hepatopancreas extracts from marine periwinkle snails (*Littorina littorea*) decreased by 50% within just 30 min when snails were transferred from aerobic to anoxic conditions *(56)*. This indicates that protein synthesis inhibition is a proactive response by cells that is an integral part of metabolic arrest rather than a reactive response to ATP limitation. Indeed, ATP content and energy charge did not change significantly over the course of 72 h of anoxia exposure in this facultative anaerobe species.

Two factors could contribute to protein synthesis inhibition during hypometabolism: (a) mRNA substrate availability, and (b) specific inhibition of the ribosomal translational machinery. The former possibility has been largely discounted by multiple studies that found no change in either global mRNA content or the transcript levels of various constitutively active genes during transitions to or from hypometabolic states *(10)*. This makes intuitive sense because it preserves the cellular pool of mRNA transcripts so that messages are immediately available for translation when organisms arouse from hypometabolism.

The primary mechanism of translational control in hypometabolism is again RPP, this time regulating the activities of ribosomal initiation and elongation factors. The mechanisms identified to date are the same as the RPP controls on translation that have been described for the regulation of mammalian protein synthesis in response to various stresses (e.g., amino acid limitation, hypoxia/ischemia, heat shock, infection) *(57)*. Because most hypometabolic states are also typically states of prolonged starvation, it is perhaps not unexpected that mechanisms for controlling protein synthesis with respect to nutritional status (starved vs fed) are also employed to achieve long-term suppression of biosynthesis during hypometabolism.

Recent studies have documented consistent inhibition of the eukaryotic initiation factor-2 (eIF2) in hypometabolic states. This factor introduces initiator methionyl-tRNA into the 40S ribosomal subunit and phosphorylation of the α-subunit inhibits this function (Fig. 3). The content of phosphorylated, inactive eIF2α increases strongly during hibernation; for example, the figure shows Western blot assessments of total eIF2α protein (no change) vs eIF2α phosphopeptide content (a strong increase in hibernation) in ground squirrel kidney *(44)*. Similarly, in ground squirrel brain, phospho-eIF2α content was less than 2% of the total in euthermia but rose to 13% in hibernation *(43)*. Anoxia exposure had the same effect

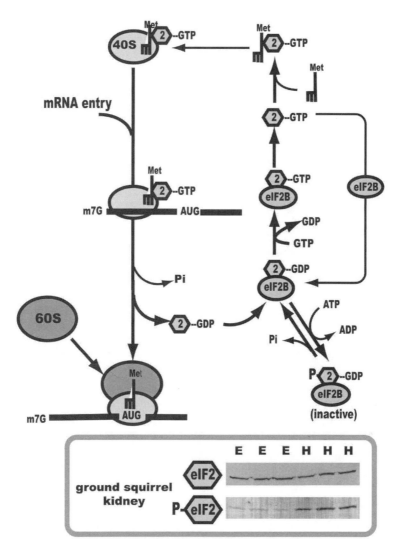

Fig. 3. Reversible phosphorylation control of the eukaryotic initiation factor 2 (eIF2) regulates translation initiation by restricting the availability of methionine tRNA to the 40S ribosomal subunit. Phosphorylation of the α-subunit of eIF2 keeps the protein bound in an inactive complex with the guanine nucleotide exchange factor, eIF2B, and prevents eIF2α recycling between successive rounds of peptide synthesis. The inset shows Western blots of ground squirrel kidney extracts crossreacted with antibodies that recognize total eIF2α versus the phosphopeptide segment of eIF2α. Total eIF2α protein content did not change during hibernation (H) compared with euthermia (E) but the amount of phosphorylated eIF2α rose sharply. Western blot data from Hittel and Storey *(44)*.

on eIF2α in the marine snail, *L. littorea*; total eIF-2α content in hepatopancreas was unaffected over a cycle of anoxia and aerobic recovery but the amount of phospho-eIF2α rose approx 15-fold in anoxic animals, compared with aerobic controls *(56)*. However, when oxygen was reintroduced, phospho-eIF2α fell to control levels or below within 1 h. Inhibitory control of protein synthesis also occurs at other loci. Phosphorylation-mediated inhibition of the eukaryotic elongation factor-2 (eEF2) raised mean transit times for polypeptide elongation by ribosomes by threefold in extracts of hibernator brain as compared with euthermic ground squirrels *(43)*. Regulation was traced to both a 50% higher activity of eEF2 kinase in hibernator tissues and a 20 to 30% decrease in protein phosphatase-2A activity (that opposes eEF2 kinase) *(58)*. EF2 kinase is, in turn, subject to phosphorylation and activation by several of the major cellular protein kinases including PKA, MAPKs, p90^{RSK1}, and p70 S6 kinase, one or more of which may mediate the hibernation response.

Other ribosomal initiation factors are also involved in the suppression of translation under situations such as starvation in mammals and, although they have not yet been investigated as elements of hypometabolism, it is likely that they will prove to be involved. For example, the eukaryotic initiation factor-5 (eIF5), a GTPase-activating protein that promotes GTP hydrolysis within the 40S initiation complex, is also regulated by RPP. Similarly, the eukaryotic initiation factor-4E binding protein (4E-BP1) is subject to RPP and when dephosphorylated it binds to and inhibits eIF4E. Another mechanism, proteolytic fragmentation, controls subunit G of eIF4. Fragmentation occurs under stress (e.g., ischemia) and changes the type of mRNA that can be translated because intact eIF4G is needed to allow eIF4E-bound m^{7}G-capped mRNAs (the vast majority of cellular mRNAs) to bind to the 40S ribosomal subunit (Fig. 3) *(57)*. Without intact eIF4G, message selection favors only those messages that contain an internal ribosome entry site (IRES) *(59)*. Such messages often code for proteins involved in apoptosis but the mRNA transcripts of several stress-responsive proteins also contain IRES elements so that these can be translated under stress conditions (e.g., hypoxia, amino acid limitation) that normally inhibit protein synthesis. For example, the mRNA for HIF-1 α-subunit contains an IRES that allows enhanced synthesis of HIF-1α to occur under hypoxic conditions when overall protein synthesis is suppressed *(60)*. In turn, enhanced levels of the α-subunit, when combined with the more stable ß-subunit, can elevate overall HIF levels and stimulate the expression of HIF-1 regulated genes whose protein products function to alleviate hypoxia stress. The presence of an IRES may also be key to the translation of the protein products of the selected few genes that are upregulated during entry into hypometabolic states including hibernation, anaerobiosis and freezing (see section on gene expression).

5.4. Regulation of Protein Synthesis by Changes in Ribosome Assembly

The activity state of protein synthesis in a cell can generally be inferred from the state of ribosome assembly—active translation occurs on polysomes (aggregates of ribosomes moving along a strand of mRNA), whereas monosomes are translationally silent. Polysome dissociation is a recognized cellular response to stress. For example, stresses such as hypoxia, starvation, and diabetes all trigger polysome dissociation in rat tissues. Recent work has also shown that polysome dissociation is a mechanism of MRD in stress-tolerant organisms.

The situation is well- illustrated by recent studies with hibernators. To assess the state of ribosomal assembly, tissue extracts are separated on a sucrose gradient and fractions collected. Polysomes migrate into denser fractions whereas monosomes are found in lighter fractions; ribosomal RNA presence is detected by absorbance at 254 nm, ethidium bromide staining, or, as shown in Fig. 4, by Northern blotting for 18 S rRNA. When tissue extracts from euthermic ground squirrels (Tb = 37°C) were assessed, most ribosomes were present in the higher density polysome fraction as was a high proportion of the mRNA for constitutively active genes as illustrated by the distribution of cytochrome c oxidase subunit 4 (*Cox4*) mRNA in the figure, as detected by Northern blotting *(44)*. When animals entered hibernation, however, tissues showed consistent polysome disassembly with movement of a high percentage of 18S rRNA and *Cox4* mRNA into the monosome fractions. Hence, the principle here is that an overall suppression of protein synthesis during hibernation is achieved by the dissociation of active polysomes and the storage of mRNA transcripts in the translationally silent monosome fraction. During arousal the reverse transition occurs and allows protein synthesis to be rapidly reinitiated without a need for *de novo* gene transcription. Note also that by this mechanism an effective "life extension" of mRNA transcripts is achieved.

However, transcripts of genes that are specifically upregulated during entry into hibernation behaved differently. Transcript levels of fatty acid binding protein (*fabp*) increase several-fold in most ground squirrel tissues during hibernation *(30)* and, in addition, they remain associated with polysomes, ensuring their active translation and leading to a strong increase in FABP protein content in hibernation versus euthermia as illustrated by the Western blots in Fig. 4. By contrast, COX4 protein levels were unchanged or declined slightly during hibernation. This illustrates another principle of metabolic control—the rate of translation of individual mRNA species can be altered by differential distribution of transcripts between translationally active and inactive ribosomes. The polysomes remaining in hibernator tissues contain disproportionately higher numbers of those mRNAs (such as *fabp*) that are crucial to the hibernation phenotype, whereas mRNA species that are not needed during hibernation are relegated into the translationally silent monosome fractions.

The same mechanisms characterize the response to a very different stress (anoxia) in a very different species (the marine snail, *L. littorea*) and, together with a number of other studies, show that polysome disassembly is a common feature of MRD across phylogeny. Anoxia-induced MRD in *L. littorea* was also accompanied by a movement of rRNA and the mRNA for constitutive genes (α-tubulin in this case) into the monosome fractions, whereas mRNA for genes that were upregulated in anoxia (e.g., ferritin) stayed in the high-density fractions that contained the few remaining polysomes *(56,61)*. The result was a twofold rise in ferritin protein levels in anoxic snails; by increasing iron storage, elevated ferritin is believed to contribute to minimizing oxidative stress during the transition back from anoxic to aerobic life. However, when snails were returned to aerobic conditions, the control situation was reestablished within 6 h with a return to a high polysome content and a high percentage of all mRNA localized with the polysomes.

The trigger for polysome disaggregation is not known with certainty but in hibernators there is evidence that temperature is a factor. When the distribution of rRNA (monitoring ribosomes) and actin mRNA (monitoring constitutive transcripts) was assessed in liver samples taken at different Tb values a distinct shift occurred when core Tb reached 18°C *(62)*. In euthermia, actin mRNA was localized mainly in the polysomes and remained there during entry into torpor until animals cooled to 18°C. Below 18°C a large portion of the transcripts, as well as rRNA, suddenly shifted to the monosome fraction and remained there throughout torpor. Conversely, during arousal, polysome reassembly was first evident when Tb rose to 18°C. Whether this temperature effect derives from a passive influence of temperature on polysome assembly or is due to temperature-stimulated regulation of one or more ribosomal proteins is not yet known.

Another variation on translational control has also been illustrated from our studies with hibernators. A prominently upregulated gene in kidney of hibernating ground squirrels is the organic cation transporter type 2 (*Oct2*); *Oct2* transcript levels were two to threefold higher

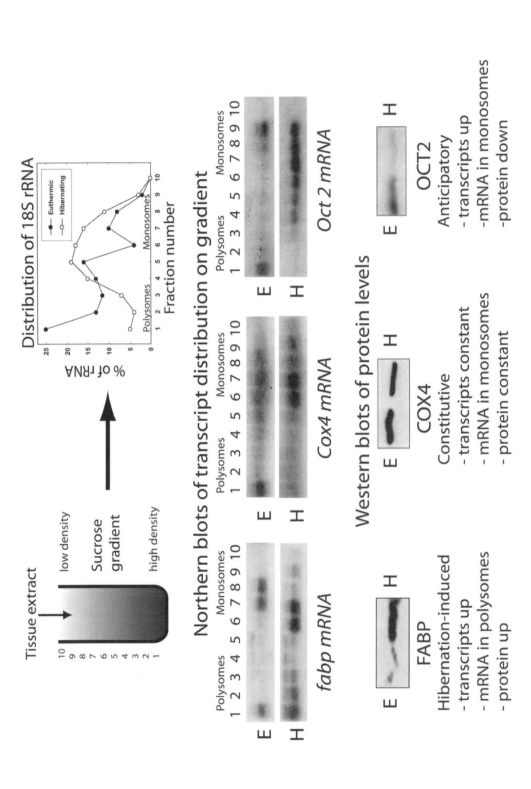

in torpid animals compared with euthermic controls *(44)*. However, despite this, OCT2 protein content actually decreased during hibernation. Why would this be? The answer came from an analysis of *Oct2* transcript distribution on polysome profiles. Although transcript levels were much higher in hibernation, they were largely sequestered into the translationally silent monosome fraction during hibernation (Fig. 4). We suggested that the reason for this is to allow OCT2 protein to be produced very rapidly from existing *Oct2* transcripts as soon as torpor is broken. Kidney function is virtually shut down during hibernation and it is possible that this includes an actual degradation of OCT2 protein (rather than a reversible inactivation) since immunoblotting showed much lower levels of OCT2 in hibernator kidney, compared with euthermia (Fig. 4). In such a situation, resumption of transporter action to support renewed kidney function during interbout arousals would require the rapid synthesis of OCT2, which could be potentiated using the high levels of *Oct2* transcripts already present. Hence, this suggests another possible principle of translational control—anticipatory upregulation of selected transcripts during hibernation can support a rapid activation of protein synthesis during arousal that does not depend on enhanced gene expression.

Hence, multiple mechanisms of translational control are available for use in association with MRD and these can accomplish a variety of specific goals including a general suppression of protein translation (via inhibition of ribosomal factors and mRNA sequestering into the monosome fraction), the specific upregulation of selected transcripts (e.g., IRES-mediated translation, preferential transcript presence in polysomes), and anticipatory upregulation with delayed translation. In combination, all of these mechanisms (and probably others yet to be uncovered) contribute to the overall suppression of protein synthesis as part of the general conservation strategy of MRD while still providing the means to regulate the stress-induced production of selected proteins that are needed for the long-term survival of the organism in the hypometabolic state.

6. ADVANCES IN BIOCHEMICAL ADAPTATION THROUGH GENE DISCOVERY

One of the major mechanisms of animal response to extreme environmental stress is the activation of gene expression to synthesize new protein products. Stress responses, particularly in situations where MRD is also employed, are typically selective with a relatively small group of genes upregulated against the background of an overall suppression of protein synthesis in the energy-limited state. For many years, the approach to finding proteins that aided biochemical adaptation worked from physiological phenotype backward; that is, gross differences between stress-tolerant and intolerant species were identified (e.g., a major

---

Fig. 4. *(continued from opposite page)* Effect of hibernation on the distribution of ribosomes and mRNA between translationally active polysomes and translationally silent monosomes. Tissue extracts were fractionated on a sucrose gradient. The graph shows 18S ribosomal RNA content in the fractions, documenting the shift in rRNA presence from predominance in the polysome fractions in euthermia (solid circles) to the monosome fractions in hibernation (open circles). Northern blots of RNA extracts of each fraction show changes in abundance and position on the gradient of mRNA for three genes in euthermia (E) versus hibernation (H): fatty acid binding protein (*fabp*) and cytochrome *c* oxidase subunit 4 (*Cox4*) from brown adipose tissue and the organic cation transporter type 2 (*Oct2*) from kidney. Western blots show corresponding changes in protein contents. Data compiled from Hittel and Storey *(44)*.

hysteresis between freezing and melting points of the blood of coldwater marine fish, nonshivering thermogenesis by arousing hibernators) and then the underlying protein adaptations were traced. For instance, in the above examples this led to the discovery of novel antifreeze proteins in marine fish and of UCP1 in brown adipose of hibernating mammals. Such an approach is excellent for exploring adaptations that have an obvious "footprint" but is less good for elucidating more subtle changes in gene/protein expression that may be of equal importance to overall survival of the organism.

New advances in gene-screening technology are revolutionizing the exploration of biochemical adaptation. Broad-based screening techniques can produce an unbiased assessment of the genes that are upregulated in response to stress and are identifying many genes, protein products and cell functions that have never before been linked with the response to a particular environmental stress. For example, in our first use of cDNA library screening to evaluate freeze-induced gene upregulation in wood frog liver, we expected to find increased expression of genes associated with previously identified adaptations for freeze tolerance such as cryoprotectant biosynthesis. What we found instead was strong upregulation of the genes encoding fibrinogen subunits and the mitochondrial ADP–ATP translocase (AAT) (63,64) and this led us to two previously unrecognized facets of freezing survival. One is the need for improved plasma-clotting capacity, provided by the synthesis and export of fibrinogen (and perhaps other clotting factors as well), as a mechanism of dealing with any physical injuries caused by ice expansion within the delicate microvasculature of organs. Indeed, such damage, which causes a loss of vascular integrity upon thawing, is a major reason for the current failure of solid organ cryopreservation in medicine. The rationale for AAT upregulation is still not confirmed but interestingly, the stress-induced upregulation of membrane transporters is proving to be a common theme in organs of freeze tolerant frogs with two more examples recently documented: the mitochondrial inorganic phosphate transporter and the monocarboxylic acid transporter (38,65). Similarly, another recurring theme that has been highlighted from gene screening, but never previously considered, is the stress-induced upregulation of mitochondrially encoded subunits of respiratory chain proteins that we have now identified in anoxia-tolerant, freeze-tolerant, and hibernating animals. This includes subunits 2, 4, and 5 of NADH ubiquinone-oxidoreductase (complex I), cytochrome b (from complex III), subunit 1 of cytochrome c oxidase (COX, complex IV) and ATPase subunits 6 and 8 (from the F_1F_0ATPase, complex V) (54,66–69). The rationale for upregulation of these mitochondrial-encoded proteins has not yet been adequately explained although, clearly, this is a common theme among stress-tolerant animals that potentially represents a new principle of MRD. All of the mitochondrial respiratory chain complexes are large multisubunit enzymes with only a few subunits encoded on the mitochondrial genome and, interestingly, when we assessed the responses of nuclear-encoded genes of the same proteins these were not stress-induced (e.g., no changes were seen in transcript levels of subunit 4 of COX and ATPα, a nuclear encoded subunit complex V) (69). Hence, the phenomenon is peculiar to the mitochondrial genome and our working hypothesis is that selective changes to protein components of the respiratory complexes help to stabilize mitochondrial energetics and preserve viable organelles in stress situations.

Gene-screening techniques have also made major inroads in identifying many more adaptations that support mammalian hibernation. Studies by my lab and others have identified hibernation-responsive upregulation a variety of genes including of α_2-macroglobulin in liver, moesin in intestine, isozyme 4 of pyruvate dehydrogenase kinase (PDK4) and pancreatic lipase in heart, isoforms of UCP and FABP in multiple tissues, the ventricular isoform of myosin light-chain 1 (MLC1$_v$) in heart and skeletal muscle, OCT2 in kidney, the melatonin receptor, and four genes on the mitochondrial genome *(10,41)*. Although the genes identified to date are a disparate group, once again principles of adaptation are emerging that are providing new directions for study. For example, the upregulation of MLC1$_v$ in ground squirrel heart *(67)*, when combined with studies of hamster heart that show changes in the proportions of myosin heavy-chain isoforms during hibernation *(70)*, suggests that myosin restructuring occurs during hibernation. This would presumably provide an optimal mix of myosin isoforms to adapt the contractile apparatus of the heart to the new workload and thermal conditions of the torpid state. Other studies suggest that adjustments are made to minimize the risk of thrombosis in the microvasculature under the very low blood flow (ischemic) conditions during torpor. Up-regulation and export of α_2-macroglobulin (which inhibits proteases of the clotting cascade) from the liver, reduced platelet numbers (sequestered into the spleen) and reduced levels of several clotting factors all support a decreased clotting capacity during torpor *(71)*. Indeed, these results illustrate the natural solution to the clotting problems that are typically noted during organ ischemia caused by heart attack or stroke.

6.1. Gene Discovery From cDNA Library Screening

cDNA library screening is one of two major methods of gene discovery that have been used to make major advances in our understanding of biochemical adaptation to extreme environments. The other is cDNA array screening (discussed later). The construction and screening of cDNA libraries is often difficult and time-consuming and favors the detection of genes that have abundant transcripts. However, because the library is made from the organism under study, it has one key advantage and that is its ability to detect the presence of novel species-specific genes. For example, we have discovered two novel genes that are upregulated in response to anoxia exposure in the marine snail, *L. littorea*, and three novel genes in freeze-tolerant frogs *(72–76)*; none of these show similarity to any gene/protein sequences present in international databases. Each of the protein products encoded by these genes may have a key role to play in stress tolerance and a wide range of technologies can be applied to elucidate their functions. Among others, we have used (a) proteomics programs to identify regionalities, functionalities and structural elements within the proteins; (b) Northern blotting or polymerase chain reaction (PCR) techniques to assess time-, organ- and stress-specific changes in mRNA transcript levels; (c) Western blotting using peptide antibodies raised against segments of the putative amino acid sequence to assess comparable time-, organ- and stress-specific changes in protein levels as well as subcellular location of the protein; (d) in vitro tissue incubations to test the influences of external stresses, hormones, and second messengers; and (e) techniques of transgenics and cloning to insert the novel gene into cell lines for high yield manufacture of the gene product for use in protein chemistry or analysis of the effects of the protein on stress survival of the transgenic cell. For example, such techniques when applied to analyzing the snail anoxia-responsive protein (SARP)-19, in *L. littorea* correlated the presence of two EF-hand Ca^{2+}-binding domains in the protein sequence with the upregulation of *sarp* transcript levels in hepatopancreas explants incubated with calcium ionophore A23187 or with phorbol 12-myristate 13-acetate, a stimulator of the Ca^{2+} and phospholipid dependent protein kinase C (PKC) *(73)*. With the further information from Northern blots that showed a progressive rise in *sarp* transcripts over several days of anoxia exposure but a rapid reversal within 1 h of aerobic recovery, we proposed an important function for SARP-19 in Ca^{2+} signaling under anaerobic conditions.

Our investigations of three novel freeze-responsive genes identified by cDNA library screening of wood frog liver are also providing fascinating results. FR10, Li16, and FR47 encode proteins of 10, 13, and 47 kD, respectively *(74–76)*. They share no structural features in common other than the presence of a hydrophobic region of 21 amino acids in length in each; in FR10 and Li16 this region is N-terminal but it is near the C-terminus in FR 47. Hydrophobic regions often represent transmembrane segments and this suggests that all three proteins may associate with membranes. Transcripts of all three are elevated by three to five-fold in liver after 24 h of freezing at $-2.5°C$ and Western blotting showed a comparable rise in Li16 and FR47 protein in frozen frogs (FR10 has not been tested) reaching a maximum of 8.4- and 3.5-fold higher than control values after 2 h of thawing at 5°C (before subsequently declining) *(see* Fig. 5 for *li16)*. The occurrence of maximum protein levels in 2 h thawed frogs (that still have substantial internal ice, no heart beat and visibly shrunken organs) with a strong reduction by 8 h thawed (when heart beat and breathing have resumed and liver appears visibly restored to normal size) suggests that both proteins play roles in dealing with the ischemic or cell volume stresses associated with freezing and are no longer needed once recovery is well advanced. However, despite similar patterns of transcript and protein changes during freeze–thaw, the three proteins show very different organ distributions and responses to other stimuli, which argues for very different functions for each. For example, transcripts of *fr47* are found only in liver, *li16* occurs in liver, heart, and gut, whereas *fr10* was found in all organs that we tested. Furthermore, FR47 protein was detected in liver of two other freeze-tolerant frog species *(76)* but not in intolerant species, which is a strong indication of a freeze-specific function.

We gained further information about these three proteins by testing their responses to stimuli both in vivo and in vitro. Two main components of freezing stress are (a) ischemia that results from plasma freezing, and (b) cellular dehydration that results from water out-flow into extracellular ice masses. In early studies of the control of cryoprotectant (glucose) synthesis in wood frogs we found that dehydration of frogs was just as effective as freezing in stimulating the hyperglycemic response whereas anoxia exposure (mimicking ischemia) had no effect on plasma glucose levels *(37)*. This suggested that glucose synthesis as a colligative cryoprotectant is triggered or regulated by changes in cell volume. We used the same strategy to assess the expression of the three novel proteins and found that transcripts of *fr10* were strongly upregulated by dehydration (suggesting a role for FR10 in cell volume regulation), whereas both *li16* and *fr47* transcripts responded strongly to anoxia exposure (suggesting links to ischemia resistance) *(74–76)*. Li16 protein levels increased strongly under anoxia and also increased somewhat in 40% dehydrated frogs (Fig. 5); note, however, that high dehydration values also impose a hypoxic stress owing to high blood viscosity and low blood volume that impairs tissue oxygenation. Other studies used in vitro incubation of liver slices with different second messenger molecules to derive information about the signal transduction pathways that regulate the *li16* and *fr47* genes. The cryoprotectant response in frogs is regulated by β-adrenergic receptors and a cyclic AMP (cAMP)-dependent activation of liver glycogenolysis *(37,38)* but neither *li16* nor *fr47* responded to tissue incubations with dibutyryl cAMP. However, *li16* transcript levels were stimulated about twofold by dibutyryl cyclic GMP, whereas *fr47* responded to phorbol 12-myristate 13-acetate, indicating PKC involvement *(75,76)*. The response of *li16* to both anoxia and cGMP is very interesting because adenosine receptor signaling is mediated intracellularly by cGMP and the

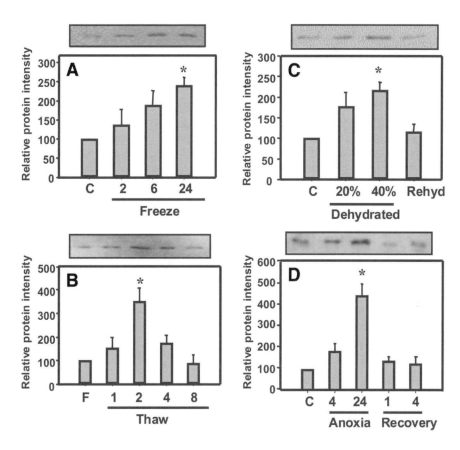

Fig. 5. Changes in the levels of the novel freeze-responsive protein, Li16, in liver of wood frogs, *Rana sylvatica*, under different conditions: (**A**) freezing at –2.5°C for 2, 6, or 24 h, (**B**) thawing at 5°C after 24 h frozen for 1, 2, 4, or 8 h; note that on this graph data are expressed relative to protein levels in 24 h frozen (F) frogs, (**C**) Dehydration at 5°C to 20 or 40% of total body water lost followed by 24 h rehydration, and (**D**) Anoxia exposure under a nitrogen gas atmosphere at 5°C for 4 or 24 h followed by aerobic recovery for 1 or 4 h. Shown are representative Western blots with histograms showing mean values for relative band intensities, $n = 3$ trials. C– control at 5°C. *, significantly different from zero time value, $p < 0.01$. (Data complied from McNally et al. *[75]*).

adenosine A1 receptor is one of freeze-responsive genes that have been identified from cDNA array screening in wood frogs *(54)*. This suggests a role for *li16* in ischemia resistance during freezing. FR47, by contrast, seems to be regulated by PKC and in other studies we have shown that levels of the PKC second messenger, inositol 1,4,5-trisphosphate (IP_3), rise over time during freezing or anoxia exposures in wood frog liver *(77)*. Overall, then, it appears that the three novel proteins are regulated by different signal transduction pathways and probably serve quite different, although as yet unknown, functions in the cell under freezing stress. Interestingly, one of the key conclusions from these studies is that natural freezing survival involves multiple pro-active responses by cells that include the upregulation of both known and unknown genes and regulation via at least three different signal transduction pathways.

6.2. Gene Discovery From cDNA Array Screening

The use of cDNA arrays to screen for stress-induced gene expression is revolutionizing the study of biochemical adaptation. State-of-the-art glass microarrays now have thousands of nonredundant cDNAs bound to them and offer one-step screening for stress-responsive genes. Critical advantages of array screening over other technologies include (a) a capacity to detect and quantify transcripts of different genes in a single sample that can vary as much as 1000-fold in abundance, (b) most of the genes are identified so screening allows evaluation of a huge number of genes involved in a wide array of cell functions, and (c) the ability to analyze the responses of functional groupings of genes to an imposed stress (e.g., families of transmembrane transporters, enzymes of different signal transduction cascades, enzymes in specific metabolic pathways, transcription factors, etc.). An illustration of this latter point came when we used rat macroarrays (Clontech ATLAS™) to screen for hibernation-responsive genes in ground squirrel skeletal muscle. The results showed that several genes that encode components of the small and large ribosomal subunits were consistently downregulated during hibernation including L19, L21, L36a, S17, S12 and S29 *(78)*. This further implicates control of the ribosomes as critical to the inhibition of protein synthesis in hibernation. Array screening of liver and kidney samples from both ground squirrels and bats also showed consistent upregulation of genes associated with antioxidant defense during hibernation. Transcript levels of superoxide dismutase, glutathione peroxidase, and glutathione-*S*-transferase were elevated by twofold or more in hibernator kidney, whereas these genes plus peroxiredoxin and metallothionein were upregulated in liver *(78)*. It is well known that hibernators elevate antioxidant defenses in brown adipose tissue as a means of dealing with high rates of oxygen-free radical generation during thermogenesis *(79)* but our screening data now suggests that the improvement of antioxidant defenses is widespread in multiple tissues. This would aid all organs in defense against oxidative damage during arousal when oxygen consumption can rise by 10- to 20-fold within minutes as the animal rewarms to 37°C.

In other studies, we used 19,000 gene human microarrays (Ontario Cancer Institute) to screen for genes that were upregulated in wood frog heart during freezing (comparing 5°C acclimated controls with frogs frozen for 24 h at –3°C) *(54)*. More than 200 genes were designated as putatively upregulated during freezing by at least 1.5-fold, some by as much as four to sevenfold. These included a variety of protein kinases and protein phosphatases that further stresses the point made earlier that the cells of freeze-tolerant organisms respond actively to the encroachment of extracellular ice around them with responses by a wide variety of cellular systems, triggered and coordinated by multiple signal transduction cascades. Furthermore, the key advantages of array screening for identifying functionally related groups of genes as well as genes that had not previously been associated with freeze tolerance was again apparent. Putative upregulation of genes involved with multiple metabolic functions was documented including antioxidant defense (e.g., glutathione-*S*-transferase, metallothionein, thioredoxin), transmembrane carriers and ion motive ATPases (monocarboxylic acid transporter, adenine nucleotide translocator, Na^+-K^+ ATPase), signal reception (e.g., adenosine A1 receptor, atrial natriuetic peptide [ANP] receptor), glucose production and transport (e.g., glucose transporter type 4, glucose-6-phosphatase), defense against high glucose damage (e.g., receptor for advanced glycosylation end products [RAGE]), and hypoxia-related proteins (HIF-1 α-subunit, F_o subunit c of the F_oF_1 ATPase complex *(54)*. In particular, the putative upregulation of receptors has provided key new clues about freeze tolerance that we are currently pursuing. For example, adenosine A1 receptor upregulation suggests that mechanisms of hypoxia-ischemia-induced MRD are activated during freezing. Up-regulation of the ANP receptor has implicated ANP, a cardiac hormone that regulates intravascular volume, in regulating the major changes in fluid dynamics of the cardiovascular system that occur during freeze–thaw. Up-regulation of RAGE suggests that advanced glycosylation end-products may accumulate in frogs as a result of nonenzymatic glycation of proteins under the extreme hyperglycemia of the frozen state. How frogs deal with this problem may suggest solutions for human diabetics because nonenzymatic glycation of long-lived proteins is the cause of diabetic vasculopathies and cataract.

The use of cDNA arrays for screening in comparative animal systems raises the critical issue of heterologous probing—the use of arrays containing immobilized cDNAs from one species to screen the mRNA populations of another species. Clearly, gene-sequence differences between species will prevent 100% cross-reaction between sample and array in any heterologous pairing. Predictably, our trials with 19K human cDNA arrays show that cross-hybridization falls off rapidly with phylogenetic distance. After optimization, we achieved cross-hybridization of 85 to 90% for human arrays hybridized with cDNA from hibernating mammals (ground squirrels or bats), 60 to 80% for frogs, and only approx 18% for hepatopancreas of *L. littorea (22,78)*. The latter value for snails may seem very low especially when only 10.6% of the genes that hybridized were scored as putatively upregulated this still provided us with more than 300 anoxia-responsive candidate genes for future work. Among the anoxia-responsive genes identified by this screening were protein phosphatases and kinases, mitogen-activated protein kinase interacting factors, translation factors, antioxidant enzymes, and nuclear receptors *(22)*. Heterologous screening would clearly have problems if the experimental goal was to evaluate qualitative and quantitative responses to stress by specific genes or groups of genes. However, our use of array screening has been as a general screening tool to find any genes that are stress-responsive and, in this mode, heterologous screening can be highly successful, providing the researcher with tens or hundreds of positive "hits" that provide multiple new directions for future research. Indeed, as noted above for frogs, array screening has provided numerous new directions to follow (e.g., receptors, antioxidants, glucose-related proteins, etc.) in our exploration of vertebrate freeze tolerance.

Another issue that has caused concern with heterologous probing is the possibility of false-positives (sample cDNA binding to cDNA on the array that is not its homologue) but in our work, to date, we have not encountered any case of a false-positive result. We treat all results from heterogeneous array screening with caution until they are verified by other techniques (e.g., RT-PCR or Northern blotting) using homologous probes. Indeed, our current protocol for the follow-up of genes that are highlighted as putatively upregulated from array screening is as follows. We search Genbank to gather sequences for the gene of interest from multiple sources, preferably from species phylogenetically close to the one under study, and then use those to identify a consensus sequence from which we design a PCR probe. The probe is then used to isolate the species-specific PCR product from a total RNA mix. We then sequence the species-specific product, confirm its identity, and use it to assess relative mRNA expression levels under multiple conditions via quantitative PCR. With the further use of peptide antibodies designed from the translated species-specific amino acid sequence we can also assess stress-induced changes in the accompanying protein levels. Thus, when combined with appropriate follow-up techniques, heterologous probing of cDNA arrays can provide an amazingly powerful search tool for gaining insights into the genes/proteins that underlie biochemical adaptation to extreme environmental stress.

In addition to gene-screening technologies, a variety of other new molecular methods are allowing enormous advances to be made in the field of biochemical adaptation. Proteomics approaches are rapidly developing, combining two-dimensional gel electrophoresis for protein isolation with analysis of tryptic peptides via liquid chromatography and mass spectrometry to identify stress upregulated proteins. The development of peptide and phosphopeptide antibodies for dozens of protein kinases, transcription factors and other proteins has greatly improved the ability to trace stress-activated signal transduction pathways to identify the stimulators of gene expression and of enzyme phosphorylation. For example, we used this technology to trace the upregulation of FABP in hibernator organs to stimulation by the γ isoform of the peroxisome proliferator-activated receptor (PPARγ) transcription factor and its co-activator, PGC-1 *(80)*. Techniques for evaluating protein–protein interactions and the roles of targeting proteins are allowing researchers to decipher the three-dimensional organization and compartmentation of metabolism *(81)*. These technologies and many more are allowing researchers to elucidate the unifying principles of biochemical adaptation that allow organisms to exploit and endure every extreme environment on Earth.

ACKNOWLEDGMENTS

We thank the many graduate students and postdoctoral fellows who have contributed to the research that is summarized in this article. Research in the Storey lab is supported by a discovery grant from the Natural Sciences and Engineering Research Council of Canada. KBS holds the Canada Research Chair in Molecular Physiology.

REFERENCES

1. Hochachka, P.W. and Somero, G.N. (2002) *Biochemical Adaptation: Mechanism and Process in Physiological Evolution*. Oxford University Press, Oxford, UK.
2. Schmidt-Nielsen, K. (1997) *Animal Physiology: Adaptation and Environment*, 5th ed. Cambridge University Press, Cambridge.
3. Ashcroft, F. (2000) *Life at the Extremes: The Science of Survival*. HarperCollins, London, UK.
4. Margesin, R. and Schinner, F. (eds.) (1999) *Cold-Adapted Organisms: Ecology, Physiology, Enzymology and Molecular Biology*. Springer-Verlag, Berlin.
5. Willmer, P., Stone, G. and Johnston, I. (2000) *Environmental Physiology of Animals*. Blackwell Science, Oxford, UK.
6. Lutz, P.L., Nilsson, G.E. and Prentice, H.M. (2003) *The Brain Without Oxygen*. Kluwer Academic Publishers, Amsterdam.
7. Gerday, C. and Glansdorff, N. (eds.) (2003) Extremophiles. in *Encyclopedia of Life Support Systems*. EOLSS Publishers, Oxford, UK [http://www.eolss.net]
8. Trueman, R.J., Tiku, P.E., Caddick, M.X. and Cossins, A.R. (2000) Thermal thresholds of lipid restructuring and delta(9)-desaturase expression in the liver of carp (*Cyprinus carpio* L.). *J. Exp. Biol.* **203**:641–650.
9. Storey, K.B. and Storey, J.M. (1990) Facultative metabolic rate depression: molecular regulation and biochemical adaptation in anaerobiosis, hibernation, and estivation. *Quart. Rev. Biol.* **65**:145–174.
10. Storey, K.B. and Storey, J.M. (2004a) Metabolic rate depression in animals: transcriptional and translational controls. *Biol. Rev. Camb Philos Soc.* **79**:207–233.
11. Duman J.G. (2001) Antifreeze and ice nucleator proteins in terrestrial arthropods. Annu. Rev. Physiol. **63**:327–357.
12. Storey, K.B. (2002) Life in the slow lane: molecular mechanisms of estivation. Comp. *Biochem. Physiol. A* **133**:733–754.

13. Storey, K.B. (ed.) (2004) *Functional Metabolism: Regulation and Adaptation*. John Wiley and Sons, New York, NY.
14. Wenger, R.H. (2000) Mammalian oxygen sensing, signaling and gene regulation. *J. Exp. Biol.* **203**:1253–1263.
15. Hochachka, P.W. (1986) Defense strategies against hypoxia and hypothermia. *Science* **23**:234–241.
16. Hochachka, P.W. and Lutz, P.L. (2001) Mechanism, origin and evolution of anoxia tolerance in animals. *Comp. Biochem. Physiol. B* **130**:435–459.
17. Claussen, T. (1986) Regulation of active Na⁺K⁺ transport in muscle. *Physiol. Rev.* **66**:542–576.
18. Berridge, M.J., Bootman, M.D. and Roderick, H.L. (2003) Calcium signalling: dynamics, homeostasis and remodelling. *Nat. Rev. Mol. Cell Biol.* **4**:517–529.
19. Orrenius, S., Zhivotovsky, B. and Nicotera P. (2003) Regulation of cell death: the calcium–apoptosis link. *Nat. Rev. Mol. Cell Biol.* **4**:552–565.
20. Halliwell, B. and Gutteridge, J.M.C. (1999) *Free Radicals in Biology and Medicine*. 3rd Ed. Clarendon Press, Oxford.
21. Jackson, D.C. (2001) Anoxia survival and metabolic arrest in the turtle. In *Molecular Mechanisms of Metabolic Arrest* (Storey, K.B., ed.), BIOS Scientific Publishers, Oxford, pp. 103–114.
22. Larade, K. and Storey, K.B. (2002) A profile of the metabolic responses to anoxia in marine invertebrates, in *Cell and Molecular Responses to Stress* (Storey, K.B. and Storey, J.M., eds.), Elsevier Press, Amsterdam, Vol. 3, pp. 27–46.
23. Hermes-Lima, M., Storey, J.M. and Storey, K.B. (2001) Antioxidant defenses and animal adaptation to oxygen availability during environmental stress. In *Cell and Molecular Responses to Stress* (Storey, K.B. and Storey, J.M., eds.), Elsevier Press, Amsterdam, Vol. 2, pp. 263–287.
24. Hermes-Lima, M. and Zenteno-Savin, T. (2002) Animal response to drastic changes in oxygen availability and physiological oxidative stress. *Comp. Biochem. Physiol. C* **133**:537–556.
25. Wang, L.C.H. and Wolowyk, M.W. (1988) Torpor in mammals and birds. *Can. J. Zool.* **66**:133–137.
26. Rousseau, K., Atcha, Z. and Loudon, A.S. (2003) Leptin and seasonal mammals. J. Neuroendocrinol. **15**:409–414.
27. Frank, C.L., Dierenfeld, E.S. and Storey, K.B. (1998) The relationship between lipid peroxidation, hibernation, and food selection in mammals. *Am. Zool.* **38**:341–349.
28. Storey, K.B. (1997) Metabolic regulation in mammalian hibernation: enzyme and protein adaptations. *Comp. Biochem. Physiol. A* **118**:1115–1124.
29. Andrews, M.T., Squire, T.L., Bowen, C.M. and Rollins, M.B. (1998) Low-temperature carbon utilization is regulated by novel gene activity in the heart of a hibernating animal. *Proc. Natl. Acad. Sci. USA* **95**:8392–8397.
30. Hittel, D. and Storey, K.B. (2001) Differential expression of adipose and heart type fatty acid binding proteins in hibernating ground squirrels. *Biochim. Biophys. Acta* **1522**:238–243.
31. Porras, A. and Benito, M. (2002) Regulation of proliferation, differentiation and apoptosis of brown adipocytes: signal transduction pathways involved. In *Cell and Molecular Responses to Stress* (Storey, K.B. and Storey, J.M., eds.), Elsevier Press, Amsterdam, Vol. 3, pp. 269–282.
32. Alves-Guerra, M-C., Pecqueur, C., Shaw, A., et al. (2002) The uncoupling proteins family: from thermogenesis to the regulation of ROS. In *Cell and Molecular Responses to Stress* (Storey, K.B. and Storey, J.M., eds.), Elsevier Press, Amsterdam, Vol. 3, pp. 257–268.
33. Stewart, J.M., English, T.E. and Storey, K.B. (1998) Comparisons of the effects of temperature on the liver fatty acid binding proteins from hibernator and nonhibernator mammals. *Biochem. Cell Biol.* **76**:593–599.
34. Thatcher, B.J. and Storey, K.B. (2001) Glutamate dehydrogenase from liver of euthermic and hibernating Richardson's ground squirrels: evidence for two distinct enzyme forms. *Biochem. Cell Biol.* **79**:11–19.
35. Fletcher, G.L., Hew, C.L. and Davies, P.L. (2001) Antifreeze proteins of teleost fishes. *Annu. Rev. Physiol.* **63**:359–390.

36. Storey, K.B. and Storey, J.M. (1988) Freeze tolerance in animals. *Physiol. Rev.* **68**:27–84.

37. Storey, K.B. and Storey, J.M. (1996) Natural freezing survival in animals. *Ann. Rev. Ecol. Syst.* **27**:365–386.

38. Storey, K.B. and Storey, J.M. (2004) Physiology, biochemistry and molecular biology of verte-brate freeze tolerance: the wood frog, in *Life in the Frozen State* (Benson, E., Fuller, B. and Lane, N., eds.), CRC Press, Boca Raton, FL, pp. 243–274.

39. Wharton, D.A. (2003) The environmental physiology of Antarctic terrestrial nematodes: a re-view. J. Comp. Physiol. B.**173**:621–628.

40. Barros, R.C.H., Zimmer, M.E., Branco, L.G.S. and Milsom, W.K. (2001) Hypoxic metabolic response of the golden-mantled ground squirrel. *J. Appl. Physiol.* **91**:603–612.

41. Storey, K.B. (2003) Mammalian hibernation: transcriptional and translational controls. *Adv. Exp. Med. Biol.* **543**:21–38.

42. Hochachka, P.W., Buck, L.T., Doll, C.J. and Land, S.C. (1996) Unifying theory of hypoxia tolerance: molecular/metabolic defense and rescue mechanisms for surviving oxygen lack. *Proc. Natl. Acad. Sci. USA* **93**:9493–9498.

43. Frerichs, K.U., Smith, C.B., Brenner, M., DeGracia, D.J., Krause, G.S., Marrone, L., Dever, T.E. and Hallenbeck, J.M. (1998) Suppression of protein synthesis in brain during hibernation in-volves inhibition of protein initiation and elongation. *Proc. Natl. Acad. Sci. USA* **95**:14511–14516.

44. Hittel, D. and Storey, K.B. (2002) The translation status of differentially expressed mRNAs in the hibernating thirteen-lined ground squirrel (*Spermophilus tridecemlineatus*). *Arch. Biochem. Biophys.* **401**:244–254.

45. Storey, K.B. (1993) Molecular mechanisms of metabolic arrest in mollusks. In *Surviving Hy-poxia: Mechanisms of Control and Adaptation* (Hochachka, P.W., Lutz, P.L., Sick, T.J., Rosenthal, M. and Thillart, G. van den, eds.) CRC Press, Boca Raton, pp. 253–269.

46. Plaxton, W.C. and Storey, K.B. (1984) Purification and properties of aerobic and anoxic forms of pyruvate kinase from red muscle tissue of the channeled whelk, *Busycotypus canaliculatum*. *Eur. J. Biochem.* **143**:257–265.

47. Brooks, S.P.J. and Storey, K.B. (1997) Glycolytic controls in estivation and anoxia: a comparison of metabolic arrest in land and marine molluscs. *Comp. Biochem. Physiol. A* **118**:1103–1114.

48. Brooks, S.P.J. and Storey, K.B. (1992) Mechanisms of glycolytic control during hibernation in the ground squirrel *Spermophilus lateralis*. *J. Comp. Physiol. B* **162**:23–28.

49. MacDonald, J.A. and Storey, K.B. (1999) Regulation of ground squirrel Na⁺ K⁺-ATPase activity by reversible phosphorylation during hibernation. *Biochem. Biophys. Res. Commun.* **254**:424–429.

50. Malysheva, I.N., Storey, K.B., Lopina, O.D. and Rubtsov, A.M. (2001) Ca-ATPase activity and protein composition of sarcoplasmic reticulum membranes isolated from skeletal muscles of typical hibernator, the ground squirrel *Spermophilus undulatus*. *Biosci. Rep.* **21**:831–838.

51. Bickler, P.E. and Donohoe, P.H. (2002) Adaptive responses of vertebrate neurons to hypoxia. *J. Exp. Biol.* **205**:3579–3586.

52. Buck, L.T. and Bickler, P.E. (1998) Adenosine and anoxia reduce N-methyl-D-aspartate recep-tor open probability in turtle cerebrocortex. *J. Exp. Biol.* **201**:289–297.

53. Lankford, A.R., Byford, A.M., Ashton, K.J., et al. (2002) Gene expression profile of mouse myocardium with transgenic over-expression of A1 adenosine receptors. *Physiol. Genomics* **11**:81–89.

54. Storey, K.B. (2004) Strategies for exploration of freeze responsive gene expression: advances in vertebrate freeze tolerance. *Cryobiology* **48**:134–145.

55. Shiomi, H. and Tamura, Y. (2000) Pharmacological aspects of mammalian hibernation: central thermoregulation factors in hibernation cycle. *Nippon Yakurigaku Zasshi* **116**:304–312.

56. Larade, K. and Storey, K.B. (2002) Reversible suppression of protein synthesis in concert with polysome disaggregation during anoxia exposure in *Littorina littorea*. *Mol. Cell. Biochem.* **232**:121–127.

57. DeGracia, D.J., Kumar, R., Owen, C.R., Krause, G.S. and White, B.C. (2002) Molecular path-ways of protein synthesis inhibition during brain reperfusion: implications for neuronal survival or death. *J. Cereb. Blood Flow Metab.* **22**:127–141.

58. Chen, Y., Matsushita, M., Nairn, A.C., et al. (2001) Mechanisms for increased levels of phosphorylation of elongation factor-2 during hibernation in ground squirrels. *Biochemistry* **40**:11565–11570.

59. Gingras, A.C., Raught, B. and Sonenbert, N. (1999) eIF4 initiation factors: effectors of mRNA recruitment to ribosomes and regulators of translation. *Ann. Rev. Biochem.* **68**:913–963.

60. Lang, K.J.D., Kappel, A. and Goodall, G.J. (2002) Hypoxia-inducible factor-1α mRNA contains an internal ribosome entry site that allows efficient translation during normoxia and hypoxia. *Mo.l Biol. Cell* **13**:1792–1801.

61. Larade, K. and Storey, K.B. (2004) Accumulation and translation of ferritin heavy chain transcripts following anoxia exposure in a marine invertebrate. *J. Exp. Biol.* **207**:1353–1360.

62. Van Breukelen, F. and Martin, S.L. (2001) Translational initiation is uncoupled from elongation at 18°C during mammalian hibernation. *Am J Physiol* **281**:R1374–R1379.

63. Cai, Q. and Storey, K.B. (1997) Freezing-induced genes in wood frog (*Rana sylvatica*): fibrinogen upregulation by freezing and dehydration. *Am. J. Physiol.* **272**:R1480–R1492.

64. Cai, Q., Greenway, S.C. and Storey, K.B. (1997) Differential regulation of the mitochondrial ADP/ATP translocase gene in wood frogs under freezing stress. *Biochim. Biophys. Acta* **1343**:69–78.

65. De Croos, J.N.A., McNally, J.D., Palmieri, F. and Storey, K.B. (2004) Up-regulation of the mitochondrial phosphate carrier during freezing in the wood frog *Rana sylvatica*: potential roles of transporters in freeze tolerance. *J. Bioenerg. Biomemb.* **36**:229–239.

66. Cai, Q. and Storey, K.B. (1996) Anoxia-induced gene expression in turtle heart: upregulation of mitochondrial genes for NADH-ubiquinone oxidoreductase subunit 5 and cytochrome C oxidase subunit 1. *Eur. J. Biochem.* **241**:83–92.

67. Fahlman, A., Storey, J.M. and Storey, K.B. (2000) Gene upregulation in heart during mammalian hibernation. *Cryobiology* **40**:332–342.

68. Willmore, W.G., English, T.E. and Storey K.B. (2001) Mitochondrial gene responses to low oxygen stress in turtle organs. *Copeia* **2001**:628–637.

69. Hittel, D. and Storey, K.B. (2002) Differential expression of mitochondria-encoded genes in a hibernating mammal. *J. Exp. Biol.* **205**:1625–1631.

70. Morano, I., Adler, K., Agostini, B. and Hasselbach, W. (1992) Expression of myosin heavy and light chains and phosphorylation of the phosphorylatable myosin light chain in the heart ventricle of the European hamster during hibernation and in summer. *J. Muscle Res. Cell Motility* **13**:64–70.

71. McCarron, R.M., Sieckmann, D.G., Yu, E.Z., Frerichs, K. and Hallenbeck, J.M. (2001) Hibernation, a state of natural tolerance to profound reduction in organ blood flow and oxygen delivery capacity. In *Molecular Mechanisms of Metabolic Arrest* (Storey, KB., ed.) BIOS Scientific Publishers, Oxford, pp. 23–42.

72. Larade, K. and Storey, K.B. (2002) Characterization of a novel gene upregulated during anoxia exposure in the marine snail *Littorina littorea*. *Gene* **283**:145–154.

73. Larade, K. and Storey, K.B. (2004) Anoxia-induced transcriptional upregulation of sarp-19: cloning and characterization of a novel EF-hand containing gene expressed in hepatopancreas of *Littorina littorea*. *Biochem. Cell Biol.* **82**:285–293.

74. Cai, Q. and Storey, K.B. (1997) Up-regulation of a novel gene by freezing exposure in the freeze-tolerant wood frog (*Rana sylvatica*). *Gene* **198**:305–312.

75. McNally, J.D., Wu, S., Sturgeon, C.M. and Storey, K.B. (2002) Identification and characterization of a novel freezing inducible gene, *li16*, in the wood frog, *Rana sylvatica*. *FASEB J.* 10.1096/fj.02–0017fje (online) http://www.fasebj.org/cgi/content/abstract/02-0017 fjev1.

76. McNally, J.D., Sturgeon, C.M. and Storey, K.B. (2003) Freeze induced expression of a novel gene, *fr47*, in the liver of the freeze tolerant wood frog, *Rana sylvatica*. *Biochim. Biophys. Acta* **1625**:183–191.

77. Holden, C.P. and Storey, K.B. (1997) Second messenger and cAMP-dependent protein kinase responses to dehydration and anoxia stresses in frogs. *J. Comp. Physiol. B* **167**:305–312.

78. Eddy, S.F. and Storey, K.B. (2002) Dynamic use of cDNA arrays: heterologous probing for gene discovery and exploration of animal adaptations in stressful environments. In *Cell and Molecular Responses to Stress* (Storey, K.B. and Storey, J.M., eds.), Elsevier Press, Amsterdam, Vol. 3, pp. 315–325.
79. Buzadzic, B., Spasic, M.B., Saicic, Z.S., Radojicic, R., Petrovic, V.M. and Halliwell, B. (1990) Antioxidant defenses in the ground squirrel *Citellus citellus*. 1. The effect of hibernation. *Free Rad Biol Med* **9**:407–413.
80. Eddy, S.F. and Storey, K.B. (2003) Differential expression of Akt, PPAR-γ and PGC-1 during hibernation in bats. *Biochem. Cell Biol.* **81**:269–274.
81. Sullivan, D.T., MacIntyre, R., Fuda, N., Fiori, J., Barrilla, J. and Ramizel, L. (2003) Analysis of glycolytic enzyme co-localization in *Drosophila* flight muscle. *J. Exp. Biol.* **206**:2031–2038.

11

Repair and Defense Systems at the Epithelial Surface in the Lung

Pieter S. Hiemstra

1. INTRODUCTION

The vast majority of infectious agents enter the body through the mucosal surfaces, such as those of the lung and intestine. These surfaces are covered by an epithelium that forms the interface between the external environment and the internal milieu. The epithelium is an essential physical barrier between the potentially hostile outside world and the host, but is also actively involved in processes like infection, inflammation, and immunity. The epithelium is therefore considered an important component of the host defense system. Epithelial layers are frequently injured because they are exposed to environmental toxic agents such as chemicals and micro-organisms, and to mechanical injury. To prevent invasion of the underlying tissue by pathogens and toxic agents, this epithelial injury must be followed by a rapid repair response. This chapter focuses on the interplay between this epithelial repair and the host defense system at the epithelial surface, two mechanisms that appear to act in concert to protect the host from infection. Whereas many of the principle mechanisms that operate in the diverse epithelia display marked similarities, some characteristics are typical for one type of epithelium. This chapter focuses on the pulmonary epithelium.

2. THE EPITHELIUM OF THE LUNG: A SHORT INTRODUCTION

The conducting airways of the lower respiratory tract conduct air to the distal alveoli where gas exchange occurs. The surface epithelium of the conducting airways (the trachea, bronchi, and bronchioli) is a pseudostratified epithelium, that is, all of its cells rest on the basement membrane but not all cells reach the lumen of the airways (reviewed in refs. *1* and *2*). It is composed of a variety of cell types, including the five major cell types: basal cells, ciliated cells, mucous goblet cells, nonciliated bronchiolar cells (Clara cells), and neuroendocrine cells. The first approx 12 generations of the tracheal–bronchial airways also contain the submucosal glands that are the major source of bronchial secretions. These glands are located between the airway smooth muscle layer and the cartilage and are lined by distinct epithelial cell types. The serous and mucous gland cells produce a mixed secretion containing mucins, lactoferrin, lysozyme, and other defense factors. In addition, the serous cells transport secretory immunoglobulin A that has been produced by plasma cells from the basal to the apical site of the gland cell through the action of secretory component. Finally, two

From: *Integrative Physiology in the Proteomics and Post-Genomics Age*
Edited by: W. Walz © Humana Press Inc., Totowa, NJ

types of epithelial cells line the alveoli: the cuboid, surfactant producing type II cells that act as progenitor for the flattened type I cells that allow gas exchange. The type II cells are twice as numerous as the type I cells, but cover only 7% of the surface area of the alveoli *(3)*.

The airway epithelium protects the host against environmental hazards that reach the lung. It acts as an effective first line of defense by (a) providing an effective physical barrier; (b) secreting compounds that act against inhaled substances; (c) clearing the airways by the coordinated action of cilia and mucus producing secretory cells (mucociliary clearance); and (c) secreting mediators that recruit cells of the innate and adaptive immune system. Therefore, the airway epithelium is not only a passive physical barrier, but plays an active role in regulating infection and inflammation.

Epithelial injury is a hallmark of an inflammatory lung disease such as asthma, and may also result from inhalation of e.g., cigarette smoke, infection, and endotracheal intubation and mechanical ventilation. The result of epithelial injury may vary from a transient loss of the epithelial barrier function to complete denudation of the airway epithelium. Injury may initiate a rapid repair response, resulting in reconstitution of normal epithelium. It may also cause hyperplastic or metaplastic changes in the airway epithelium. These may be induced by the injurious event itself in the absence of epithelial shedding, or may occur following repair of denuded epithelium. Such epithelial changes are a characteristic feature of a variety of pulmonary disorders. This is illustrated by squamous metaplastic changes and goblet cell hyperplasia observed in e.g., patients with chronic bronchitis or chronic obstructive pulmonary disease (Fig. 1). Goblet cell hyperplasia may result in mucus hypersecretion, which is not only observed in chronic bronchitis, but also in asthma.

3. HOST DEFENSE AT THE EPITHELIAL SURFACE

3.1. The Epithelium and Its Fight Against Infections

The airway epithelium is exposed to large amounts of airborne substances, including respiratory pathogens. Yet, severe lung infections are rare. This may be explained by the presence of efficient epithelial mechanisms that deal with such pathogens. How does the airway epithelium protect the underlying tissue? As outlined above and discussed in recent reviews *(4,5)*, it acts as a physical barrier, contributes to mucociliary clearance, and provides protection by secreting antimicrobial peptides and proteins. In addition, it may amplify the host defense response by producing chemokines and cytokines that attract a range of cell types involved in innate and adaptive immunity. Its ability to sense microbial exposure by a range of cellular receptors allows it to mount a rapid response and is essential for an effective host defense. These features are discussed in more detail in the following sections.

3.2. Response to Microbial Exposure

Airway epithelial cells react to microbial exposure by a number of cellular responses, including the production of mediators such as cytokines, chemokines and antimicrobial polypeptides *(4,6)*. Cytokines and chemokines mediate the recruitment of a variety of cell types, including neutrophils and macrophages that ingest and kill the invading micro-organisms, and immune cells that may initiate an immune response *(4)*. Other cytokines and chemokines are involved in the antiviral response of the airway epithelium, and include the type I interferons (IFN-α and IFN-β) that mediate the establishment of an antiviral state of the epithelium by induction of IFN-stimulated genes *(6)*. Antimicrobial polypeptides con-

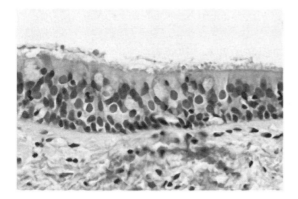

normal epithelium

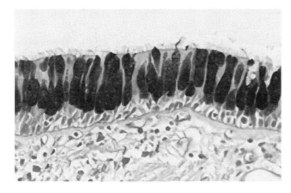

goblet cell hyperplasia

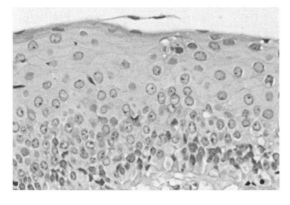

squamous metaplasia

Fig. 1. Epithelial changes in inflammatory lung disease. Micrographs of bronchial biopsies obtained from patients with chronic obstructive pulmonary disease (COPD). Top panel: hematoxylin and eosin (H&E)-stained section showing the normal epithelial structure composed of ciliated and some mucus cells; middle panel: periodic acid Schiff (PAS)—Alcian Blue-stained section showing an increased number of mucus containing goblet cells (stained dark) characteristic of goblet cell hyperplasia; lower panel: H&E-stained section showing multilayered epithelium with a top layer of flattened cells typical of squamous metaplasia; note the absence of ciliated cells.

tribute by directly killing micro-organisms and their ability to activate inflammatory and immune cells, as discussed in the next sections.

How do epithelial cells sense microbial exposure? Our knowledge of the ability of cells of the innate immune system to recognize micro-organisms has sharply increased in the last decade. This has led to the insight that the innate immune system is not aspecific in nature, but instead discriminates between exposures to different classes of micro-organisms. This is the result of recognition by so-called pattern-recognition molecules that bind conserved molecular patterns present on microbial components. An important family of the membrane-bound pattern-recognition molecules is the Toll-like receptor (TLR) family *(7)*. Ten members of the human TLR family have so far been identified, and different TLRs are expressed by almost all cell types in the body, including airway epithelial cells. The cells use these receptors to recognize a range of bacterial, viral, and fungal products that have been identified as ligands for the TLRs. Recently, also endogenous components including fragments of hyaluronic acid that are generated during inflammation, β-defensins, and surfactant proteins have been shown to signal through selected TLRs *(4,7)*. The importance of TLRs for host defense against infection was illustrated soon after the identification of TLRs in mammals. A mutation in TLR4 in mice was found to be associated with hyporesponsiveness to lipopolysaccharide (LPS), a component of the outer membrane of Gram-negative bacteria *(8)*. Subsequent studies showed that also mutations in human TLR4 are associated with a defective host defense against infection *(9)*.

3.3. Mucus Production and Mucociliary Clearance

Submucosal glands and mucus-producing goblet cells in the surface epithelium produce a range of mucins that are secreted into the lumen of the airways. Particles, including micro-organisms, that are present in inspired air are trapped in the mucus layer that resides on top of the surface epithelium and are usually cleared within 6 h *(10)*. Mucus is separated from the cells that constitute the surface epithelium by a thin layer of airway surface liquid (ASL). The cilia beat in this fluid layer, allowing them to move the gel-like mucus layer including the trapped particles by their tips towards the oral cavity, a process that is further enhanced by coughing *(10,11)*. The height of this fluid layer is adapted to the length of the cilia to allow optimal mucociliary transport. The importance of this clearance mechanism is illustrated by the severe lung infections that occur in patients in whom this mucociliary clearance is impaired. In patients with cystic fibrosis (CF), their increased susceptibility to infection may be the consequence of depletion of the volume of the ASL layer, which is a result of the abnormal ion transport owing to mutations in gene encoding the CF transmembrane regulator protein *(12)*. The thickened mucus, which can no longer be efficiently transported, forms a favorable environment for bacterial infections. Another example of the importance of mucociliary clearance in prevention of respiratory infection is primary ciliary dyskinesia, a rare genetic disorder resulting in deficient ciliary function. Secondary ciliary dyskinesia may result from dysfunction of ciliary beat activity owing to microbial toxins, for example.

3.4. Antimicrobial Polypeptides

Antimicrobial peptides are small, often cationic elements of innate immunity that are found in bacteria, plants, and animals *(13)*. These peptides display broad-spectrum antimicrobial activity against a wide range of bacteria, fungi, and viruses. The airway epithelium is a major site of synthesis of these peptides that are produced both by cells of the surface epithelium and by glandular cells. The direct antimicrobial action of antimicrobial peptides allows rapid elimination of invading micro-organisms. A variety of antimicrobial peptides have been identified, including the families of the defensins and cathelicidins. Larger antimicrobial proteins include lysozyme and lactoferrin, and the cationic proteinase inhibitors secretory leukocyte proteinase inhibitor (SLPI) and elafin *(4,5,14)*. These molecules, collectively referred to as antimicrobial polypeptides, kill a wide variety of micro-organisms by interfering with the integrity of the microbial membrane.

Defensins and cathelicidins are the major families of antimicrobial peptides in mammals. Structurally, defensins (3–5-kDa cationic peptides) are characterized by a β-sheet rich fold and six disulphide-linked cysteine residues (reviewed in refs. *15* and *16*). The two main families of human defensins, α- and β-defensins, are distinguished by the localization and pairing of the cysteine residues. Airway epithelial cells are main producers of β-defensins in the lung, whereas infiltrating neutrophils are the main cellular source of α-defensins. Four β-defensins (hBD1–4) that are produced by airway epithelial cells have been characterized in detail *(16)*, but computerized searching of the genome revealed that many more β-defensins (as many as 6 to 28) may be expressed *(17,18)*. Based on gene knockout studies, it appears that β-defensins are an important component of the antibacterial screen in the mouse lung. This is based on the observation that mice deficient in mouse β-defensin 1 (mBD1) show delayed clearance of *Haemophilus influenzae* from the lung *(19)*.

Airway surface epithelial cells and submucosal glands also produce the 4.5-kDa peptide LL-37, the only member of the human cathelicidin family of antimicrobial peptides that has been identified to date (reviewed by Zanetti *[20]*). Neutrophils are the main cellular source of hCAP-18/LL-37, but also various squamous epithelia and keratinocytes produce hCAP-18/LL-37. This molecule is produced and stored as an inactive precursor, hCAP-18, that is cleaved into two fragments: the N-terminal cathelin-like domain and the active C-terminal cationic antimicrobial peptide LL-37. The importance of hCAP-18/LL-37 in host defense in the lung is illustrated by the observation that overexpression of the gene in a human epithelial xenograft model of CF epithelium restores the defect in bacterial killing *(21)*. Furthermore, mice defective in the mouse homologue of LL-37, CRAMP, are more sensitive to skin infections than wild-type mice *(22)*.

Since their discovery in the mid-1990s, the mechanisms that regulate expression of β-defensins and cathelicidins by epithelial cells have been partly unraveled. These studies showed that the expression of hBD-1 is mainly constitutive, whereas that of hBD2, hBD3, hBD4 and hCAP-18/LL-37 is increased by a variety of pro-inflammatory and microbial stimuli. Microbial stimuli directly activate epithelial cells through pattern-recognition receptors, as illustrated by the ability of bacterial lipopeptide to increase hBD-2 expression in tracheobronchial epithelial cells through TLR2 *(23)*. In addition, cytokines such as interleukin-1β and tumor necrosis factor-α, which are produced by macrophages that have been exposed to microbial stimuli, are another stimulus of epithelial cells to produce antimicrobial peptides *(24,25)*.

In addition to the small defensins and cathelicidins, also larger antimicrobial proteins are abundant in airway secretions *(5)*. These include not only the well-known antimicrobials lysozyme and lactoferrin, but also the cationic serine proteinase inhibitors SLPI and elafin. Whereas lysozyme is a major constituent of neutrophils and macrophages, submucosal glands and the surface epithelium of the conducting airways appear to be its main cellular source in the lung. Lysozyme is 14-kDa antimicrobial protein that kills bacteria by both enzymatic and nonenzymatic mechanisms. Overexpression of lysozyme in the mouse lung was found to increase the resistance of mice against intratracheally administered *P. aeruginosa* and group B *Streptococcus (26)*. Lactoferrin is an 80-kDa antimicrobial polypeptide that displays antimicrobial activity through its ability to bind iron and by a mechanism independent of iron sequestration. Recently, it was also found to affect the formation of biofilms formed by *P. aeruginosa (27)*. This observation that components of innate immunity affect biofilm formation is an important one because bacteria in biofilms (matrix-encapsulated bacterial communities), are markedly resistant to host defence mechanisms and antibiotics *(28)*. SLPI is an 11.7-kDa cationic protein that, like elafin, was first identified based on its ability to inhibit the proteolytic activity of neutrophil elastase, and found to be expressed mainly in epithelial cells. Subsequent studies showed that both inhibitors also display defensin-like activity, by their broad-spectrum antimicrobial activity *(29–31)*. Studies by Cole and co-workers identified SLPI as one of the major antimicrobial components in nasal secretions, together with lysozyme and lactoferrin *(32)*. The in vivo role of SLPI in host defense was illustrated in a study showing its role in protection against LPS-induced septic shock in mice *(33)*. The combined antimicrobial and anti-inflammatory action of elafin in the lung in vivo was illustrated by overexpression of elafin in a mouse model of *P. aeruginosa (34)*.

Most antimicrobial polypeptides have been identified based on their ability to kill or inhibit the growth of micro-organisms. However, subsequent studies revealed that these polypeptides display many other activities. These include their ability to recruit a variety of inflammatory and immune cells (shown for α-defensins, β-defensins, LL-37, and elafin), to mediate degranulation of mast cells (α- and β-defensins, LL-37) and activate dendritic cells, macrophages and epithelial cells (α- and β-defensins, LL-37) *(4,35,36)*. Based on these findings, mainly derived from in vitro studies, antimicrobial polypeptides may turn out to be key regulators of inflammation and immunity. This conclusion is supported by in vivo studies, such as those showing the ability of defensins to increase humoral and T helper (Th)1/Th2 immune responses in mice *(37)*. As discussed elsewhere in this chapter, antimicrobial polypeptides also contribute to tissue repair by increasing epithelial repair (α-defensins, LL-37, and SLPI) and angiogenesis (LL-37).

4. EPITHELIAL REPAIR

A main function of any epithelium in the body is to act as a self-sealing barrier layer. Therefore, loss of epithelial integrity after injury of the airway epithelium requires a quick repair response. Epithelial repair is characterized by a highly coordinated series of cellular responses (for reviews, see refs. *38–40*). Interestingly, these responses do not appear to be restricted to adult life. Many of the mechanisms that act during embryonic development also play a role in epithelial repair processes after birth. Indeed, there is marked similarity between the array of signals and cytoskeletal machinery used during embryonic development and repair in adult life (reviewed in ref. *41*). Our insight in epithelial repair in the lung is based on a variety of animal models *(42,43)*, in vitro cell culture models *(44)*, and human airway xenograft models *(45)* that have provided complementary information. The epithelial repair process can be divided into three phases (Fig. 2) as described here:

1. Spreading and migration of flattened epithelial cells to cover the denuded area. Following injury, epithelial cells at the wound edge dedifferentiate into "repair cells" and migrate into the wound area to cover the denuded area *(42,43)*. Small wounds can be closed within 1 d by this process *(42)*.
2. Proliferation of epithelial cells to replace injured cells. Cell proliferation within and adjacent to the site of injury is the next step in the repair process that usually occurs after closure of the wound by spreading and migration *(42,44)*. The largest degree of cell proliferation is observed in cells located 160 to 400 μm from the wound edge *(44)*.

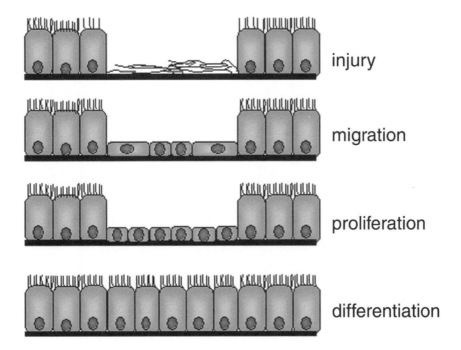

Fig. 2. Schematic presentation of the different stages of epithelial repair. Following injury, surrounding epithelial cells migrate into the wound area. Next these cells start to proliferate and finally will differentiate to form a fully reconstituted epithelial layer.

3. Differentiation leading to reconstitution of the original epithelial architecture. After coverage of the epithelium by undifferentiated cells, cell proliferation decreases and is followed by a phase of differentiation, that ultimately results in the formation of a fully functional mucociliary epithelium *(42)*.

The mediators that regulate the different steps in the repair processes have been partially characterized, and include a variety of growth factors and matrix proteins that are produced by the epithelium itself, by structural cells in the submucosa, and by infiltrating inflammatory cells. An inflammatory reaction appears to be a normal component in the process of wound healing, and is characterized by infiltration of neutrophils, macrophages, and lymphocytes. The inflammatory cells not only ingest invading micro-organisms and cause tissue injury, but may also contribute to repair by releasing factors that stimulate epithelial proliferation as shown for monocytes, neutrophils, and eosinophils, for example *(46–49)*.

As discussed earlier in this chapter, differentiation of airway epithelium following injury normally leads to restoration of the original architecture of the airway epithelium. However, the metaplastic and hyperplastic changes shown in Fig. 1, may be the consequence of injury. Both in vitro studies using cultures of airway epithelial cells, and animal studies have shown this. Indeed, both neutrophils and cigarette smoke cause an increase in mucus producing cells in the bronchial epithelium in vitro and in vivo, an effect that is mediated via the epidermal growth factor receptor *(50,51)*. In addition, intratracheal instillation of neutrophil elastase or bacterial LPS in hamsters not only results in the development of emphysematous changes, but also increases the number of mucus-producing cells in the conducting airways *(52,53)*. In elastase-treated animals, a marked loss of mucus-containing cells in the surface epithelium was observed after acute exposure, whereas a marked increase in such cells was observed after chronic exposure *(52)*.

5. REPAIR AND HOST DEFENSE AIMING TO PROTECT AGAINST INFECTION

Innate immunity and wound repair appear to act in concert to protect the host from infection. This is illustrated by the influx of inflammatory cells during wound repair, and their release of epithelial growth factors in the wound area. As discussed here, the increased expression of antimicrobial polypeptides in the wound area and the ability of selected antimicrobial polypeptides to stimulate wound repair are other examples of the integration of defense and repair responses.

A variety of studies have shown that the local expression of antimicrobial polypeptides in the epithelium and by infiltrating inflammatory cells is a characteristic of the repair response. Biochemical analysis of human wound fluid showed that it contains high concentrations of various antimicrobial peptides, that may be derived from inflammatory cells or from epithelial cells *(54)*. Studies in the skin of mice and humans showed that epithelial injury and wound healing increases the local epithelial expression of cathelicidins *(55,56)*, β-defensins *(57)*, SLPI *(58)*, and elafin *(59)*. In addition, keratinocytes in the skin of patients with the hyperproliferative skin disease, psoriasis, display high levels of expression of a variety of antimicrobial polypeptides *(60)*. In fact, members of the β-defensin family of antimicrobial peptides were first isolated from lesional psoriatic scales collected from such patients *(61,62)*. A possible explanation for these findings was provided by the observation that growth factors that are involved in wound repair (insulin-like growth factor-I and transforming growth factor-α) increase the expression of hCAP-18/LL-37, human β-defensin-3, neutrophil gelatinase-associated lipocalin, and SLPI in cultured human keratinocytes *(63)*. This clearly illustrates that repair and defense mechanisms act in a highly integrated fashion.

Antimicrobial polypeptides may also directly accelerate wound closure or affect epithelial differentiation. This is illustrated by the ability of neutrophil-derived lactoferrin and α-defensins to increase epithelial proliferation *(46,47,64,65)*. Neutrophil defensins, when present at high concentrations, may cause injury to epithelial cells, but increase epithelial wound closure at low concentrations in cultured bronchial epithelial cells (Fig. 3 *[64]*). Interestingly, neutrophil defensins affect various phases of the repair process because they not only display mitogenic activity, but also act as chemoattractants for epithelial cells and stimulate mucin production. The ability of neutrophil defensins to increase epithelial cell migration, proliferation, and mucin production, indicates that defensins may contribute to all stages of the epithelial repair process. Whether chronic exposure to neutrophil defensins also contributes to mucus hypersecretion and squamous metaplastic changes is unknown. However, we did observe an association between the presence of neutrophils defensins and squamous metaplastic changes in bronchial epithelium *(66)*. This was demonstrated by immunohistochemistry, showing a higher number of neutrophil-defensin positive cells and proliferating epithelial cells in areas with squamous cell metaplasia, as compared to areas with intact or damaged epithelium. Another neutrophil-derived antimicrobial peptide that may affect tissue repair following injury is hCAP-18/LL-37. This is based on the observation that LL-37 and its mouse homologue CRAMP display angiogenic activity and increase wound healing in an ex vivo human skin model *(56,67)*. The observation in bronchial epithelial cell cultures that LL-37 causes transactivation of the epidermal growth factor receptor *(68)*, a

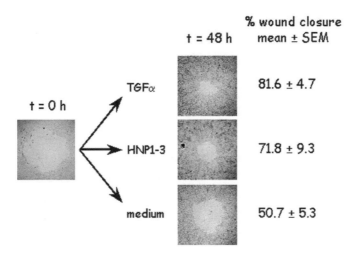

Fig. 3. An in vitro model for epithelial repair using the bronchial epithelial cell line NCI-H292: increased wound closure induced by the growth factor TGF-α and neutrophil defensins. A mechanical wound is created in a monolayer of NCI-H292 cells, that is subsequently incubated with neutrophil defensins (a mixture of HNP 1-3), TGF-α, or medium alone. Percentage wound closure was calculated from digital images obtained after 48 h. TGF, transforming growth factor; HNP, human neutrophil peptide; SEM, standard error of the mean. (*See* ref. *64* for details.)

receptor involved in various processes including epithelial repair, may partially explain the contribution of LL-37 to wound repair. Finally, hBD-1 may participate in differentiation processes in keratinocytes *(69)*.

The involvement of SLPI in epithelial repair was first demonstrated in mice rendered deficient in SLPI. These mice showed impaired repair of cutaneous wounds *(70)*. Subsequent in vitro studies revealed several mechanisms by which SLPI may promote growth of epithelial cells, including modulation of processing of the epithelial growth factor proepithelin *(71,72)*. In addition to its direct effect on epithelial cells, SLPI also stimulates human lung fibroblasts to release hepatocyte growth factor *(73)*, a growth factor that increases airway epithelial repair *(74)*. These results show that like many other antimicrobial polypeptides, SLPI contributes to host defense in various ways, including inhibition of neutrophil-derived serine proteinases, antimicrobial activity, and wound repair.

It appears that both growth factors and antimicrobial polypeptides contribute to prevention of infection of the wound area and to epithelial wound repair. It is therefore interesting to note that animals, by licking their wounds, provide the wound area with an additional supply of saliva-derived growth factors and antimicrobial peptides *(75,76)*.

6. CONCLUDING REMARKS

How does local expression of antimicrobial polypeptides, either by the epithelial cells themselves or by infiltrating inflammatory cells, contribute to protection of epithelial surfaces? A model for the role of antimicrobial polypeptides in host defense and repair in epithelium is presented in Fig. 4. First, antimicrobial polypeptides produced during wound repair increase local defense by killing invading micro-organisms. Second, because of their chemotactic activity for a variety of immune and inflammatory cells, including dendritic cells, T cells, and granulocytes, these polypeptides recruit professional phagocytes and

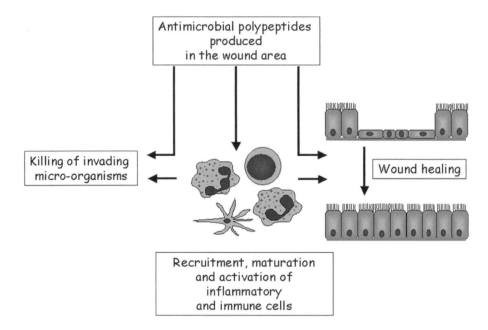

Fig. 4. Expression of antimicrobial polypeptides and their contribution to epithelial repair and host defense. Antimicrobial polypeptides in the wound area may be derived from epithelial cells or from inflammatory cells recruited to the wound area. These polypeptides cause removal or growth restriction of pathogens in the wound area by their direct antimicrobial activity, or indirectly by recruiting phagocytes and immune cells. These peptides further may protect the underlying tissue from infection by promoting wound repair directly, or indirectly by recruitment of cells that release growth factors. Note: effects of antimicrobial polypeptides on host cells are clearly concentration-dependent, with high concentrations of many peptides showing cytotoxicity.

immune cells to deal with the invading micro-organisms. Finally, antimicrobial substances such as neutrophil α-defensins and SLPI may facilitate epithelial wound closure directly.

ACKNOWLEDGMENTS

The author would like to thank Annemarie van Schadewijk and Jamil Aarbiou for help in preparation of Figs. 1 and 3, and Sandra van Wetering for critical reading of the manuscript. Studies in this research area in the author's laboratory were supported by grants from the Netherlands Asthma Foundation and the Netherlands Organization for Scientific Research.

REFERENCES

1. Corrin, B. (2000) Normal lung structure, In *Pathology of the lungs* (Corrin, B., ed), Churchill Livingstone, London, pp. 1–34.
2. Danel, C. J. (1996) Morphological characteristics of human airway structures: diversity and unity, in *Environmental impact on the airways. From injury to repair* (Chretien, J. and Dusser, D., eds), Marcel Dekker, New York, pp. 19–42.
3. Crapo, J. D., Barry, B. E., Gehr, P., Bachofen, M. and Weibel, E. R. (1982) Cell number and cell characteristics of the normal human lung. *Am. Rev. Resp. Dis.* **126**:332–337.
4. Bals, R. and Hiemstra, P. S. (2004) Innate immunity in the lung: how epithelial cells fight against respiratory pathogens. *Eur. Respir. J.* **23**:327–333.

5. Ganz, T. (2004) Antimicrobial polypeptides. *J. Leukoc. Biol.* **75**:34–38.
6. Message, S. D. and Johnston, S. L. (2004) Host defense function of the airway epithelium in health and disease: clinical background. *J. Leukoc. Biol.* **75**:5–17.
7. Takeda, K., Kaisho, T. and Akira, S. (2003) Toll-like receptors. *Annu. Rev. Immunol.* 21:335–376.
8. Poltorak, A., He, X., Smirnova, I., andet al. (1998) Defective LPS signaling in C3H/HeJ and C57BL/10ScCr mice: mutations in Tlr4 gene. *Science* **282**:2085–2088.
9. Agnese, D. M., Calvano, J. E., Hahm, S. J., andet al. (2002) Human toll-like receptor 4 mutations but not CD14 polymorphisms are associated with an increased risk of gram-negative infections. *J. Infect. Dis.* **186**:1522–1525.
10. Wanner, A., Salathe, M. and O'Riordan, T. G. (1996) Mucociliary clearance in the airways. *Am. J. Respir. Crit Care Med.* **154**:1868–1902.
11. Sleigh, M. A., Blake, J. R. and Liron, N. (1988) The propulsion of mucus by cilia. *Am. Rev. Respir. Dis.* **137**:726–741.
12. Boucher, R. C. (2004) New concepts of the pathogenesis of cystic fibrosis lung disease. *Eur. Respir. J* **23**:146–158.
13. Zasloff, M. (2002) Antimicrobial peptides of multicellular organisms. *Nature* **415**:389–395.
14. Sallenave, J. M. (2000) The role of secretory leukocyte proteinase inhibitor and elafin (elastase-specific inhibitor/skin-derived antileukoprotease) as alarm antiproteinases in inflammatory lung disease. *Respir. Res.* **1**:87–92.
15. Ganz, T. (2003) Defensins: Antimicrobial peptides of innate immunity. *Nature Rev. Immunol.* **3**:710–720.
16. Schutte, B. C. and McCray, P. B., Jr. (2002) [beta]-defensins in lung host defense. *Annu. Rev. Physiol* **64**:709–748.
17. Schutte, B. C., Mitros, J. P., Bartlett, J. A.,et al. (2002) Discovery of five conserved beta -defensin gene clusters using a computational search strategy. *Proc. Natl. Acad. Sci. U. S. A* **99**:2129–2133.
18. Kao, C. Y., Chen, Y., Zhao, Y. H. and Wu, R. (2003) ORFeome-Based Search of Airway Epithelial Cell-Specific Novel Human β-Defensin Genes. *Am. J. Resp. Cell Mol. Biol.* **29**:71–80.
19. Moser, C., Weiner, D. J., Lysenko, E., Bals, R., Weiser, J. N. and Wilson, J. M. (2002) beta-Defensin 1 contributes to pulmonary innate immunity in mice. *Infect. Immun.* **70**:3068–3072.
20. Zanetti, M. (2004) Cathelicidins, multifunctional peptides of the innate immunity. *J. Leukoc. Biol.* **75**:39–48.
21. Bals, R., Weiner, D. J., Meegalla, R. L. and Wilson, J. M. (1999) Transfer of a cathelicidin peptide antibiotic gene restores bacterial killing in a cystic fibrosis xenograft model. *J. Clin. Invest* **103**:1113–1117.
22. Nizet, V., Ohtake, T., Lauth, X., Trowbridge, J.,et al. (2001) Innate antimicrobial peptide protects the skin from invasive bacterial infection. *Nature* **414**:454–457.
23. Hertz, C. J., Wu, Q., Porter, E. M.,et al. (2003) Activation of toll-like receptor 2 on human tracheobronchial epithelial cells induces the antimicrobial peptide human beta defensin-2. *J. Immunol.* **171**:6820–6826.
24. Liu, L., Roberts, A. A. and Ganz, T. (2003) By IL-1 signaling, monocyte-derived cells dramatically enhance the epidermal antimicrobial response to lipopolysaccharide. *J. Immunol.* 170:575–580.
25. Tsutsumi-Ishii, Y. and Nagaoka, I. (2003) Modulation of human beta-defensin-2 transcription in pulmonary epithelial cells by lipopolysaccharide-stimulated mononuclear phagocytes via proinflammatory cytokine production. *J. Immunol.* **170**:4226–4236.
26. Akinbi, H. T., Epaud, R., Bhatt, H. and Weaver, T. E. (2000) Bacterial killing is enhanced by expression of lysozyme in the lungs of transgenic mice. *J. Immunol.* **165**:5760–5766.
27. Singh, P. K., Parsek, M. R., Greenberg, E. P. and Welsh, M. J. (2002) A component of innate immunity prevents bacterial biofilm development. *Nature* **417**:552–555.
28. Costerton, J. W., Stewart, P. S. and Greenberg, E. P. (1999) Bacterial Biofilms: A Common Cause of Persistent Infections. *Science* **284**:1318.
29. Hiemstra, P. S., Maassen, R. J., Stolk, J., Heinzel-Wieland, R., Steffens, G. J. and Dijkman, J. H. (1996) Antibacterial activity of antileukoprotease. *Infect. Immun.* **64**:4520–4524.

30. McNeely, T. B., Dealy, M., Dripps, D. J., Orenstein, J. M., Eisenberg, S. P. and Wahl, S. M. (1995) Secretory leukocyte protease inhibitor: a human saliva protein exhibiting anti-human immunodeficiency virus 1 activity in vitro. *J. Clin. Invest.* **96**:456–464.

31. Simpson, A. J., Maxwell, A. I., Govan, J. R., Haslett, C. and Sallenave, J. M. (1999) Elafin (elastase-specific inhibitor) has anti-microbial activity against gram-positive and gram-negative respiratory pathogens. *FEBS Lett.* **452**:309–313.

32. Cole, A. M., Liao, H. I., Stuchlik, O., Tilan, J., Pohl, J. and Ganz, T. (2002) Cationic Polypeptides Are Required for Antibacterial Activity of Human Airway Fluid. *J. Immunol.* **169**:6985.

33. Nakamura, A., Mori, Y., Hagiwara, K., et al. (2003) Increased susceptibility to LPS-induced endotoxin shock in secretory leukoprotease inhibitor (SLPI)-deficient mice. *J. Exp. Med.* **197**:669–674.

34. Simpson, A. J., Wallace, W. A., Marsden, M. E., et al. (2001) Adenoviral augmentation of elafin protects the lung against acute injury mediated by activated neutrophils and bacterial infection. *J. Immunol.* **167**:1778–1786.

35. Yang, D., Chertov, O. and Oppenheim, J. J. (2001) Participation of mammalian defensins and cathelicidins in anti- microbial immunity: receptors and activities of human defensins and cathelicidin (LL-37). *J. Leukoc. Biol.* **69**:691–697.

36. Davidson, D. J., Currie, A. J., Reid, G. S. D., et al. (2004) The Cationic Antimicrobial Peptide LL-37 Modulates Dendritic Cell Differentiation and Dendritic Cell-Induced T Cell Polarization. *J. Immunol.* **172**:1146–1156.

37. Lillard, J. W., Jr., Boyaka, P. N., Chertov, O., Oppenheim, J. J. and McGhee, J. R. (1999) Mechanisms for induction of acquired host immunity by neutrophil peptide defensins. *Proc. Natl. Acad. Sci. U. S. A* **96**:651–656.

38. Erjefalt, J. S. and Persson, C. G. A. (1997) Airway epithelial repair: breathtakingly quick and multipotentially pathogenic. *Thorax* **52**:1010–1012.

39. Puchelle, E. and Zahm, J.-M. (1996) Repair processes of the airway epithelium, in *Environmental impact on the airways. From injury to repair.* (Chretien, J. and Dusser, D., eds), Marcel Dekker, New York, pp. 157–182.

40. Rennard, S. I. (1996) Repair mechanisms in asthma. *J. Allergy Clin. Immunol* **98**:S278–S286.

41. Jacinto, A., Martinez-Arias, A. and Martin, P. (2001) Mechanisms of epithelial fusion and repair. *Nature Cell Biology* **3**:E117–E123.

42. Keenan, K. P., Combs, J. W. and McDowell, E. M. (1982) Regeneration of hamster tracheal epithelium after mechanical injury. I. Focal lesions: quantitative morphologic study of cell proliferation. *Virchows Arch. B Cell Pathol. Incl. Mol. Pathol.* **41**:193–214.

43. Erjefalt, J. S., Erjefalt, I., Sundler, F. and Persson, C. G. (1995) In vivo restitution of airway epithelium. *Cell Tissue Res.* **281**:305–316.

44. Zahm, J. M., Kaplan, H., Herard, A. L.,et al. (1997) Cell migration and proliferation during the in vitro wound repair of the respiratory epithelium. *Cell Motil. Cytoskeleton* **37**:33–43.

45. Puchelle, E. and Peault, B. (2000) Human airway xenograft models of epithelial cell regeneration. *Resp.Res.* **1**:125–128.

46. Murphy, C. J., Foster, B. A., Mannis, M. J., Selsted, M. E. and Reid, T. W. (1993) Defensins are mitogenic for epithelial cells and fibroblasts. *J. Cell. Physiol.* **155**:408–413.

47. Aarbiou, J., Ertmann, M., van Wetering, S.,et al. (2002) Human neutrophil defensins induce lung epithelial cell proliferation in vitro. *J. Leukoc. Biol.* **72**:167–174.

48. Calafat, J., Janssen, H., Stahle-Backdahl, M., Zuurbier, A. E., Knol, E. F. and Egesten, A. (1997) Human monocytes and neutrophils store transforming growth factor-alpha in a subpopulation of cytoplasmic granules. *Blood* **90**:1255–1266.

49. Egesten, A., Calafat, J., Knol, E. F., Janssen, H. and Walz, T. M. (1996) Subcellular localization of transforming growth factor-alpha in human eosinophil granulocytes. *Blood* **87**:3910–3918.

50. Takeyama, K., Jung, B., Shim, J. J.,et al. (2001) Activation of epidermal growth factor receptors is responsible for mucin synthesis induced by cigarette smoke. *Am. J. Physiol. Lung Cell Mol. Physiol.* **280**:L165–L172.

51. Takeyama, K., Dabbagh, K., Jeong, S. J.,et al. (2000) Oxidative stress causes mucin synthesis via transactivation of epidermal growth factor receptor: role of neutrophils. *J. Immunol.* **164**:1546–1552.

52. Breuer, R., Christensen, T. G., Lucey, E. C., Stone, P. J. and Snider, G. L. (1987) An ultrastructural morphometric analysis of elastase-treated hamster bronchi shows discharge followed by progressive accumulation of secretory granules. *Am. Rev. Respir. Dis.* **136**:698–703.

53. Stolk, J., Rudolphus, A., Davies, P.,et al. (1992) Introduction of emphysema and bronchial mucus cell hyperplasia by intratracheal instillation of lipopolysaccharide in the hamster. *J. Pathol.* **167**:349–356.

54. Frohm, M., Gunne, H., Bergman, A.-C.,et al. (1996) Biochemical and antibacterial analysis of human wound and blister fluid. *Eur. J. Biochem.* **237**:86–92.

55. Dorschner, R. A., Pestonjamasp, V. K., Tamakuwala, S.,et al. (2001) Cutaneous injury induces the release of cathelicidin anti-microbial peptides active against group A Streptococcus. *J. Invest. Dermatol.* **117**:91–97.

56. Heilborn, J. D., Nilsson, M. F., Kratz, G.,et al. (2003) The cathelicidin anti-microbial peptide LL-37 is involved in re- epithelialization of human skin wounds and is lacking in chronic ulcer epithelium. *J. Invest. Dermatol.* **120**:379–389.

57. Schmid, P., Grenet, O., Medina, J., Chibout, S. D., Osborne, C. and Cox, D. A. (2001) An intrinsic antibiotic mechanism in wounds and tissue-engineered skin. *J. Invest. Dermatol.* **116**:471–472.

58. Wingens, M., van Bergen, B. H., Hiemstra, P. S.,et al. (1998) Induction of SLPI (ALP/HUSI-I) in epidermal keratinocytes. *J. Invest. Dermatol.* **111**:996–1002.

59. van Bergen, B. H., Andriessen, M. P., Spruijt, K. I., van de Kerkhof, P. C. and Schalkwijk, J. (1996) Expression of SKALP/elafin during wound healing in human skin. *Arch. Dermatol. Res.* **288**:458–462.

60. Frohm, M., Agerberth, B., Ahangari, G.,et al. (1997) The expression of the gene coding for the antibacterial peptide LL-37 is induced in human keratinocytes during inflammatory disorders. *J. Biol. Chem.* **272**:15258–15263.

61. Harder, J., Bartels, J., Christophers, E. and Schröder, J.-M. (1997) A peptide antibiotic from human skin. *Nature* **387**:861.

62. Harder, J., Bartels, J., Christophers, E. and Schroder, J. M. (2001) Isolation and characterization of human beta -defensin-3, a novel human inducible peptide antibiotic. *J. Biol. Chem.* **276**:5707–5713.

63. Sorensen, O. E., Cowland, J. B., Theilgaard-Monch, K., Liu, L., Ganz, T. and Borregaard, N. (2003) Wound healing and expression of antimicrobial peptides/polypeptides in human keratinocytes, a consequence of common growth factors. *J. Immunol.* **170**:5583–5589.

64. Aarbiou, J., Verhoosel, R. M., van Wetering, S.,et al. (2004) Neutrophil defensins enhance lung epithelial wound closure and mucin gene expression in vitro. *Am. J. Respir. Cell Mol. Biol.* **30**:193–201.

65. Hagiwara, T., Shinoda, I., Fukuwatari, Y. and Shimamura, S. (1995) Effects of lactoferrin and its peptides on proliferation of rat intestinal epithelial cell line, IEC-18, in the presence of epidermal growth factor. *Biosci. Biotechnol. Biochem.* **59**:1875–1881.

66. Aarbiou, J., Van Schadewijk, A., Stolk, J.,et al. (2004) Human neutrophil defensins and secretory leukocyte proteinase inhibitor in squamous metaplastic epithelium of bronchial airways. *Inflamm. Res.* **53**:230–238.

67. Koczulla, R., von Degenfeld, G., Kupatt, C., Krotz, F., Zahler, S., Gloe, T., Issbrucker, K., Unterberger, P., Zaiou, M., Lebherz, C., Karl, A., Raake, P., Pfosser, A., Boekstegers, P., Welsch, U., Hiemstra, P. S., Vogelmeier, C., Gallo, R. L., Clauss, M. and Bals, R. (2003) An angiogenic role for the human peptide antibiotic LL-37/hCAP-18. *J. Clin. Invest.* **111**, 1665–1672.

68. Tjabringa, G. S., Aarbiou, J., Ninaber, D. K.,et al. (2003) The Antimicrobial Peptide LL-37 Activates Innate Immunity at the Airway Epithelial Surface by Transactivation of the Epidermal Growth Factor Receptor. *J. Immunol.* **171**:6690–6696.

69. Frye, M., Bargon, J. and Gropp, R. (2001) Expression of human beta-defensin-1 promotes differentiation of keratinocytes. *J Mol. Med.* **79**:275–282.

70. Ashcroft, G. S., Lei, K., Jin, W.,et al. (2000) Secretory leukocyte protease inhibitor mediates non-redundant functions necessary for normal wound healing. *Nat. Med.* **6**:1147–1153.

71. Zhang, D., Simmen, R. C., Michel, F. J., Zhao, G., Vale-Cruz, D. and Simmen, F. A. (2002) Secretory leukocyte protease inhibitor mediates proliferation of human endometrial epithelial cells by positive and negative regulation of growth-associated genes. *J. Biol. Chem.* **277**:29,999–30,009.

72. Zhu, J., Nathan, C., Jin, W.,et al. (2002) Conversion of proepithelin to epithelins: roles of SLPI and elastase in host defense and wound repair. *Cell* **111**:867–878.

73. Kikuchi, T., Abe, T., Yaekashiwa, M., et al. (2000) Secretory leukoprotease inhibitor augments hepatocyte growth factor production in human lung fibroblasts. *Am. J. Respir. Cell Mol. Biol.* **23**:364–370.

74. Zahm, J. M., Debordeaux, C., Raby, B., Klossek, J. M., Bonnet, N. and Puchelle, E. (2000) Motogenic effect of recombinant HGF on airway epithelial cells during the in vitro wound repair of the respiratory epithelium. *J. Cell. Physiol.* **185**:447–453.

75. Mogi, M., Inagaki, H., Kojima, K., Minami, M. and Harada, M. (1995) Transforming growth factor-alpha in human submandibular gland and saliva. *J. Immunoassay* **16**:379–394.

76. Murakami, M., Ohtake, T., Dorschner, R. A. and Gallo, R. L. (2002) Cathelicidin antimicrobial peptides are expressed in salivary glands and saliva. *J Dent. Res.* **81**:845–850.

12

The Zebrafish As an Integrative Physiology Model

Alicia E. Novak and Angeles B. Ribera

1. INTRODUCTION

The complexity of nervous system structure and function presents considerable challenges for understanding the output of the nervous system (behavior) in terms of the molecular and cellular properties of component parts (neurons). For integrated studies of nervous system physiology, a model system needs to meet several requirements. First, the model system should have a well-defined anatomy that ideally is simple and minimally complex. Furthermore, the ability to identify neurons of interest for physiological analysis and perturbation of known cellular processes augments the power of the system. Finally, the system should allow elimination of the unique functional contribution of specific neurons either by targeted ablation methods or genetic and molecular perturbations.

1.1. The Developing Zebrafish As a Model

A classic model that is being used for integrative physiology and that has the above characteristics is the zebrafish embryo (Fig. 1). Zebrafish (*Danio rerio*) develop externally from the mother's body and are transparent, allowing nonobtrusive observation of anatomical and developmental processes *(1)*. Both imaging and behavioral assays can be performed at multiple time points without perturbing the embryo.

The anatomy of this vertebrate's nervous system is not only relatively simple but also accessible to cellular and molecular perturbations. The neurons that are present in the central nervous system (CNS) and peripheral nervous system (PNS) during embryonic stages have been extensively studied both anatomically and molecularly (*see* refs. *2–9*). Characterization of the molecular determinants expressed in differential cell types by techniques such as *in situ* hybridization and immunocytochemistry allow biochemical assessment of the neuron's functional role. In addition, electrophysiological characterization of CNS neurons including spinal cord neurons (sensory, interneurons, and motoneurons) and retinal ganglion cell (RGC) neurons has been done *(10–15)*. When examined, zebrafish neurons display a developmental program for excitability that is similar to that for neurons in other well-characterized models (e.g., *Xenopus laevis [11,16]*).

The genetics offered by the zebrafish system are one of its greatest strengths. The forward genetics mutagenesis approach has successfully been applied to the zebrafish to isolate genes

From: *Integrative Physiology in the Proteomics and Post-Genomics Age*
Edited by: W. Walz © Humana Press Inc., Totowa, NJ

Fig. 1. The zebrafish from embryo to adult. (Left) A bright field image of a 12 somite (14 h post fertilization) zebrafish embryo at early stages of neural development. The animal is wrapped around the round structure, the yolk sac, with the presumptive head at the top right (asterisk), developing spinal cord and trunk muscle at the bottom, and tail at the top left (arrowhead). Scale bar: 0.25 mm. (Right) An adult zebrafish with its characteristic stripes at home in a laboratory tank. Scale bar: 1 cm.

that are essential for embryogenesis *(17–20)*. Large-scale mutagenesis screens have been performed using either a chemical mutagen (ENU) or retroviral vectors *(19–21)*. More than 500 genes with essential developmental roles have been isolated using a variety of anatomical and behavioral screens *(19–21)*.

A practical consideration of classic genetic screens is that they typically require three generations to bring recessive mutations to homozygosity. A large number of embryos and adult animals must be raised and the generation time and breeding space limit the number of organisms that can be analyzed. Zebrafish embryos are easy to raise and maintain and have a relatively short generation time. Embryonic development occurs quickly and processes that occur within the first 2 days post-fertilization (dpf) of zebrafish development take between 6 and 20 d in chick, rat, and mouse models *(22)*. Furthermore, a large amount of fish can be housed in a small tank. Breeding and collection of zebrafish embryos is straightforward, and willing pairs of adult fish will produce more than 200 progeny per breeding *(19)*. In fact, *The Wall Street Journal* recently stated that the zebrafish is one of the most cost-effective models used in research today *(23)*.

Forward genetics approaches have also been successful in the zebrafish because of the large amount of sequence information that is available to identify mutated genes. The zebrafish genome is being rapidly sequenced and extensive databases are available on the Internet. Expressed sequence tags (EST), yeast, bacterial, and P1 artificial chromosome sequence databases also are available, as are the sequences of many cloned genes *(24–26)*. The 25 zebrafish chromosomes, also referred to as linkage groups (LG), are rapidly being mapped *(27,28)*. Markers, such as simple-sequence length polymorphisms, random-amplified polymorphic DNAs, ESTs, microsatellites, and cloned genes are being placed on the chromosomal maps as anchors *(29,30)*. Thus, candidate gene analysis can be undertaken to discover the identity of mutated genes.

The creation of zebrafish genomic maps has also allowed for the comparison of gene and chromosomal organization between species. An analysis of synteny between 523 zebrafish and human orthologous genes showed that more than 80% map to syntenic regions *(31)*. However, although many syntenic relationships exist between human and zebrafish chromosomes, there is no one-to-one correspondence of chromosomes or genes. Zebrafish have 25 haploid chromosomes, whereas humans only have 23 *(27,28)*. Furthermore, zebrafish genes found on a single chromosome show synteny to different human chromosomes *(31,32)*. Recombination may have split up genes that were once situated close to each other. Further-more, identifying syntenic regions between the human and zebrafish genome is complicated because zebrafish often have more than one copy of analogous human genes.

In addition to the genetic advantages, the zebrafish is quite amenable to pharmacological and toxicological experimentation. Because the embryo lives in an aqueous media, it is quite easy to administer toxins to the developing fish. For example, the effect of an ion channel blocker, tricaine (MS222), on the development of spinal sensory neurons has been examined *(33)*. Other studies have examined the effects of nicotine and thyroid hormone on develop-ment of nervous system structures and other organ systems *(34,35)*.

Finally, the zebrafish provides a very interesting look at genetic evolution. In recent years, fish have become popular models for studying development and genomics. With the accu-mulation of genomic data, it has become apparent that many gene families in fish have more members than mammals *(36)*. For example, several developmental gene families, such as *Hox, distal-less, hedgehog,* and *engrailed* contain more isotypes in zebrafish than they do in mammals *(37–40)*. How did this occur? In zebrafish and other closely related fish (Teleo-sts), it has been proposed that a large-scale genome duplication occurred based on the in-creased number of *Hox* clusters and other closely associated genes *(40,41,41a)*. *Hox* genes are transcription factors that are found in well-defined clusters in the genome, and whose mRNAs are expressed along the anterior-posterior axis of the embryo during development (for review *see* ref. *42*). There are four *Hox* clusters in mammals, each on a different chro-mosome and containing 9 to 11 orthologous genes (with respect to both nucleotide sequence and genomic position *[42]*). A hypothesized full-genome duplication is proposed to have taken place after the divergence of modern fish from mammals, but before the divergence of fish lineages *(40,41a)*. Thus, most fish species should share a similar abundance of genes. This duplication event is likely to have increased the number of *Hox* gene clusters from four to eight during evolution of the zebrafish genome. Today, only seven *Hox* clusters remain in the zebrafish *(40,41)*.

However, the concept of large-scale genome duplication remains controversial. Although many species of Teleost fish do have duplicated genes, phylogenetic comparison with their mammalian and invertebrate orthologues suggests that some gene duplications occurred after the divergence of fish lineages (i.e., zebrafish from *Fugu*). This may not support the hypoth-esis of large-scale genome duplication *(43–45)*. A database called Wanda is currently keep-ing track of orthologous genes from human, mouse, chicken, and their duplicates in *Xenopus* and ray-finned fish *(46)*. As more genes and gene families are characterized, the evolution of Teleost genomes may be elucidated.

In summary, the zebrafish offers many advantages as a model system for study of integra-tive physiology. These characteristics (e.g., simple neuroanatomy, accessibility of cells of interest) facilitate cellular physiology and developmental analyses. In addition, behaviors can be analyzed in detail at genetic, morphological, and molecular levels to provide an inte-grated understanding of nervous system development.

2. INTEGRATIVE PHYSIOLOGY APPLICATION

In the following sections, we focus on how the integrative approach can and is being applied to investigate mechanisms that underlie nervous system function. The two major topics covered include motor output (e.g., swimming and the escape response) and sensory input (e.g., auditory, vestibular, lateral line, visual, olfactory and gustatory, and tactile stimuli). These systems are interconnected (e.g. stimuli through sensory input lead to motor output and behavior). Thus, the zebrafish system can also be used to analyze the many physiological levels that contribute to organismal behavior. We highlight how integrative physiology approaches in zebrafish are providing important new information about the nervous system, and consequently do not summarize the vast amount of material that has been gathered on each topic.

2.1. Motor Activities

Here we focus on the motor behaviors displayed by developing zebrafish embryos and larva and the mechanisms that underlie their generation. Reviews of motor behavior that focus on early and larval behaviors, anatomy, and molecular determinants can be found in refs. *47–51*. Our focus in this section is on the integration of these elements in the developing embryo and the output motor behaviors.

2.1.1. Anatomy of the Neurons Involved in Motor Behaviors

The developing zebrafish relies on neurons in both the hindbrain and spinal cord to elicit the complex movements required for locomotion and behavior *(13)*. Within the hindbrain, reticulospinal interneurons are the largest population of descending signals used in motor behaviors *(52)*. These neurons receive synaptic inputs from the brain and project to spinal motor and interneurons *(53–55)*. Well over 100 neurons in 21 cell categories reside in the larval reticular formation and participate in locomotive behaviors *(49,55)*. The first-born and largest of the reticulospinal neurons is the Mauthner (M) cell *(56,57)*. Two other large reticulospinal neurons, MiD2cm and MiD3cm, are thought to be segmental homologues of the M cell based on their size, temporal specification and location *(52,53,57–59)*. Descending reticulospinal neuron axons have extensive cell-specific axonal arborizations that enable contact with other spinal interneurons, the central pattern generator, and motoneurons *(53,55)*.

Motor behaviors are also controlled by interneurons and motoneurons in the spinal cord that form complex distant and local interactions. These cells can be stereotypically identified by their location, size, and time of development (Fig. 2 *[3,60,61]*). The earliest born motoneurons, also known as primary motoneurons, are easily identified by the stereotypic locations of their cell body and axons in the ventral region of each spinal segment (Fig. 3 *[60,62]*). Three motoneuron types—caudal primary (CaP), middle primary (MiP), and rostral primary (RoP)—extend their axons into the ventral, dorsal, and medial muscle, respectively *(60)*. A fourth motoneuron type, variable primary (VaP), has similar axonal trajectories to its neighbor CaP, but is only found in about 50% of segments *(63)*. While motoneurons tend to be classified based on the muscle they innervate and their time of development, interneurons are much more complex. In larval fish, eight classes of interneurons have been identified using retrograde labeling techniques *(2,3,61)*. Interneurons are classified based on (a) the direction that the axon projects from the cell body, (b) whether the axon is commissural or ipsilateral to the cell body, and (c) special characteristics of the cell body and dendrites *(3,61)*. As with other spinal neurons, each of the eight classes is thought to have unique functions that are related to their anatomical and physiological properties *(61)*. While the specific function of each spinal interneuron class is just beginning to be understood, future imaging and physiological studies will elucidate the role of interneurons in the motor network.

DORSAL

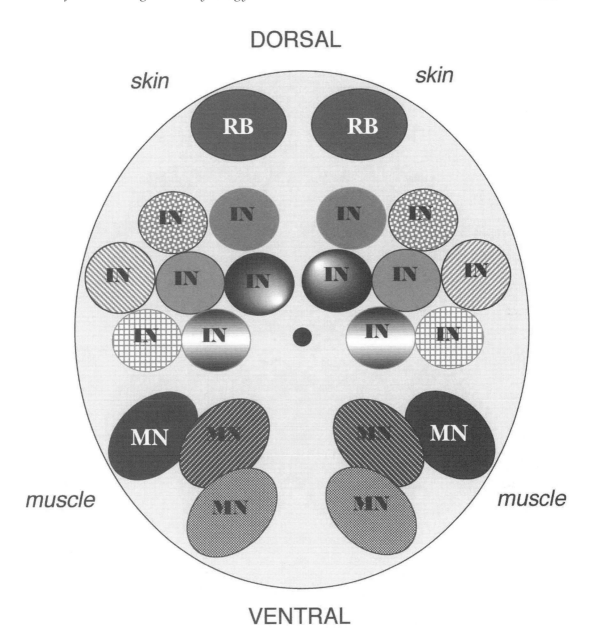

VENTRAL

Fig. 2. Primary spinal neurons have sterotypic positions in the spinal cord. There are three populations of primary spinal neurons found in the early developing embryo. The most dorsally located are the mechano-sensory Rohon-Beard (R-B) neurons. In the medial portion of the spinal cord, several types of interneurons (IN) reside and motoneurons (MN) are found most ventrally. Primary neurons function in swimming, escape, and hatching behaviors, and later are joined by secondary neurons (MN and IN) or replaced by dorsal root ganglion neurons (R-B).

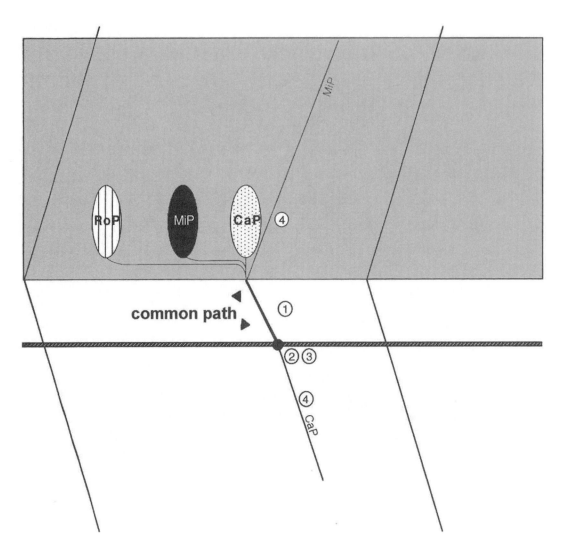

Fig. 3. Schematic of primary motoneuron axogenesis. The rostral primary (RoP), middle primary (MiP), and caudal (CaP) motoneurons lie in stereotypic locations within the spinal cord (gray) and extend their axons into the adjacent muscle (white and lateral to the gray spinal cord; Adapted from ref. *67*). One cluster of neurons is found per segment (indicated by the slanting boundary lines). During axogenesis, the motoneurons extend their axons along the spinal cord to the ventral root, where they exit the spinal cord and travel ventrally along the common pathway in the medial myotome (*see* ref. *60*). At the end of the common pathway, the reach a somitic 'choice point' at the horizontal myoseptum (dark gray circle; ref. *60*). Motoneurons in the *diwanka* mutant have errors in pathfinding along the common path (1), whereas those in *unplugged* and *stumpy* traverse the common path correctly, but do not make correct pathfinding decisions with their ventrally projecting CaP axons (paths 2 and 3). Finally, Smn morphants have truncated and highly branched axons into the dorsal and ventral muscle (path 4).

2.1.2. Development of Motor Circuitry

Here we discuss how neurons involved in motor behaviors form the necessary integrated circuitry. In most cases, the cues and expressed proteins that are used by neurons to reach their target cells and form proper synapses involved in these motor pathways have not been completely identified. Evidence exists that cues from the external environment play a major role in axonal pathfinding (*see* ref. *64* for review).

Genetic studies have identified mutants that have defects in pathfinding in neurons needed for motor activities *(65,66)*. The mutant *diwanka* was isolated in a motility screen because of its accordion like swimming movements *(65)*. Mutant *diwanka* and wild-type embryos were examined histochemically using an antibody that labels axonal projections of the CaP, MiP and RoP neurons *(67)*. Wild-type primary motoneurons extended their axons along the common path and out to their myotome-specific areas by 23 h post-fertilization (hpf) *(67)*. In *diwanka* embryos, primary motoneurons were able to extend their axons along the spinal cord to the ventral root. However, there were defects in growth of axons along the common path to the choice point *(67)*. The mutant *unplugged* also shows pathfinding defects. However, *unplugged* mutants display defects in the axonal targeting of only CaP and RoP primary neurons and their secondary motoneuron counterparts *(68,69)*. Furthermore, in contrast to *diwanka*, targeting of *unplugged* axons is aberrant after the choice point where the axons do not extend to their appropriate myotomal segments *(68,69)*. Chimeric studies with both *unplugged* and *diwanka* mutants indicate that each gene encodes a protein that is expressed by adaxial muscle precursor cells within the horizontal myoseptum *(66,67,69)*. Transplantation of wild-type adaxial cells into mutant *diwanka* or *unplugged* embryos restores proper axonal trajectories of the primary motoneurons *(67,68)*. A third mutant, *stumpy*, also has defects in pathfinding of motoneurons, with the most severe defects seen in CaPs *(66)*. Similarly to *unplugged* and *diwanka*, *stumpy* is a factor that is expressed in adaxial muscle precursor cells in the horizontal myoseptum, which gives cell type-specific information to the motoneurons that pass them *(66)*. Thus, the *diwanka*, *unplugged* and *stumpy* genes contribute to pathfinding of specific motoneuron axons at different stages of outgrowth.

Determinants of motoneuron trajectories have also been examined in "morphant" embryos in which a transient knockdown of survival motor neuron (Smn) protein has been made with a morpholino *(70)*. A decrease in the expression levels of the Smn protein has been implicated in the mortality of motoneurons in the spinal cord, a leading cause of hereditary death in infants *(71–73)*. The *Smn* gene is thought to encode an intracellular protein, associated with ribosomal processing proteins *(74,75)*. A single *Smn* gene in zebrafish is expressed ubiquitously in all cell types at a time when motoneurons are undergoing axogenesis *(70)*. However, reduction of the expression levels of Smn in morphant embryos results in a motor axon specific deficit in both primary and secondary motoneurons *(70)*. Axonal projections to the choice point do not seem to be affected in the morphant embryos. However, past the choice point, both dorsally and ventrally extending morphant motoneuron axons show a dose-dependent increase in truncation of the axonal projections with a corresponding increase in branching of the axons *(70)*. Although the mechanism that underlies the function of the *Smn* gene remains elusive, the morphant phenotype provides another clue for understanding the molecules needed for correct axonal outgrowth of motoneurons.

2.1.3. Electrical Activity in Neurons Involved in Motor Behaviors

During the time that the motor behavior circuitry is being formed, neurons are also becoming electrically excitable and forming their first synapses, thus allowing simple behaviors. By examining the developmental acquisition of electrical excitability, the role of specific cell types in early behaviors has been examined. Furthermore, electrophysiological analysis allows examination of molecular components, such as voltage-gated ion channels, expressed during early development. The physiology of several cell types involved in motor activities have been characterized with whole-cell patch electrophysiology. The M cell is electrically excitable as early as 24 hpf *(10,12)*. Spontaneous action potentials and synaptic activity, comprised of inhibitory and excitatory input associated with the spontaneous firing of other hindbrain neurons, can be observed as early as 26 hpf *(10,12)*. Primary motoneurons have also been characterized electrophysiologically in both embryos and larva *(12)*. Both synaptic and action potentials were observed in embryonic motoneurons, but most notable was a periodic excitatory drive responsible for bursting needed during contractions of the muscle *(12)*. Larval motoneurons were constantly receiving synaptic input including γ-aminobutyric acid, glycine, and glutamatergic input *(12,76)*. Interneurons were also characterized electrophysiologically at early times (17–27 hpf). Some interneuron populations have bursting activity reminiscent of the motoneurons at these early stages *(12)*. Dorsal interneurons show synaptic currents from early (24 hpf) to later (96 hpf) stages *(11)*. In summary, many cell types in the developing reticular formation and spinal cord acquire electrical activity during early embryogenesis. Furthermore, as evidenced by both bursting activity and synaptic currents, these neurons actively form synapses with other neurons in the motor circuit.

2.1.4. Synapse Formation

The components required for the formation of functional neuromuscular synapses have been studied in great detail in many systems (for reviews, *see* refs. *77–80*). Interestingly, several zebrafish embryonic mutations have been found that affect aspects of synapse formation and have revealed new functions for these proteins. The *nic1* and *sofa potato* mutants, identified by their immotile phenotypes, have defects in neuromuscular transmission *(65,81–83)*. Spinal motoneurons in both mutants extend their axons to their appropriate muscle targets at the right time. However, neither mutant displays a muscular response, a phenotype thought to be mediated by the loss of acetylcholine receptor (AChR) functionality in the postsynaptic muscle *(82–84)*. In both *nic1* and *sofa potato* mutants, AChRs fail to cluster at the neuromuscular junction and thus proper synaptic transmission cannot occur. The *nic1* gene has been identified as the AChR α-subunit *(82)*. The *nic1* mutation lies within an intron and results in improper splicing of the mRNA and loss of function of the AChR *(82)*. Similar phenotypes are seen in morphant embryos whose AChR ion channel expression is knocked down with morpholinos *(85)*. Analysis of *sofa potato* mutants has shown that the mutation most likely lies in a gene whose product is responsible for AChR clustering at the neuromuscular junction, such as the AChR β-subunit *(83)*. Studies of the neuromuscular junction of *sofa potato* mutants have shown that the *sofa potato* gene may also be necessary for localization of other proteins, such as rapsyn, to the postsynaptic membrane *(83)*. Interestingly, rapsyn proteins have a well-documented function of clustering AChRs to the subsynaptic membrane *(86–88)*. Mutations in the zebrafish *rapsyn* gene cause decreases in AChR clustering at the neuromuscular junction, as seen in the *twitch once* mutant *(65,89)*. Behaviorally, *twitch once* mutants do not exhibit spontaneous swimming nor do they mount a full escape response to mechanical stimulation *(89)*. Mutant *twitch once* larva show signs of synaptic depression at the neuromuscular junction, which may be mediated by receptor desensitization or other mechanisms that have not been elucidated *(89)*.

2.1.5. Early Motor Behaviors

The earliest motor activity of the zebrafish is a spontaneous slow lateral bending of the trunk beginning at about 17–18 hpf, at a time before motor circuitry is completed *(13,54)*. At this time, CaP motoneuron axons are beginning to enter the myotomes and these contractions may be caused by the release of ACh from the motoneurons *(90)*. Electrophysiological characterization of primary motoneurons at this time suggests that bursts of activity responsible for these contractions are mediated through gap junctions with a population of adjacent interneurons *(13,14)*. These spinal interneurons form an early and local circuit with motoneurons that functions independent of chemical transmission, until a few hours later *(13,14)*. The peak in the frequency of spontaneous contractions is reached at about 19 hpf, after which the embryo acquires a stimulus-evoked response to touch *(47,90)*.

By 27 hpf, the embryo is able to evoke a fast escape response mediated by the M cell *(47,91)*. The escape response is initiated by a fast bend that turns the embryo 180 degrees from the stimuli, called a C-start bend *(49,59,92)*. Confocal calcium imaging of the M, MiD2cm, and MiD3cm cells during evoked escape responses has revealed that all three cells function together during escape responses *(91)*. Furthermore, the response level of each cell is dependent on the placement of the stimulus. For example, stimuli placed at the head of the larva produces activation of all three cells, whereas stimulation of the tail activates only the M cell *(91)*. However, other hindbrain neurons appear to be able to take over the escape response when the M cell and its homologues are laser ablated *(59)*. Interestingly, calcium imaging used to determine which hindbrain neurons are still active after M cell ablation reveals that a substantial proportion of descending neurons exhibit calcium signals in response to a stimuli *(93)*. This reveals that the escape response involves perhaps more cells than once thought.

A third motor activity that the zebrafish embryo acquires is the ability to swim. Swimming is a sequence of alternating contractions of the tail at low amplitude *(47)*. A swimming episode often follows an escape response, but is usually slower and uses smaller body angles *(49)*. High-speed digital imaging has been able to discern different movements used in escape responses, slow and fast swimming episodes and prey capture *(49,94)*. The highly specific movements associated with each of these behaviors suggest that each has unique sets of motor circuits that control them. Data gleaned from calcium imaging studies of two types of interneurons confirm this hypothesis *(95)*. Circumferential descending (CiD) interneurons are active during escape responses, but quiescent during fictive swimming when Multipolar Commisural descending (McoD) interneurons are active *(95)*. Because specificity of the cell types is needed for each behavior, this may provide a reason why hundreds of neurons contribute to spinal motor circuitry *(49)*.

2.2. Sensory Systems

In this section, we discuss how components of the auditory, vestibular, lateral line, visual, olfactory, gustatory, and tactile systems develop in the zebrafish. Furthermore, we show how an integrative approach has been invaluable for illuminating the complex developmental processes relevant to these sensory systems (for review, *see* ref. 96). Interestingly, comparitively more advances have been made in auditory, vestibular, lateral line,

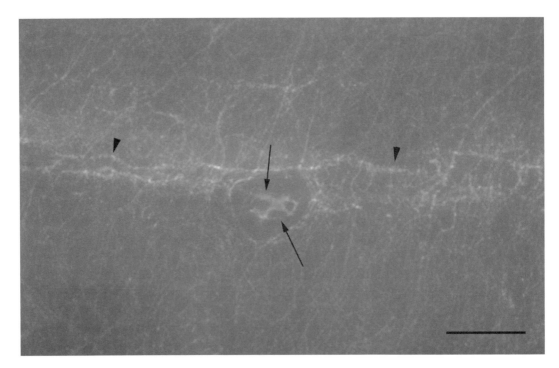

Fig. 4. Neuromasts in the head and trunk sense water pressure and displacement. An image of a neuromast in the trunk of a 48 h post fertilization embryo as revealed by the zn-12 antibody. Hair cells in the center of the neuromast (arrows) are surrounded by support cells. Processes in the lateral line (arrowheads) are also labeled by the zn-12 antibody, which recognizes a carbohydrate moiety found on cell surfaces thought to be involved in axon guidance and cell recognition *(101a,157)*. Scale bar: 25 µM.

visual, and touch systems than in olfaction and gustation. Possible reasons for this difference are discussed.

2.2.1. Auditory and Vestibular Function

As for mammals, the zebrafish inner ear holds organs responsible for audition (hearing) and vestibular (balance) functions. Complex developmental pathways allow differentiation of the proper cell types needed for each structure (for reviews, *see* refs. *97–100*). Hair cells, the neural component of the ear, are formed from sensory epithelia at specified locations *(99)*. Areas of sensory cells overlain by otoliths form separate auditory and vestibular structures, the saccule and utricle, respectively *(99)*. Additional sensory cells are found in other vestibular structures, (1) the three semicircular canals (that detect movement), and (2) in the the utricle (detects gravity) *(99)*. A third detector in the zebrafish that senses changes in pressure in the water is the lateral line system *(101)*. Lateral line organs, called neuromasts, are found in the head and trunk and are comprised of hair and support cells like those in the ear (Fig. 4 *[102,103]*).

In comparison to motor behaviors, the acquisition of both hearing and vestibular function occurs late in development. Although hair cells can be detected in presumptive maculae of the ear as early as 24 hpf by phalloidin staining, neurons in the stereoacoustic ganglia do not innervate their targets until almost 2 d later *(104)*. Lateral line hair primordia finish differentiating into hair cells by 48 hpf and are innervated by their sensory neurons shortly afterward *(101,103,105)*. However, behavioral tests show that mature auditory, vestibular, and lateral line responses do not appear until 96 hpf *(104,106)*.

Mutants with morphological defects of the ear that lead to deafness and often times loss of vestibular function have been isolated *(107,108)*. In addition, mutations have been isolated that impair vestibular and/or auditory function despite normal or minimally perturbed morphological development of the ear *(65,99,110)*. Mutants that represent at least seven human diseases, including Usher and Waardenburg syndromes, have been identified in the zebrafish *(100,109)*. One example, *mariner*, has defects in sensory hair cell function *(65,110,111)*. Morphologically, the integrity of the hair cell stereocilia was disrupted in *mariner;* behaviorally, startle responses to acoustic/vibrational stimuli were abolished *(110)*. Micropotentials generated by bending of lateral line hair cell stereocilia were also absent. The gene responsible for the defects in the *mariner* mutants is the *myosin VIIA* gene, which is concentrated in stereocilia and regions responsible for apical endocytosis *(111)*. Another mutant, *sputnik*, displays similar phenotypes but is caused by degeneration of the stereocilia owing to a mutation in otocadherin *(99,110,112)*. Interestingly, defects in genes involved in hair cell integrity, including stereocilia morphology, and transduction affect both auditory and vestibular function, suggesting that these two sensory systems share developmental programs. However, independent programs of development are also indicated because the mutant *monolith* shows selective loss of vestibular function *(106,113)*.

Transduction events in the hair cell are mediated through the opening of mechanosensitive channels at the tip of the stereocilia *(97,114)*. A transient receptor channel, Drosophila no mechanoreceptor potential C (NompC), is vital for hair cell mechanotransduction *(114)*. Morpholino knockdowns of NompC led to an absence of microphonic currents in the morphologically normal stereocilia, as well as a loss of the acoustic startle response *(114)*. A group of three other mutants, *orbiter*, *mercury,* and *gemini*, also have normal stereocilia morphology but diminished or absent microphonic potentials, indicating that additional genes are also critical for proper formation of the transduction apparatus *(65,110)*. In summary, the auditory and vestibular systems are intimately associated. They appear to use similar mechanisms for transduction of sound/vestibular stimuli and mutations in genes that affect hair cells tend to disrupt both sensory processes. Furthermore, using an integrative approach, zebrafish mutants identified by their abnormal behaviors can be studied anatomically, molecularly, and physiologically to understand the basis of their defects.

2.2.2. The Visual System

Using an integrative physiology approach, advances have been made in understanding the development of the zebrafish visual system. The zebrafish retina is composed of seven different cell types responsible for transducing and processing light signals from the eye to the brain *(see* refs. *115–117* for review). These cell types have been studied anatomically and molecularly in the zebrafish and are very similar to those in other vertebrates *(115,116,118,119)*. However, the zebrafish system has allowed for advances in both developmental and disease models owing to the accessibility of neurons and characterization of cellular markers at early stages *(115,119,120)*. The zebrafish retina is also amenable to many physiological assays including whole-cell recordings, electroretinograms, and behavioral

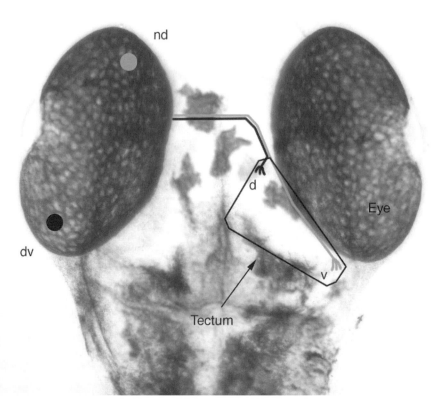

Fig. 5. Lipophilic dye injection reveals topographic specificity of retinal projections. Injection of dyes into the nasodorsal (nd) region of the retina labels axons that project to the contralateral ventral tectum (v), while dyes injected at the temporoventral (dv) region, label axons that project to the contralateral dorsal (d) tectum (figure modified from ref. *135*). Both the projection pathways and arborizations within the termination fields are highly regulated and perturbation of either of these processes can lead to fish that are blind. More than 100 mutants have been identified by dye labeling screens that have perturbations in these processes *(123,134,135)*.

tests including two that measure eye movements in response to visual stimuli, the optokinetic and optomotor response *(121–123)*.

As with the auditory system, the visual system develops late. While RGCs become postmitotic as early as 28 hpf, functional synapses only begin to form within the retina by 120 hpf (5 dpf), and reach maturity by 10 dpf *(115,124,125)*. Formation of the retina is dependent on the expression of many genes involved in both differentiation of individual cells types as well as formation of the functional layers. For example, specification and differentiation of RGCs is dependent on a number of genes, including *atonal*, a basic helix-turn-helix transcription factor, and *sonic hedgehog (shh)*, a secreted signaling factor *(126–128)*. Disruptions in the *atonal* gene, as identified in the *lakritz* mutant, prevent the specification of RGCs, with a corresponding increase in amacrine, bipolar and Müller cell number *(126,129)*. Similarly, a loss of *shh*, as seen in the *sonic-you* mutant, perturbs RGC neurogenesis, leading to a reduced population of RGCs, many of which die *(128,130)*.

The RGCs are the first retinal cell type to undergo axogenesis, as early as 30 hpf *(124)*. Axons traverse the optic stalk and project to the contralateral tectum forming a topographic map corresponding to their locations on the retina *(15,131,132)*. The precision of this mapping has been elegantly examined by injection of lipophilic dyes into desired regions of retina (Fig. 5 *[131,133,135]*). Furthermore, this method has served as a screen for identification of mutant embryos that have improper RGC axonal connections *(15,123,133–135)*. Mutants have been found with an array of phenotypes, with defects in both pathfinding within the retina as well as with processes to correct topographical locations within the tectum. The *astray* mutant has one of the more severe phenotypes in which RGC axons make incorrect decisions at the midline, leading to aberrant projections and defasciculation of the retinal projection *(134,136)*. A mutation in an axon guidance receptor, robo2, causes the *astray* phenotype, and is thought to be involved in correction of pathfinding errors of the RGC neurons *(136,137)*. Mutants have also been identified with defects in mapping within the tectum *(15,133–135)*. The topographical map within the retina is formed twice, once during early development when axons first reach the tectum, and later, at 4–6 dpf when it is reorganized as more RGCs send their axons to the tectum *(15)*. Morphologically, defects in the retinal axon rearrangements at 4–6 dpf can be identified by enlarged retinotectal projection fields, as seen in retina of the mutant *macho (15,135)*. Interestingly, the maturation of the retinotectal map has been found to be activity dependent *(15,138)* . RGCs in the *macho* embryo do not fire overshooting action potentials at 4–6 dpf and their phenotype can be copied by raising wild-type zebrafish embryos in solution with a sodium channel blocker, tetrodotoxin *(15,138)*. Deficits in sodium current can also be seen in other sensory cells in the *macho* mutant, and are discussed in the section on sensory mechanotransduction *(11)*.

During development of the retinotectal connections, axo- and synaptogenesis are also occurring between photoreceptor and second-order cells within the retinal outer plexiform layer (OPL) *(125)*. Histological analysis of synapses in the OPL indicates that the structural components for a functional synapse are present at an early stage of development *(125)*. Synaptic proteins SNAP-25 (synaptosomal-associated protein of 25kDA), syntaxin 3 and SV2, a synaptic vesicle transporter protein, are first revealed by immunocytochemistry in the OPL at 2.5 dpf, and increase to adult levels by 10 dpf *(125)*. The onset of synaptic transmission is thought to begin later, between 3.5 and 5 dpf, when synaptic structures at photoreceptor and secondary cell junctions mature *(125)*.

In summary, there are many developmental events that occur during embryogenesis that underlie the formation of the visual system. These include specification and differentiation of retinal cell types as well as creating ordered layers and functional synapses. Forward genetics has enabled the discovery of many molecules involved in these processes, as well as those that are important for the continued survival of retinal cell types in larval and adult models (*see* ref. *139* for review).

2.2.3. Olfaction and Gustation

Next, we consider the olfactory (smell) and gustatory (taste) systems in the developing zebrafish. The organs and cells responsible for smell and taste perception in the zebrafish have been examined anatomically and resemble those found in mammals, with some exceptions *(140–146)*. Furthermore, olfactory sensory neurons have been characterized physiologically *(147–149)*. However, forward genetics screens have not been able to identify many mutants with defects that elucidate the molecular mechanisms underlying both development and function of these systems *(150,151)*. One reason may be that neurons in the tongue and nose have highly regulated gene expression patterns that code for many different and unique olfactory and gustatory cell types. For example, olfactory sensory neurons express one of 100 olfactory receptors and the mutation of just one type may be difficult to detect *(152–154)*. Behavioral assays that can screen for changes to these systems are time intensive and impractical for large-scale screens. Another reason may be that receptor cells in the taste bud have very short life spans, which has hindered detailed analysis of both wild type and mutant states *(155)*. In the future, targeted mutagenesis or knockdown technology may provide a glimpse of the importance of the many genes required for the development of these two senses.

2.2.4. Mechanosensation

The last sensory system that we discuss is that of mechanosensation, specifically with respect to tactile sensation. Mechanosensitive cells reside in both the head and the trunk of the embryo, where they extend peripheral processes into the skin *(3,156)*. During early development sensory neurons in the spinal cord, the Rohon-Beards (R-Bs), and trigeminal ganglia (TG) relay mechanosensitive information from the trunk and head, respectively (Fig. 6 *[3,157]*). The R-Bs are a transient population of cells that mediate touch sensitivity from 27 hpf until the time that they are eliminated by programmed cell death (PCD) by 72–120 hpf *(13,33,47,90,158,159)*. R-B function is taken over by dorsal root ganglia (DRG) neurons *(160)*. At 45 hpf, the DRG contains one to three sensory neurons, which steadily increases to more than 100 neurons over the next month *(161)*. Although DRG and R-B neurons have the same function, it is thought that they are derived from evolutionarily related, but different populations of neural precursors *(162)*. For example, R-Bs are completely absent in the *narrowminded* mutant, whereas DRG neurons are still present, however in a decreased amount *(162)*. Thus, DRG and R-B neurons are under similar developmental influences that differently affect their specific cell types *(162)*. Both R-B and TG neurons are born in the primary wave of neurogenesis, however, TG sensory neurons do not die during development, but remain functional throughout adulthood *(163)*.

The mechanosensory neurons feed in to the complex circuitry responsible for eliciting the escape response, as discussed in Subheading 2.1 *(47)*. Thus, perturbations in the touch response, as assayed by changes in locomotion, are an excellent marker for disruptions in mechanosensory neurons *(65)*. Indeed, a number of mutants have been isolated that have a reduction in the touch response such as the mutant *macho*, which has both defects in the touch response and retinal pathfinding *(11,15,33,65,123)*. As discussed earlier (Subheading 2.2.2), RGCs in *macho* do not acquire developmentally regulated over shooting action potentials, preventing maturation of the topographical map on the tectum *(15)*. Interestingly, R-B neurons, responsible for carrying mechanosensing information to locomotor circuitry, also do not fire overshooting action potentials *(11)*. Electrophysiological data show that in both RGCs and R-Bs, there is a marked reduction of sodium current amplitude, preventing normal action potential generation *(11,15)*. Thus it is likely that the *macho* gene is essential for proper formation of a functional sodium channel *(11,33)*.

R-Bs in *macho* also do not die as early in development, implicating a role for electrical activity in the regulation of PCD *(33)*. R-B PCD has also been shown to be dependent on the expression of the neurotrophin receptor TrkC1, which mediates NT-3 dependent cell survival *(160)*. Endogenous application of NT-3 promotes R-B survival on spinal cord explants *(160)*. Developmentally, *macho* embryos do not respond to a tap on the head, nor do they acquire a touch response later in development, indicating that this mutation also affects TG

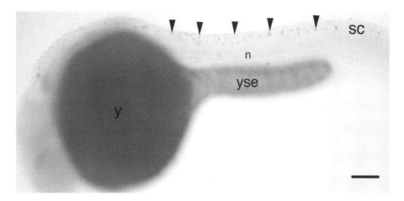

Fig. 6. Rohon-Beard (R-B) neurons mediate the touch response of the zebrafish embryo. Lateral view of a 24 hpf zebrafish embryo, with the head to the left and yolk sac (y) and its extension (yse) rostral and ventral to the spinal cord (sc) and notochord (n). R-B neurons (arrows) can be seen occupying the most dorsal region of the spinal cord of a zebrafish embryo, as seen by *in situ* hybridization detecting voltage-gated sodium channel α gene transcripts. By 120 hpf, most R-Bs have undergone programmed cell death and their function has been replaced by dorsal root ganglia neurons. Scale bar: 100 μM.

and DRG sensory neurons *(33,158)* . In contrast, mutant *fakir* embryos have decreased touch sensitivity at early times, but regain a normal touch response by 120 dpf *(65)*. This suggests that *fakir* affects R-Bs but not DRGs, whereas *macho* acts on both cell types *(65)*.

In summary, the ability to respond to tactile stimuli in the developing zebrafish is dependent on a number of cell types with slightly different overlapping developmental programs. TG neurons are born early and remain functional in the adult fish, whereas R-Bs die during the first 3 d of life, to be replaced by DRG neurons. Defects in mechanosensation can be assayed by a behavioral response to touch during different times of development *(65)*. Mutants with reduced touch responses during early development have been isolated and helped elucidate factors essential for proper mechanosensation *(11,15,33,65,123)*. In addition, mutants such as *macho* have aided in the understanding of the very diverse functions of a single gene.

3. SUMMARY AND FUTURE DIRECTIONS

This chapter has given some examples of how the integrative physiology approach has been used in the zebrafish. As a model, the zebrafish has many traits that allow a comprehensive look at systems in the developing fish using anatomical, molecular, behavioral, and genetic approaches. Integration of these data allow an in-depth understanding of the pathways and processes underlying both normal and perturbed states. The zebrafish nervous system is optimal for understanding both sensory and motor behaviors, with well-defined and simple anatomical structures and circuitry. We have reviewed the sensory systems responsible for hearing, balance, vision, smell, taste, and touch in the developing zebrafish (*see* ref. *96* for review). We have also looked at motor output, which is responsible for the locomotive behaviors of the fish. For each system, a unique set of processes have been elucidated based on the studies that have been undertaken. However, in most cases data has been obtained defining cellular specification, axogenesis, electrical properties, and behaviors in wild-type and mutant embryos and larva. These systems can also be analyzed at a higher level (i.e. separate sensory modalities interact with motor behaviors). For example, both visual cues and tactile stimulation can trigger locomotion *(47,115)*.

The future of the zebrafish developmental model looks bright. With more than 300 mutants representing genes essential for development, the forward genetics approach is allowing the discovery of undescribed genes. Furthermore, new retroviral techniques are being used that allow much easier identification of these mutated genes *(20)*. Traditional reverse genetics approaches are also allowing targeted disruption of genes of interest using morpholino technology *(85,164)*. Cellular physiology is being elucidated with transgenic models and calcium imaging *(165–167)*. The analyses of these experiments are being made possible with the plethora of DNA sequences that are available. In turn, this information can be used to identify novel or orthologous genes to which unique functions can be ascribed. Thus, the zebrafish model is able encompass many roles, from elucidating human disease models with mutant fish, to providing an evolutionary model that describes the divergence of fish and mammals. Finally, the ability to apply integrative approaches to this developmental model has allowed important insights into the development of function of vertebrate sensory and motor systems.

ACKNOWLEDGMENTS

This work was supported by National Institutes of Health grants T32-NS07083 and NS38937 (A.B.R.).

REFERENCES

1. Mullins, M.C. and C. Nüsslein-Volhard (1993) Mutational approaches to studying embryonic pattern formation in the zebrafish. *Curr. Op. Gen. Devel.* **3**:648–654.
2. Kuwada, J.Y. and R.R. Bernhardt (1990) Axonal outgrowth by identified neurons in the spinal cord of zebrafish embryos. *Exp. Neurol.* **109**(1):29–34.
3. Bernhardt, R.R., A.B. Chitnis, L. Lindamer and J.Y. Kuwada. (1990) Identification of spinal neurons in the embryonic and larval zebrafish. *J. Comp. Neurol.* **302**(3):603–616.
4. Kimmel, C.B. (1993) Patterning the brain of the zebrafish embryo. *Ann. Rev. Neurosci.* **16**:707–732.
5. Inoue, A., M. Takahashi, K. Hatta, Y. Hotta and H. Okamoto. (1994) Developmental regulation of islet-1 mRNA expression during neuronal differentiation in embryonic zebrafish. *Devel. Dyn.* **199**:1–11.
6. Martin, S.C., G. Heinrich and J.H. Sandell (1998) Sequence and expression of glutamic acid decarboxylase isoforms in the developing zebrafish. *J. Comp. Neurol.* **396**(2):253–66.
7. Nguyen, V.H., J. Trout, S.A. Connors, P. Andermann and E. Weinberg. (2000) Dorsal and intermediate neuronal cell types of the spinal cord are established by a BMP signaling pathway. *Devel.* **127**(6):1209–1220.
8. Hivert, B., Z. Liu, C.Y. Chuang, P. Doherty and V. Sundaresan. (2002) Robo1 and Robo2 are homophilic binding molecules that promote axonal growth. *Mol. Cell. Neurosci.* **21**(4):534–545.
9. Kurita, R., H. Sagara, Y. Aoki, B.A. Link, K. Arai and S. Watanabe. (2003) Suppression of lens growth by alphaA-crystallin promoter-driven expression of diptheria toxin results in disruption of retinal cell organization in zebrafish. *Devel. Biol.* **255**(1):113–127.
10. Legendre, P.A. (1998) A reluctant gating model of glycine receptor channels determines the time course of inhibitory miniature synaptic events in zebrafish hindbrain neurons. *J. Neurosci.* **18**:2856–2870.
11. Ribera, A.B. and C. Nüsslein-Volhard (1998) Zebrafish touch-insensitive mutants reveal an essential role for the developmental regulation of sodium current. *J. Neurosci.* **18**(22):9181–9191.

12. Drapeau, P., D.W. Ali, R.R. Buss and L. Saint-Amant. (1999) *In vivo* recording from identifiable neurons of the locomotor network in the developing zebrafish. *J. Neurosci. Meth.* **88**:1–13.
13. Saint-Amant, L. and P. Drapeau (2000) Motoneuron activity patterns related to the earliest behavior of the zebrafish embryo. *J. Neurosci.* **20**(11):3964–3972.
14. Saint-Amant, L. and P. Drapeau (2001) Synchronization of an embryonic network of identified spinal interneurons solely by electric coupling. *Neuron* **31**:1035–1046.
15. Gnuegge, L., S. Schmid and S.C.F. Neuhauss (2001) Analysis of the activity-deprived zebrafish mutant *macho* reveals an essential requirement of neuronal activity for the development of a fine-grained visuotopic map. *J. Neurosci.* **21**(10):3542–3548.
16. Spitzer, N.C. and J.E. Lamborghini (1976) The development of the action potential mechanism of amphibian neurons isolated in culture. *Proc. Natl. Acad. Sci.* **73**(5):1641–1645.
17. Mullins, M.C. and C. Nüsslein-Volhard (1993) Mutational approaches to studying embryonic pattern formation in the zebrafish. *Curr. Op. Genet. Devel.* **3**:648-654.
18. Currie, P.D. (1996) Zebrafish genetics:mutant cornucopia. *Curr. Biol.* **6**(12):1548-1552.
19. Haffter, P., M. Granato, M. Brand, et al. (1996) The identification of genes with unique and essential functions in the development of the zebrafish, *Danio rerio*. *Devel.* **123**:1–36.
20. Golling, G., A. Amsterdam, Z. Sun, et al. (2002) Insertional mutagenesis in zebrafish rapidly identifies genes essential for early vertebrate development. *Nature Genet.* **31**:135–140.
21. Driever, W., L. Solnica-Krezel, A.F. Schier, et al. (1996) A genetic screen for mutations affecting embryogenesis in zebrafish. *Devel.* **123**:37–46.
22. Linares, A.E. (2002) Characterization of voltage-gated sodium channel genes expressed in zebrafish embryos. Ph.D. thesis in Neuroscience. University of Colorado Health Sciences Center:Denver.
23. Callahan, P. (2003) Building a better lab mouse - The zebrafish. In *Wall Street J.*
24. Roush, W. (1997) A Zebrafish Genome Project? *Science* **275**:923.
25. Amemiya, C.T., T.P. Zhong, G.A. Silverman, M.C. Fishman and L.I. Zon. (1999) Zebrafish YAC, BAC and PAC genomic libraries. *Methods Cell Biol.* **60**:235–258.
26. Lo, J., S. Lee, M. Xu, F, et al. (2003) 15000 unique zebrafish EST clusters and their future use in microarray for profiling gene expression patterns during embryogenesis. *Genome Res.* **13**(3):455–466.
27. Endo, A. and T.H. Ingalls (1968) Chromosomes of the zebrafish. A model for cytogenetic, embryologic and ecologic study. *J. Hered.* **59**:382–384.
28. Daga, R.R., G. Thode and A. Amores (1996) Chromosome complement C-banding, Ag-NOR and replication banding in the zebrafish *Danio rerio*. *Chrom. Res.* **6**(1):29–32.
29. Knapik, E.W., A. Goodman, M. Ekker, et al. (1998) A microsatellite genetic linkage map for zebrafish (*Danio rerio*). *Nature Gen.* **18**(4):338–343.
30. Shimoda, N., E.W. Knapik, J. Ziniti, et al. (1999) Zebrafish genetic map and 2000 microsatellite markers. *Genomics* **58**(3):219–232.
31. Barbazuk, W.B., I. Korf, C. Kadavi, et al. (2000) The syntenic relationship of the zebrafish and human genomes. *Gen. Res.* **10**:1351–1358.
32. Postlethwait, J., M. Ekker, K. Frazer, M. Mullins and M. Westerfield. (2000) To orthologue or not to orthologue, that is the question. *Zebrafish Sci. Monitor* **7**(1).
33. Svoboda, K.R., A.E. Linares and A.B. Ribera (2001) Activity regulates programmed cell death of zebrafish Rohon-Beard neurons. *Devel.* **128**:3511–3520.
34. Svoboda, K.R., S. Vijayaraghavan and R.L. Tanguay (2002) Nicotinic receptors mediate changes in spinal motoneuron development and axonal pathfinding in embryonic zebrafish exposed to nicotine. *J. Neurosci.* **22**(24):10731–10741.
35. Liu, Y.-W. and W.-K. Chan (2002) Thyroid hormones are important for embryonic to larval transitory phase in zebrafish. *Diff.* **70**:36-45.
36. Wittbrodt, J., A. Meyer and M. Schartl (1998) More genes in fish? *Bioessays* **20**:511–515.
37. Holland, P.W. and N.A. Williams (1990) Conservation of engrailed-like homeobox sequences during vertebrate evolution. *FEBS Letts.* **277**(1):250–252.
38. Ekker, S.C., A.R. Ungar, P. Greenstein, et al. (1995) Patterning activities of vertebrate hedgehog proteins in the developing eye and brain. *Curr. Biol.* **5**(8):944–955.

39. Stock, D.W., D.L. Ellies, Z. Zhao, M. Ekker, F.H. Ruddle and K.M. Weiss. (1996) The evolution of the vertebrate Dlx family. *Proc. Natl. Acad. Sci.* **93**:10858–10863.

40. Amores, A., Force, Y.-L. Yan, L. (1998) Zebrafish *hox* clusters and vertebrate genome evolution. *Science* **282**:1711–1714.

41. Myer, A. and E. Malaga-Trillo (1999) Vertebrate genomics: more fishy tales about *Hox* genes. *Curr. Biol.* **9**:R210–R213.

41a. Jaillon, O., J.-M. Aury, F. Brunet, et al. (2004) Genome duplication in the teleost fish *Tetraodon nigroviridis* reveals the early vertebrate proto-karyotype. *Nature* **431**:946–957.

42. Ruddle, F.H., J.L. Bartels, K.L. Bentley, C. Kappen, M.T. Murtha and J.W. Pendleton. (1994) Evolution of *Hox* genes. *Annu. Rev. Genet.* **28**:423–442.

43. Skrabanek, L. and K.H. Wolfe (1998) Eukaryote genome duplication - where's the evidence? *Curr. Op. Gen. Devl.* **8**:694–700.

44. Hughes, M.K. and A.L. Hughes (1993) Evolution of duplicate genes in a tetraploid animal, *Xenopus laevis. Mol. Biol. Evol.* **10**:1360–1369.

45. Robinson-Rechavi, M., O. Marchand, H. Escriva, P.-L. Bardet, S. Hughes and V. Laudet. (2001) Euteleost fish genomes are characterized by expansion of gene families. *Genome Res.* **11**:781–788.

46. Van de Peer, Y., J.S. Taylor, J. Joseph and A. Myer. (2002) Wanda:a database of duplicated fish genes. *Nuc. Acids Res.* **30**(1):109–112.

47. Saint-Amant, L. and P. Drapeau (1998) Time course of the development of motor behaviors in the zebrafish embryo. *J. Neurobiol.* **37**:622–632.

48. Bate, M. (1999) Development of motor behavior. *Curr. Opin. Neurobiol.* **9**:670–675.

49. Budick, S.A. and D.M. O'Malley (2000) Locomotor repertoire of the larval zebrafish: Swimming, turning and prey capture. *J. Exp. Biol.* **203**:2565-2579.

50. Drapeau, P., L. Saint-Amant, R.R. Buss, M. Chong, J.R. McDearmid and E. Brustein. (2002) Development of the locomotor network in zebrafish. *Prog. Neurobiol.* **68**:85–111.

51. Lewis, K.E. and J.S. Eisen (2003) From cells to circuits:development of the zebrafish spinal cord. *Prog. Neurobiol.* **69**(6):419–449.

52. Metcalfe, W.K., B. Mendelson and C.B. Kimmel (1986) Segmental homologies among reticulospinal neurons in the hindbrain of the zebrafish larva. *J. Comp. Neurol.* **251**:147–159.

53. Kimmel, C.B. (1982) Reticulospinal and vestibulospinal neurons in the young larva of a teleost fish, *Brachydanio rerio. Prog. Brain. Res.* **57**:1–23.

54. Kimmel, C.B., K. Hatta and J.S. Eisen (1991) Genetic control of primary neuronal development in zebrafish. *Devel.* **Suppl**(2):47–57.

55. Gahtan, E. and D.M. O'Malley (2003) Visually guided injection of identified reticulospinal neurons in zebrafish:a survey of spinal arborization patterns. *J. Comp. Neurol.* **459**:186–200.

56. Mendelson, B. (1986) Development of reticulospinal neurons of the zebrafish. I. Time of origin. *J. Comp. Neurol.* **251**:160–171.

57. Mendelson, B. (1985) Soma position is correlated with time of development in three types of identified reticulospinal neurons. *Devel. Biol.* **112**:489–493.

58. Mendelson, B. and C.B. Kimmel (1986) Identified vertebrate neurons that differ in axonal projection develop together. *Devl. Biol.* **118**(1):309–313.

59. Liu, K. and J.R. Fetcho (1999) Laser ablation reveals functional relationships of segmental hindbrain neurons in zebrafish. *Neuron* **23**:325–335.

60. Myers, P.Z., J.S. Eisen and M. Westerfield (1986) Development and axonal outgrowth of identified motoneurons in the zebrafish. *J. Neurosci.* **6**(8):2278–2289.

61. Hale, M.E., D.A. Ritter and J.R. Fetcho (2001) A confocal study of spinal interneurons in living larval zebrafish. *J. Comp. Neurol.* **437**:1–16.

62. Eisen, J.S. (1991) Motoneuronal development in the embryonic zebrafish. *Devel.* **Suppl**(2):141–147.

63. Eisen, J.S., S.H. Pike and B. Romancier (1990) An identified motoneuron with variable fates in embryonic zebrafish. *J. Neurosci.* **10**(1):34–43.

64. Bernhardt, R.R. (1999) Cellular and Molecular bases of axonal pathfinding during embryogenesis of the fish central nervous system. *J. Neurobiol.* **38**:137–160.

65. Granato, M., F.J. van Eeden, U. Schach, T., et al. (1996) Genes controlling and mediating locomotion behavior of the zebrafish embryo and larva. *Devel.* **123**:399–413.

66. Beattie, C.E. (2000) Control of motor axon guidance in the zebrafish embryo. *Brain Res. Bull.* **53**(5):489–500.

67. Zeller, J. and M. Granato (1999) The zebrafish diwanka gene controls an early step of motor growth cone migration. *Devel.* **126**:3461–3472.

68. Zhang, J. and M. Granato (2000) The zebrafish unplugged gene controls motor axon pathfinding selection. *Devel.* **127**:2099–2111.

69. Zhang, J., S. Malayaman, C. Davis and M. Granato. (2001) A dual role for the zebrafish unplugged gene in motor axon pathfinding and pharyngeal development. *Devel. Biol.* **240**:560–573.

70. McWhorter, M.L., U.R. Monani, A.H. Burghes and C.E. Beattie. (2003) Knockdown of the survival motor neuron (Smn) protein in zebrafish causes defects in motor axon outgrowth and patterning. *J. Cell Biol.* **162**(5):919–931.

71. Roberts, D.F., J. Chavez and S.D.M. Court (1970) The genetic component in child mortality. *Arch. Dis. Child* **45**:33–38.

72. Crawford, T.O. and C.A. Pardo (1996) The neurobiology of childhood spinal muscular atrophy. *Neurobiol.* **3**:97–110.

73. Melki, J. (1997) Spinal muscular atrophy. *Curr. Opin. Neurol.* **10**(381–385).

74. Terns, M.P. and R.M. Terns (2001) Macromolecular complexes:SMN-the master assembler. *Curr. Biol.* **11**:R862–R864.

75. Rossoll, W., A.K. Kroning, U.M. Ohndorf, C. Steegborn, S. Jablonka and M. Sendtner. (2002) Specific interaction of Smn, the spinal muscular atrophy determining gene product, with hnRNP-R and gry-rbp/hnRNP-Q:a role for Smn in RNA processing in motor axons? *Hum. Mol. Genet.* **11**:93–105.

76. Buss, R.R., C.W. Bourque and P. Drapeau (2003) Membrane properties related to the firing behavior of zebrafish motoneurons. *J. Neurophysiol.* **89**:657–664.

77. Grinnell, A.D. (1995) Dynamics of nerve-muscle interaction in developing and mature neuromuscular junctions. *Physiol. Rev.* **75**(4):789–834.

78. Cohen-Cory, S. (2002) The developing synapse:construction and modulation of synaptic structures and circuits. *Science* **298**(5594):770–776.

79. Hoch, W. (2003) Molecular dissection of neuromuscular junction formation. *Trends Neurosci.* **26**(7):67–77.

80. Chakkalakal, J.V. and B.J. Jasmin (2003) Localizing synaptic mRNAs at the neuromuscular junction:it takes more than transcription. *Bioess.* **25**(1):25–31.

81. Westerfield, M., D.W. Liu, C.B. Kimmel and C. Walker. (1990) Pathfinding and synapse formation in a zebrafish mutant lacking functional acetylcholine receptors. *Neuron* **4**(6):867–874.

82. Sepich, D.S., J. Wegner, S. O'Shea and M. Westerfield. (1998) An altered intron inhibits synthesis of the acetylcholine receptor alpha subunit in the paralyzed zebrafish mutant *nic1*. *Genetics* **148**:361–372.

83. Ono, F., S. Higashijima, A. Shcherbatko, J.R. Fetcho and P. Brehm. (2001) Paralytic zebrafish lacking acetylcholine receptors fail to localize rapsyn clusters to the synapse. *J. Neurosci.* **21**(15):5439–5448.

84. Sepich, D.S., R.K. Ho and M. Westerfield (1994) Autonomous expression of the *nic1* acetylcholine receptor mutation in zebrafish muscle cells. *Devl. Biol.* **161**:84–90.

85. Brent, L.J. and P. Drapeau (2002) Targeted 'knockdown' of channel expression in vivo with an antisense morpholino oligonucleotide. *Neurosci.* **114**(2):275–278.

86. Burden, S.J., R.L. DePalma and G.S. Gottesman (1983) Crosslinking of proteins in acetylcholine receptor-rich membranes:association between the beta-subunit and the 43 kd subsynaptic protein. *Cell* **35**:687–692.

87. Froehner, S.C., C.W. Luetje, P.B. Scotland and J. Patrick. (1990) The postsynaptic 43K protein clusters muscle nicotinic acetylcholine receptors in *Xenopus* oocytes. *Neuron* **5**:403–410.

88. Gautam, M., P.G. Noakes, J. Mudd, et al. (1995) Failure of postsynaptic specialization to develop at neuromuscular junctions of rapsyn-deficient mice. *Nature* **377**:232–236.

89. Ono, F., A. Shcherbatko, S. Higashijima, G. Mandel and P. Brehm. (2002) The zebrafish motility mutant twitch once reveals new roles for rapsyn in synaptic function. *J. Neurosci.* **22**(15):6491–6498.

90. Grunwald, D.J., C.B. Kimmel, M. Westerfield, C. Walker and G. Streisinger. (1988) A neural degeneration mutation that spares primary neurons in the zebrafish. *Devl. Biol.* **126**(1):115–128.

91. O'Malley, D.M., Y.-H. Kao and J.R. Fetcho (1996) Imaging the functional organization of zebrafish hindbrain segments during escape behaviors. *Neuron* **17**:1145–1155.

92. Kimmel, C.B., R.C. Eaton and S.L. Powell (1980) Decreased fast-start performance of zebrafish larvae lacking Mauthner neurons. *J. Comp. Physiol.* **140**:343–350.

93. Gahtan, E., N. Sankrithi, J.B. Campos and D.M. O'Malley. (2002) Evidence for a widespread brain stem escape network in larval zebrafish. *J. Neurophysiol.* **87**:608–614.

94. Borla, M.A., B. Palecek, S. Budick and D.M. O'Malley. (2002) Prey capture by larval zebrafish:Evidence for fine axial motor control. *Brain Behav. Evol.* **60**:207–229.

95. Ritter, D.A., D.H. Bhatt and J.R. Fetcho (2001) *In vivo* imaging of zebrafish reveals differences in the spinal networks for escape and swimming movements. *J. Neurosci.* **21**(22):8956–8965.

96. Moorman, S.J. (2001) Development of sensory systems in zebrafish (*Danio rerio*). *ILAR* **42**(4):292–298.

97. Bryant, J., R.J. Goodyear and G.P. Richardson (2002) Sensory organ development in the inner ear: molecular and cellular mechanisms. *Brit. Med. Bull.* **63**:39–57.

98. Bever, M.M. and D.M. Fekete (2002) Atlas of the developing inner ear in zebrafish. *Devl. Dyn.* **223**(4):536–543.

99. Whitfield, T.T., B.B. Riley, M.Y. Chiang and B. Phillips. (2002) Development of the zebrafish inner ear. *Devl. Dyn.* **223**(4):427–458.

100. Whitfield, T. (2002) Zebrafish as a model for hearing and deafness. *J. Neurobiol.* **53**:157–171.

101. Metcalfe, W.K., C.B. Kimmel and E. Schabtach (1985) Anatomy of the posterior lateral line system in young larvae of the zebrafish. *J. Comp. Neurol.* **233**(3):377–389.

101a. Trevarrow, B., D.L. Marks and C.B. Kimmel (1990) Organization of hindbrain segments in the zebrafish embryo. *Neuron* **4**(5):669–679.

102. Kornblum, H.I., J.T. Corwin and B. Trevarrow (1990) Selective labeling of sensory hair cells and neurons in auditory, vestibular and lateral line systems by a monoclonal antibody. *J. Comp. Neurol.* **301**:162–170.

103. Gompel, N., N. Cubedo, C. Thisse, B. Thisse, C. Dambly-Chaudiere and A. Ghysen. (2001) Pattern formation in the lateral line of zebrafish. *Mech. Devl.* **105**(1-2):69–77.

104. Haddon, C. and J. Lewis (1996) Early ear development in the embryo of the zebrafish, *Danio rerio*. *J. Comp. Neurol.* **365**(1):113–128.

105. Gompel, N., C. Dambly-Chaudiere and A. Ghysen (2001) Neuronal differences prefigure somatotopy in the zebrafish lateral line. *Devl.* **128**:387–393.

106. Riley, B.B. and J. Moorman (2000) Development of utricular otoliths, but not saccular otoliths, is necessary for vestibular function and survival in zebrafish. *J. Neurobiol.* **43**:329–337.

107. Malicki, J., A.F. Schier, L. Solnica-Krezel S, et al. (1996) Mutations affecting development of the zebrafish ear. *Development* **123**:275–83.

108. Whitfield, T.T., M. Granato, F.J.M. van Eeden., et al. (1996) Mutations affecting development of the inner ear and lateral line. *Development* **123**:241–54.

109. Price, E.R. and D.E. Fisher (2001) Sensorineural deafness and pigmentation genes:melanocytes and the Mitf transcriptional network. *Neuron* **30**:15–18.

110. Nicolson, T., A. Rusch, R.W. Friedrich, M. Granato, J.P. Ruppersberg and C. Nüsslein-Volhard. (1998) Genetic analysis of vertebrate sensory hair cell mechanosensation: the zebrafish *circler* mutants. *Neuron* **20**:271–283.

111. Ernest, S., G.J. Rauch, P. Haffter, R. Geisler, C. Petit and T. Nicolson. (2000) *Mariner* is defective in myosin VIIA:a zebrafish model for human hereditary deafness. *Hum. Mol. Genet.* **9**(14):2189–2196.

112. Sollner, C., G.J. Rauch, J. Sieman, et al. Mutations in cadherin 23 affect tip links in zebrafish sensory hair cells. *Nature* **428**:955–959.

113. Riley, B.B. and D.J. Grunwald (1996) A mutation in zebrafish affecting a localized cellular function required for normal ear development. *Devl. Biol.* **179**:427–435.

114. Sidi, S., R.W. Friedrich and T. Nicolson (2003) NompC TRP channel required for vertebrate sensory hair cell mechanotransduction. *Science* **301**(5629):96–99.

115. Easter, S.S. and G.N. Nicola (1996) The development of vision in the Zebrafish (*Danio rerio*). *Devel. Biol.* **180**:646–663.

116. Schmitt, E.A. and J.E. Dowling (1999) Early retinal development in the zebrafish *danio rerio*:Light and electron microscopic analyses. *J. Comp. Neurol.* **404**:515–536.

117. Malicki, J. (1999) Development of the retina. *Methods Cell Biol.* **59**:273–299.

118. Hu, M. and S.S. Easter (1999) Retinal neurogenesis:the formation of the initial central patch of postmitotic cells. *Dev Biol* **207**(2):309–321.

119. Marc, R.E. and D. Cameron (2001) A molecular phenotype atlas of the zebrafish retina. *J. Neuorcytol.* **30**:593–654.

120. Vihtelic, T.S. and D.R. Hyde (2002) Zebrafish mutagenesis yields eye morphological mutants with retinal and lens defects. *Vision Res.* **42**:535–540.

121. Brockerhoff, S.E., J.B. Hurley, U. Janssen-Bienhold, S.C.Neuhauss, W. Deriever and J.E. Dowling. (1995) A behavioral screen for isolating zebrafish mutants with visual system defects. *Proc. Natl. Acad. Sci.* **92**:10545–10549.

122. Li, L. and J.E. Dowling (1997) A dominant form of inherited retinal degeneration caused by a non-photoreceptor cell-specific mutation. *Proc. Natl. Acad. Sci.* **94**:11645–11650.

123. Neuhauss, S.C.F., O. Biehlmaier, M.W. Seeliger, et al. (1999) Genetic disorders of vision revealed by a behavioral screen of 400 essential loci in zebrafish. *J. Neurosci.* **19**(19):8603–8615.

124. Nawrocki, L.W. (1985) Development of the neural retina in the zebrafish, *Brachydanio rerio*. Ph.D. Thesis - University of Oregon:Eugene.

125. Biehlmaier, O., S.C.F. Neuhauss and K. Kohler (2003) Synaptic plasticity and functionality at the cone terminal of the developing zebrafish retina. *J. Neurobiol.* **56**(3):222–236.

126. Kay, J.N., *K.C.* Finger-Baier, T. Roeser, W. Staub and H. Baier. (2001) Retinal ganglion cell genesis requires lakritz, a zebrafish atonal homologue. *Neuron* **30**:725–736.

127. Schauerte, H.E., F.J. van Eeden, C. Fricke, J. Odenthal, U. Strahle and P. Haffter. (1998) Sonic hedgehog is not required for the induction of medial floor plate cells in the zebrafish. *Development* **125**(15):2983–2993.

128. Stenkamp, D.L., R.A. Frey, D.E. Mallory and E.E. Shupe. (2002) Embryonic retinal gene expression in *sonic-you* mutant zebrafish. *Devl. Dyn.* **225**:344–350.

129. Kelsh, R.N., M. Brand, Y.J. Jiang, et al. (1996) Zebrafish pigmentation mutations and the processes of neural crest development. *Development* **123**:369–389.

130. Neumann, C.J. and C. Nüsslein-Volhard (2000) Patterning of the zebrafish retina by a wave of sonic hedgehog activity. *Science* **289**:2137–2139.

131. Stuermer, C.A.O. (1988) Retinotopic organization of the developing retinotectal projection in the zebrafish embryo. *J. Neurosci.* **8**:4513–4530.

132. Burrill, J.D. and S.S. Easter (1994) Development of the retinofugal projections in the embryonic and larval zebrafish (*Brachydanio rerio*). *J. Comp. Neurol.* **346**:583–600.

133. Baier, H., *S.* Klostermann, T. Trowe, R.O. Karlstrom, C. Nüsslein-Volhard and F. Bonhoeffer. (1996) Genetic dissection of the retinotectal projection. *Devel.* **123**:415–425.

134. Karlstrom, R.O., T. Trowe, S. Klostermann, et al. (1996) Zebrafish mutations affecting retinotectal axon pathfinding. *Devel.* **123**:427–438.

135. Trowe, T., *S.*. Klostermann, H. Baier, et al. (1996) Mutations disrupting the ordering and topographic mapping of axons in the retinotectal projection of the zebrafish, *Danio rerio*. *Development* **123**:439–450.

136. Fricke, C., J.S. Lee, S. Geiger-Rudolph, F. Bonhoeffer and C.B. Chien. (2001) *astray*, a zebrafish roundabout homologue required for retinal axon guidance. *Science* **292**(5516):507–510.

137. Hutson, L.D. and C.B. Chien (2002) Pathfinding and error correction by retinal axons:the role of astray robo2. *Neuron* **33**(2):205–217.

138. Stuermer, C.A.O., B. Rohrer and H. Münz (1990) Development of the retinotectal projection in zebrafish embryos under TTX-induced neural impulse blockade. *J. Neurosci.* **10**(11):3615–3626.

139. Malicki, J. (2000) Harnessing the power of forward genetics - analysis of neuronal diversity and patterning in the zebrafish retina. *Trends Neurosci.* **23**:531–541.

140. Baier, H., S. Rotter and S. Korsching (1994) Connectional topography in the zebrafish olfactory system:Random positions but regular spacing of sensory neurons projecting to an individual glomerulus. *Proc. Natl. Acad. Sci.* **91**:11646–11650.

141. Byrd, C.A. and P.C. Brunjes (1995) Organization of the olfactory system in the adult zebrafish:histological, immunohistochemical and quantitative analysis. *J. Comp. Neurol.* **358**(2):247–259.

142. Friedrich, R.W. and S.I. Korsching (1997) Combinatorial and chemotopic odorant coding in the zebrafish olfactory bulb visualized by optical imaging. *18* (737–752).

143. Vogt, R.G., S.M. Lindsay, C.A. Byrd and M. Sun. (1997) Spatial patterns of olfactory neurons expressing specific odor receptor genes in 48-hour old embryos of zebrafish Danio Rerio. *J. Exp. Biol.* **200**:433–443.

144. Yoshihara, Y., H. Nagao and K. Mori (2001) Sniffing out odors with multiple dendrites. *Science* **291**(5505):835–837.

145. Hansen, A., K. Reutter and E. Zeiske (2002) Taste bud development in the zebrafish, *danio rerio*. *Devl. Dyn.* **223**:483–496.

146. Yoshida, T., A. Ito, N. Matsuda and M. Mishina. (2002) Regulation by protein kinase A switching of axonal pathfinding of zebrafish olfactory sensory neurons through the olfactory-placode bulb boundary. *J. Neurosci.* **22**(12):4964–4972.

147. Corotto, F.S., D.R. Piper, N. Chen and W.C. Michel. (1996) Voltage and Ca2+-gated currents in zebrafish olfactory receptor neurons. *J. Exp. Biol.* **199**:1115–1126.

148. Fuss, S.H. and S.I. Korsching (2001) Odorant feature detection:activity mapping of structure response relationships in the zebrafish olfactory bulb. *J. Neurosci.* **21**(21):8396–8407.

149. Friedrich, R.W. and G. Laurent (2001) Dynamic optimization of odor representations by slow temporal patterning of mitral cell activity. *Science* **291**:889–894.

150. Li, L. and J.E. Dowling (2000) Disruption of the olfactoretinal centrifugal pathway may relate to the visual system defect in *night blindness b* mutant zebrafish. *J. Neurosci.* **20**(5):1883–1892.

151. Sanders, L.H. and K.E. Whitlock (2003) Phenotype of the zebrafish *masterblind* (*mbl*) mutant is dependent on genetic background. *Devl. Dyn.* **227**:291–300.

152. Ngai, J., M.M. Dowling, L. Buck, R. Axel and A. Chess. (1993) The family of genes encoding odorant receptors in the channel catfish. *Cell* **72**:657–666.

153. Barth, A.L., N.J. Justice and J. Ngai (1996) Asynchronous onset of odorant receptor expression in the developing zebrafish olfactory system. *Neuron* **16**:23–34.

154. Korsching, S.I., S. Argo, H. Campenhausen, R.W. Friedrich, A. Rummrich and F. Weth. (1997) Olfaction in zebrafish:what does a tiny teleost tell us. *Sem. Cell Devel. Biol.* **8**:181–187.

155. Nicholls, J.G., A.R. Martin and B.G. Wallace (1992)*From Neuron to Brain*. (3 ed. , Sunderland, MA:Sinauer Associates, Inc.

156. Kimmel, C.B. and M. Westerfield, *Primary neurons of the zebrafish.*, in *Signals and Sense:Local and Global Order in Perceptual Maps*, G.M. Edelman, W.E. Gall and W.H. Cowan, Editors. 1990, John Wiley and Sons:New York. p. 561–588.

157. Metcalfe, W.K., P.Z. Myers, B. Trevarrow, M.B. Bass and C.B. Kimmel. (1990) Primary neurons that express the L2/HNK-1 carbohydrate during early development in the zebrafish. *Development* **110**:491–504.
158. Kimmel, C.B., W.W. Ballard, S.R.Kimmel, B. Ullamann and T.F. Schilling. (1995) Stages of embryonic development of the zebrafish. *Devel. Dyn.* **203**:253–310.
159. Cole, L.K. and L.S. Ross (2001) Apoptosis in the developing zebrafish embryo. *Devel. Biol.* **240**:123–142.
160. Williams, J.A., A. Barrios, C. Gatchalian, L. Rubin, S.W. Wilson and N. Holder. (2000) Programmed cell death in zebrafish Rohon Beard neurons is influenced by TrkC1/NT-3 signaling. *Devl. Biol.* **226**:220–230.
161. An, M., R. Luo and P.D. Henion (2002) Differentiation and maturation of zebrafish dorsal root and sympathetic ganglion neurons. *J. Comp. Neurol.* **446**:267–275.
162. Artinger, K.B., A.B. Chitnis, M. Mercola and W. Driver. (1999) Zebrafish *narrowminded* suggests a genetic link between formation of neural crest and primary sensory neurons. *Devl.* **126**(18):3969–3979.
163. Lazarov, N.E. (2002) Comparative analysis of the chemical neuroanatomy of the mammalian trigeminal ganglion and mesencephalic trigeminal nucleus. *Prog. Neurobiol.* **66**:19–59.
164. Nasevicius, A. and S.C. Ekker (2000) Effective targeted gene 'knockdown' in zebrafish. *Nature Genetics* **26**(2):216–220.
165. Park, H.-C., C.-H. Kim, Y.-K. Bae, et al. (2000) Analysis of upstream elements in the HuC promoter leads to establishment of transgenic zebrafish with fluorescent neurons. *Devl. Biol.* **227**:279–293.
166. Higashijima, S.-I., M.A. Masino, G. Mandel and J.R. Fetcho. (2003) Imaging neuronal activity during zebrafish behavior with a genetically encoded calcium indicator. *J. Neurophysiol.* **90**(6):3986–3997.
167. O'Malley, D.M., Q. Zhou and E. Gahtan (2003) Probing neural circuits in the zebrafish:a suite of optical techniques. *Methods* **30**:49–63.

13

Curriculum Design for Integrative Physiology

Dee U. Silverthorn and Penelope A. Hansen

1. INTRODUCTION

As research in the 21st century shifts its focus from molecular biology to functional genomics, proteomics, and integrated system functions, physiologists have an opportunity to position integrative physiology as the organizing discipline of biology. The challenge is twofold: to educate practicing scientists about the importance of integrative physiology in the future of biological research, and to excite and recruit students to become the next generation of integrative physiologists. The preceding chapters in this book are designed to achieve the first objective; this chapter outlines some considerations for the second.

To inspire students to become integrative physiologists, we must teach them well. Fortunately, thanks to a growing body of research in learning science, we now have a better understanding of what comprises good teaching and of how good teaching takes advantage of what we know about how people learn. The ideal curriculum for training adaptive physiologists will tap into what students already know about biology and help them pull disparate facts together to create a "big picture" organized around the fundamental principles that govern the functioning of living organisms. Ideally, the integrative physiology curriculum will also include activities that challenge students to develop the ability and confidence to become adaptive experts *(1,2)*.

Unfortunately, the ideal curriculum is not reflected by the current state of most physiology education, at least in the United States. In the last two decades of the 20th century, the focus on molecular biology research and the shift away from integrative studies had a negative impact on the teaching of physiology, particularly in undergraduate or pre-baccalaureate institutions. Comparative and animal physiology courses lost popularity, and vertebrate or mammalian physiology classes came to be viewed as "service courses," necessary for the training of premedical students and other health professionals, but peripheral to the training of future research scientists. Indeed, a study panel of the US National Academy of Sciences, funded by the National Science Foundation (NSF) and Howard Hughes Medical Institute, found that by the year 2000, physiology had disappeared from the core curriculum for undergraduate biology students at 104 institutions surveyed *(3,4)*.

This shift in the importance of physiology in the biology curriculum, from a core course in the 1970s to an elective subject in the 21st century, in part may be owing to the traditional organ-systems-based approach to teaching physiology, which has led some to regard physi-

From: *Integrative Physiology in the Proteomics and Post-Genomics Age*
Edited by: W. Walz © Humana Press Inc., Totowa, NJ

ology as a "mature" discipline. Human and medical physiology textbooks of the 1980s and early 1990s largely overlooked the exciting role that molecular biology and contemporary biomedical research were playing in our understanding of how living organisms work, compounding the schism between the classroom and the research laboratory.

The stagnation of physiology teaching may also reflect the crisis that physiologists were undergoing in their professional lives, trying to revamp their research programs to include molecular biology methods and technology. It was more fashionable to be a molecular biologist than a physiologist, and in an academic culture that favors research over education, the disappearance of identifiable physiologists meant a disappearance of physiology courses. In US medical schools, the crisis extended to the department level, with departments of physiology being replaced or renamed.

Now, however, as physiology regains stature in research endeavors that require an integrative perspective, physiology should find a new place in the biology core curriculum. The challenge teachers face will be to make biology courses reflect the role of integrative physiology in contemporary biomedical research. A well-designed contemporary course in physiology or introductory biology should integrate traditional organ-systems physiology and its focus on homeostasis and complex interactions with the molecular and cellular phenomena that underlie control mechanisms for the organ systems *(see,* e.g., refs. *5–9)* . It should also emphasize mathematical and biophysical principles so that students can understand the complexity and interdisciplinary nature of biological processes *(3).*

More than course content needs to be updated. Physiologists should also be applying the skills and thought processes they use in their bench research to improving and monitoring learning in their classrooms *(10).* Fortunately, the academic culture is changing, and the scholarship of teaching and learning is receiving more recognition as a valid scholarly activity within the academic community (for a recent annotated bibliography on this subject, *see* ref. *11).* In the sections that follow, we present suggestions and resources for teachers who wish to revamp their curricula to incorporate integrative physiology. First, we examine some of the core issues for curriculum design. We then discuss the three steps of the design process: establishing learning objectives for the course, selecting teaching and learning strategies that most effectively allow students to meet those objectives, and designing appropriate assessment. All three steps should be dynamic processes, constantly subject to scrutiny and revision. In the final section of this chapter we address the importance of disseminating the results of local curriculum revisions. Just as in the world of bench research, the results of curriculum change should be critically reviewed by competent peers who can provide insight, and published so that colleagues can benefit from each others' work.

2. APPROACHES TO CURRICULUM DESIGN

Physiologists should approach their teaching with the same scholarly strategies that they use in their bench research. We need first to decide what we want to achieve in a particular class, think about how we will know when we have achieved it, then decide what content and strategies we should use to meet the learning objectives we have set. This approach is known as "backward design" and is explained in detail by G. Wiggins and J. McTighe in their 1998 book, *Understanding by Design (12).*

Backward design is the opposite of the traditional method for developing a course. In a traditional scheme, the teacher first decides what content to convey to the students, then develops activities for conveying that content, and finally, almost as an afterthought, writes examinations for testing whether students have learned the content. One major flaw in this approach is that it fails to take into account the "hidden curriculum," the skills, attitudes, and behaviors that we expect students to acquire along with content knowledge *(13)*. Another problem with the "content-first" design is that given the vast amount of scientific knowledge available today, deciding what to teach and what to ignore can become paralyzing *(14)*. It is impossible for teachers to convey everything they know to students within the time constraints of an academic class. Perhaps it is time to step back and use a "less-is-more" approach in our classes by asking ourselves what essentials our students must know to become independent learners. But how can we decide what information is essential and what can be ignored?

The utility of the backward design approach is illustrated by a curriculum development project for undergraduate biomedical engineering (BME) that is taking place under the auspices of the NSF-sponsored Vanderbilt-Northwestern-Texas-Harvard (VaNTH) Engineering Research Center (www.vanth.org/curriculum). One of the early tasks for those working on the VaNTH physiology domain was to develop a taxonomy of the physiology content that biomedical engineers need to know. It soon became apparent that a compendium of physiology that spanned all levels of organization, all organ systems, and all living organisms contained far more information than any single course could cover. When we examined what was actually being taught in existing BME courses, we discovered that most courses limited their coverage to four or five organ systems rather than the eight to ten that a traditional systems-based course would cover *(15)*. The BME courses, therefore, focus on the key concepts that are most important for a well-trained biomedical engineer, using the organ systems to provide examples of concepts.

To implement backward-design curriculum revision, teachers must first answer several important questions.

2.1. What Are the Aims and Requirements of My Specific Course and of the Program(s) Within Which My Course Is Taught?

Integrative physiology is relevant to graduate, undergraduate, and health professional programs, but each of these student populations has unique goals and attributes that will influence all areas of course design. Thinking about what students need to know and do after they complete a course is a key element in establishing learning objectives for a class.

2.2. What Level of Achievement Do Students in This Course Need to Master?

Student achievement should be directly related to the program's aims. A student planning a career in medical record-keeping does not need the same depth of understanding of physiology as a future physician. One scheme for classifying level of student achievement is called Bloom's taxonomy of educational objectives *(16)*. Bloom described six levels of achievement divided into three domains of learning: cognitive, affective, and psychomotor. The six levels for the cognitive domain, from basic to complex, were knowledge, comprehension, application, analysis, synthesis, and evaluation. In 2001, Bloom's taxonomy was reorganized into *A Taxonomy for Learning, Teaching, and Assessing (17)*. This revision of Bloom's taxonomy recognizes four major categories of knowledge: factual, conceptual, procedural, and metacognitive. Metacognitive knowledge can be loosely defined as understanding how people in general learn and how you as an individual learn best. Each knowledge category is then divided into Bloom's six levels, but the levels are now described with verbs: remember, understand, apply, analyze, evaluate, and create. Often, students work only at the lowest level of comprehension, believing they "know" (remember) the assigned material but failing to recognize that they do not understand it. A well-designed course also should challenge students to work at the highest levels of achievement by asking them apply, analyze, evaluate, and create.

2.3. What Aspects of the Hidden Curriculum Should This Course Promote?

The hidden curriculum comprises what students learn—desirable or not—that may not be recognized in course objectives or tested by examinations. The hidden curriculum consists of cognitive skills, habits, and attitudes that students develop because of the way a teacher organizes, conducts, and assesses activities in a course. Revealing the hidden curriculum enables both teacher and students to optimize its beneficial aspects and ameliorate its undesirable ones *(13)*.

Ethics of animal and human experimentation, aspects of professionalism such as intellectual honesty and modesty, literacy and numeracy skills—all these and more have a place in physiology courses, and yet often are not made explicit by being included in learning objectives, are not modeled by the teacher, and are not encouraged by using specially targeted teaching and learning activities and assessments.

Revealing the hidden curriculum may be especially important in integrative physiology courses. For example, the special cognitive skills and attitudes needed to be a successful student or practitioner of integrative science might include the ability to create concept maps relating physiological information across species or from diverse body systems or at different levels of organization. Students need to develop the habits and skills of inquiry that will lead them to read widely in the primary or secondary literature *(18)* and to recognize models or concepts when they occur in various contexts *(19)*.

2.4. What Resources Are Available and How Can They Be Used to Support This Class?

When considering recommendations for changes in instructional methods, it is important to understand that there can be no "one-size-fits-all" solution owing to wide variability in student populations and resources of different institutions. In some developing countries, students may not have individual copies of the required textbook, much less access to computers. In these situations, the teacher becomes the primary information source, a requirement that places certain constraints on what happens in the classroom. At the other extreme, students in technology-rich environments can be made responsible for learning some core content independently, thus freeing up valuable class time for activities other than straight lecturing. The appropriate selection of learning objectives, teaching activities, and assessment depends on the resources as well as on the other constraints of the course, such as the length and number of class periods, and the availability of laboratory activities. A combination of active learning activities for the classroom and inquiry-based laboratory exercises is ideal for understanding physiology, which is after all, a quintessential experimental science requiring an inquiring mind and the ability to think critically.

2.5. What Knowledge, Skills, and Attitudes Do Students Have When They Begin This Course?

David Ausubel, in his seminal 1968 textbook on educational psychology *(20)*, said "The most important single factor influencing learning is what the learner already knows." This statement has been extensively documented by research studies and underlies the constructivist theory of learning, which says that learners construct knowledge and understanding by linking new information to their pre-existing knowledge *(21–23)*. But how do we know what our students know when they begin our courses?

Understanding students' prior knowledge is a critical element in effective teaching, yet surprisingly, teachers, even very experienced ones, are probably not in touch with what their students know *(24)*. In one study of undergraduate physiology classes in a variety of institutions, the teachers all underestimated their students' content knowledge and overestimated students' ability to think conceptually and solve problems *(25)*. This study underscores the fact that as teachers, we cannot assume that passing a prerequisite course guarantees that students understood or remember the material. Thus, it behooves us to build multiple forms of assessment into our curricula *(26)*.

One strategy to uncover the scope and level of knowledge and skills students have acquired as they begin our classes is the diagnostic pretest. These tests should emphasize conceptual understanding of the course's important topics as well as skills, such as the ability to read graphs and interpret data. The physics community has been a leader in developing tests of conceptual understanding, and a useful set of examples can be found in Eric Mazur's book, *Peer Instruction: A User's Manual (27)*. Once we know what our students understand as they begin our course, we have established a basis on which we can build learning objectives.

3. ESTABLISHING LEARNING OBJECTIVES

The first step in planning an integrated physiology curriculum is to determining what students should have achieved by the end of the course. Establishing learning objectives is essentially a creative process of selecting appropriate knowledge, skills, attitudes, and behaviors from all those currently encompassed by the discipline of physiology. The process requires challenging and intellectually satisfying scholarly work and can be a major task for an integrative physiology course, given the breadth and depth of our scientific understanding.

Many universities and academic organizations have provided guidance for writing learning objectives on their web sites (e.g., www.aacsb.edu/resource_centers/assessment/ov-process-define.asp and http://alpha.confex.com/alpha/learningobjectives.htm). These guides include advice about expressing objectives in terms of measurable achievement by using particular action-descriptive verbs in statements of objectives. Objectives should be organized hierarchically, as described above for Bloom's taxonomy, to help students aim for and achieve higher order objectives.

As mentioned in the previous section, learning objectives will differ according to the student population. A graduate physiology course might emphasize development of professional skills such as critiquing primary literature and formulating research questions pertaining to integrative topics. Objectives for undergraduate courses will vary depending on whether the course is for non-majors or biological science majors. Non-major course objectives could focus on integrating physiology with everyday life situations that students are likely to encounter, with the goal of creating a scientifically literate lay populace. Courses for biology majors should help students integrate physiology with other life science courses so that they can make informed decisions about their career path. Classes for health professional students might integrate physiology with social and clinical aspects of the students' future professions.

For example, with a class of medical or undergraduate students, one course goal might be to give the students an appreciation for the significance of integrative physiology in biomedical research. The matching written objective could state, "Students completing the course should be able to explain and give examples of the importance and limitations of using model organisms in research." One teaching activity to illustrate the difficulties inherent in integrative physiology research might focus on the renin–angiotensin–aldosterone system (RAAS) pathway *(28–30)*. When physiologists described the blood pressure-raising effects of angiotensin II (ANG II), pharmacologists scrambled to develop angiotensin-converting enzyme (ACE) inhibitors, drugs to block formation of ANG II. ACE inhibitors were predicted to become the ultimate drug for treating hypertension in humans. Animal studies indicated that ACE inhibitors effectively lowered blood pressure in rodent models, but in human clinical trials the drugs were not as effective in some patients or caused undesirable side effects *(31)*. Further research in humans revealed the presence of an alternate pathway for ANG II production that uses chymase instead of ACE *(32)*. Thus, the results from rodent models were not strictly applicable to humans, demonstrating the limitations of using model organisms.

Examples of well-designed objectives in human physiology are available through the Internet. The American Physiological Society (APS) with the Association of Chairs of Departments of Physiology has developed and endorsed a set of knowledge objectives for medical physiology (www.the-aps.org/education/MedPhysObj/medcor.htm *[33]*). A set of professional skills objectives has been developed by the same organizations and is available at http://www.the-aps.org/education/skills.htm. Lemons and Griswold *(34)* have described an effective approach to creating learning objectives that are coherently imbedded in an undergraduate anatomy and physiology curriculum. These sets of objectives provide useful examples and starting points, but as noted above, each course requires the teacher to use professional judgment to create objectives specifically tailored for the course's program and the students' needs.

4. ASSESSING LEARNING

Once the course objectives have been designed, the next step, and one that is too often left until last, is designing appropriate assessment. The teacher must ask, "How will I know when my students have met the learning objectives I have set for this class?" Many teachers think of assessment only in terms of *summative assessment* that comes at the end of a unit or course and that is used to assign a grade based on what the student should have learned. However, assessment should also be *formative*, that is, used to give helpful feedback to teacher and students about what students are learning while there is still time to make corrections and improvements. Appropriate formative assessment in the classroom has been shown to improve learning *(35,36)*. *Program evaluation* is used to monitor the success of a course or program and is discussed later in this chapter.

Formative assessment in the classroom is a powerful tool for improving student learning. Teachers looking for guidance in implementing informal assessment should turn first to the detailed handbook on *classroom assessment techniques* (CATs), written by T. A. Angelo and K. P. Cross *(37)*. This excellent volume is packed with information and worksheets to help teachers define their teaching goals and to plan and implement different types of formative assessment activities. Some specific examples of CATs are discussed in the section of this chapter on teaching and learning strategies.

Summative assessment traditionally appears as term papers or examinations given at intervals throughout a course. One challenge to teachers is to think beyond these traditional methods of summative assessment. Graded assignments can be group or individual, creative and nontraditional (posters, journals, play-acting), and can involve the students themselves through peer grading and self-rating *(38–44)*. Although the legality of peer grading was challenged in the United States by a parent who claimed that it violated her child's privacy, the US Supreme Court in 2002 ruled that peer grading was an acceptable form of assessment *(45)*.

If one course objective is to have students work effectively in teams, who better than the team members themselves to grade participation? Peer and self-assessment can be a useful method of both formative and summative assessment *(46–48)*. The authors use this method at the University of Texas in a physiology laboratory class where students are collaborating as they design and execute their own experiments, and found that students rate themselves and others quite objectively. The addition of a peer and self-rating activity to the class allowed early identification of problem students. Before adding the formal rating system, students were reluctant to "tattle" to the instructor on classmates who were not doing their share of the work.

Another innovative type of peer assessment is the web-based Calibrated Peer Review™ (CPR) system developed at the University of California with support from the NSF and the Howard Hughes Medical Institute (http://cpr.molsci.ucla.edu/). The CPR system allows an instructor to provide samples of excellent, good, and poor answers to a writing assignment (a lab report, perhaps). Students then upload their own assignments and must review and constructively critique the assignments submitted by others. The reviewer and author remain anonymous except to the instructor. CPR is currently being used at 337 US colleges and universities *(49,50)*.

Assessment is an integral component of curriculum design, and an area that is currently coming to the forefront in educational research *(51)*. The US National Research Council's Committee on the Foundations of Assessment in 2001 *(26)* recommended increased funding for research into assessment design, and the NSF responded by adding a track on Assessment of Student Achievement to its Course, Curriculum, and Laboratory Improvement Program (http://www.ehr.nsf.gov/ehr/DUE/programs/ccli). Teachers who wish to learn more about the principles of educational assessment should refer to *Knowing What Students Know: The Science and Design of Educational Assessment*, compiled by the Committee on the Foundations of Assessment *(26)*.

Matching assessment to learning objectives can be one of the more difficult tasks in curriculum design and is often an iterative process. Alignment between instructional objectives, what is being taught, and what is being tested is a benchmark of a well-designed course *(52)*.

5. TEACHING AND LEARNING STRATEGIES

The final steps in backward design of curriculum are to select the content necessary to achieve the learning objectives and to develop activities and teaching strategies that help students master the material at the desired level of achievement. Leaving these steps until last is probably the most difficult aspect of backward design for teachers who are used to beginning at this point. However, if curriculum design is being done by committee (a process sometimes equated to herding cats), leaving content until last and requiring that all content supports the learning objectives may be an effective way to minimize dissent over what is essential.

Teaching can be viewed as a continuum as shown in Fig. 1, with the minimally interactive, authority-based lecture at one end and small independent student groups with a faculty facilitator at the other. Numerous studies have shown while that the time-honored teaching

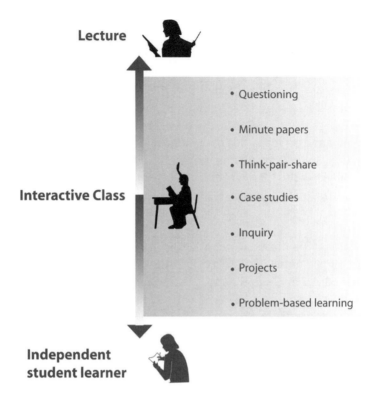

Fig. 1. Teaching and learning methods span a continuum, with the minimally interactive, authority-based lecture at one end; student-centered, faculty-facilitated activities in the middle; and small independent student groups at the other end.

practice we call the didactic lecture may be effective at conveying content, it is not an effective strategy for developing higher level skills such as analytical and creative thinking. Research on learning has shown that students retain information best when they interact with the material, either mentally, as in answering questions, or physically, as in a laboratory practical *(53–56)*.

In resource-rich environments, where students have access to books, interactive CD-ROMs, and the Internet, the teacher can place more responsibility for learning basic content on the students by providing them with a workbook of objective questions. The students can use their resources outside of class time to answer basic content questions that ask them to list, define, or describe information. This strategy frees up class time, allowing the teacher to cover difficult material and do informal assessment to test whether students understand the concepts. Even in low-resource environments, where the teacher must be the primary source of content, it is possible to enhance student learning by making the class more interactive.

Teaching techniques that promote student interaction go by an assortment of names: active learning, inquiry, discovery learning, collaborative or cooperative learning. Some critics of active learning protest that all learning is active. The key distinctions between what educators call active learning and more traditional forms of education are *when* and *where* learning takes place, and the *depth of understanding* that results. In an authority-centered lecture where the teacher talks and the students dutifully take notes with minimal interaction, relatively little learning takes place in the classroom. Learning takes place later when students study by themselves or with classmates. Even then, the learning may be superficial and consist primarily of memorization unless the teacher has provided carefully crafted home-work assignments.

5.1. Classroom Strategies

Active learning in the classroom requires interaction between teacher and students *(57, 58)*. The interaction need not be elaborate or time-consuming. Some teachers are reluctant to "try active learning" because they labor under the misconception that active learning must be elabo-rate and time-consuming. On the contrary, active learning can be as simple as questioning the students at intervals during the lecture. In large-class settings this method may be difficult to implement if only a few students are brave enough to respond, or the teacher fails to wait for a response from the students and answers the question himself. In this situation, the "think–pair–share" technique *(37)* is very effective. The teacher asks students to write down their individual answers to a question, then pair up with a neighbor and compare answers.

In large classes, the teacher can encourage participation by all students through the use of a response system, ranging from small index cards on which students write their individual answers to electronic response systems *(59–64)*. Electronic response systems are limited to true/false and multiple-choice questions. The advantage to this system is that both teacher and students get instant feedback about what the class is thinking. A low-tech version of the electronic response system uses colored index cards, with different colors corresponding to different answers (e.g., white is answer A, blue is answer B, etc.). Students simultaneously hold up the color cards that corresponds to their answer.

"Cooperative learning" is the name given to the active learning that occurs when students work in pairs or in small groups, and can easily be combined with questioning, even in large classes. Exercises for cooperative learning may be formalized and require students to turn in written reports or present oral summaries. On a less formal basis, cooperative (collaborative) learning occurs when students form study groups or study partners to review for tests *(65–73)*.

Interactive questioning between teacher and students or between student and student may cover factual information or may require higher level thinking. Problem solving and ques-tions requiring application or analytical thinking should be introduced in lecture and prac-ticed before students are tested with them. Integrative physiology contains a wealth of examples that are ideal for this type of classroom exercise.

For example, students may have learned about the osmoregulatory mechanisms that allow teleost fish to maintain low internal osmolarity in seawater despite continuous osmotic loss of water to the environment. Students could then be asked to predict what physiological and biochemical changes they would observe in anadromous fish as the fish migrate from salt-water into freshwater. The task could even include having the students design a physiologi-cal control system to monitor and regulate plasma ion concentrations. By introducing problems like this in a nonthreatening (i.e., nongraded) class situation, the teacher is able to help students uncover the weaknesses in their understanding of the mechanisms for osmo-regulation. Classroom activities provide students practice for answering similar questions that they might encounter in an examination and allow the teacher to model problem-solving techniques that the students can emulate.

Although questioning is probably the simplest and most widely used type of active learn-ing in the lecture setting, a variety of other techniques exist. Brief descriptions and resources

Table 1
Interactive Teaching Activities

Activity and brief description	References
Inquiry and project laboratories	*98–102*
Case studies and problem-based learning	*103–112*
Working with scientific literature	*113–117*
Games	*118–123*
Maps and flow charts	*124–129*
Posters	*130–132*

are listed in Table 1. The classroom assessment technique handbook described earlier is a rich source of ideas *(37)*.

Mapping (also called concept mapping) is a nonlinear way of organizing material, closely related to the flow charts commonly used to illustrate physiological processes *(74)*. A map can take a variety of forms but usually consists of terms or concepts linked by explanatory arrows *(75)*. Maps are a useful way for students to develop their conceptual understanding because construction of an accurate map requires that they understand the relationships between terms and can organize concepts into a hierarchical structure. Maps can be used for assessment as well as for studying.

Posters are a creative substitute for the traditional term paper as well as an effective way to skirt the issue of plagiarized or purchased papers. The format is similar to that for a research-based poster suitable for presentation at a scientific meeting. Students select or are assigned a topic and research it in the literature, which may require some direction *(76,77)*. They then summarize what they have learned into a poster to educate their classmates on the topic. Posters should include good visual aids (graphs, diagrams, photographs) as well as a one-page abstract and a list of references. They are organized into "poster sessions," with the posters taped to the walls of the classroom or a wide hallway. Each student presents his or her poster in one session and evaluates assigned posters in the others. Grading and feedback are simplified by use of a standardized form that assigns points based on the poster's content and appearance, appropriate use and citation of reference materials, and the author's presentation of the information. One of us (DUS) has used posters effectively in classes for both biology majors and non-majors. Students in both populations tell us that they enjoyed having an opportunity to master a subject and teach it to their peers.

Another creative activity is role-playing, in which students take the role of various physiological components and act out a process *(78)*. The classic example of role-playing is the "protein synthesis ballet" video made in 1971 by Nobel prize winner Dr. Paul Berg, in which his students choreographed an elaborate representation of transcription and translation that they executed in costume in the Stanford University football stadium *(79)*, accompanied by narration based on the Lewis Carroll poem *Jabberwocky*. Other examples of topics for role-playing include sliding filament movements in muscle contraction, the opening and closing of gates in ion channels during the action potential, renal clearance of solutes from the plasma, and oxygen transport in the blood by plasma and hemoglobin. Designing the plays requires careful thought, and allowing students to write the scripts gives them an opportunity to discover how well they understand the process being depicted.

5.2. Active Laboratories

Student laboratories or practical sessions might seem to be the epitome of active learning, but unless the protocol is carefully designed, it is possible for students to "plug and chug" their way through the exercise without ever engaging their brains *(80,81)*. All too often, thinking and learning do not take place until students are home trying to write up their reports, at which point it is too late for them to troubleshoot their mistakes.

In the laboratory, active learning can be enhanced to some degree by asking students to predict what will occur or to explain why they are carrying out a particular procedure *(82)*. The most active laboratories are those in which students design, execute, and report on their own experiments *(83,84)*. Whenever possible, the design of student laboratories should:

1. Require students to predict what results they will obtain in an experiment and explain the rationale behind their prediction.
2. Question students about why they are doing certain step in a procedure or using a piece of equipment.
3. Allow students to design, execute, and analyze their own experiments.

5.3. Incorporating Active Learning Into a Curriculum

Interactive teaching can be very intimidating for the teacher who has always used a traditional lecture technique. Angelo and Cross *(37)* have some useful guides, but the most important advice is to start small! One enthusiastic young faculty member decided to implement many different changes in his human physiology class one semester. He had the students negotiate for their grades, added peer grading and small group work in the lecture session, and asked students to write the test questions. At mid-semester, when it became obvious that the class was a disaster, he had no idea of which change or changes to discard: he had made too many at once.

One simple but effective method for incorporating some interactive time into a traditional class is for the teacher to lecture 5 to 15 min, then stop and give the class a question to answer. Another is to ask a question but to direct students to write down their individual answers in their notebooks rather than answering out loud. By changing only one aspect of a class, both the students and teacher can adapt more easily. If an activity is not successful on the first try, you should reflect on the experience to see if it can be modified to be more successful.

Any teacher changing the way he or she teaches needs flexibility, patience, and a sense of humor. Some activities may be successful on the first attempt but others may require adjustment before they achieve their intended goal. What works with one population of students may not work with others. The physical setting of the classroom can also help or hinder interactive teaching. The ideal classroom has moveable chairs and tables if small, or space for the teacher to walk between the rows if the classroom is large. Portable microphones are also helpful in large auditoriums so that the teacher can move off the stage and in among the students.

The final tips for successful interactive teaching were described earlier: tell your students what you are doing and why, and be sure your assessment matches your teaching. If you ask students to solve problems in lecture but regurgitate memorized facts on the test, they will soon become resistant to problem solving. As one colleague said, "Students are strategic—they will put their effort where they will see the greatest reward."

6. EVALUATING CURRICULUM FOR FUTURE PLANNING

One of the most difficult and least often accomplished tasks in curriculum redesign is evaluating the success of the course. Curriculum evaluation should have both subjective and objective components and should include student input as well as peer evaluation if possible *(85)*. Lechner *(86)* describes some of the difficulties of evaluating the success of teaching strategies, particularly in respect to creating what scientific methodology considers adequate controls.

One of the most common types of objective assessment is some form of written examination. Scientists are comfortable with this type of assessment as it yields numbers that can be subjected to statistical evaluation. In medical school settings, National Board examinations are often used as a benchmark for determining if curricular change has been a success or a failure. In other settings, teachers may use a content test to quantify pre- and postcurriculum change.

As with student assessment, it is essential that questions used for curriculum evaluation be consistent with the learning objectives for the class. If one goal of the course is to teach problem solving but the examination questions evaluate only whether students have memorized a large vocabulary of terms the test is not an accurate evaluation of the course design.

Subjective student feedback is also a key component of course evaluation, and can be both formative and summative. Anonymous formative evaluation can be used to monitor the progress of a class, allowing the teacher to make adjustments before students write their final course evaluations, which are documents that may be used in contract and salary negotiations or for promotion and tenure decisions. Negative student perceptions of what is happening in the classroom can be ameliorated to some degree by enlisting the students as allies from the first class day. If the teacher explains what changes are being made and the rationale for them, students are less likely to feel that the teacher is trying to trick them or be difficult. It is essential for the teacher to revisit this topic several times during the course, as students often do not assimilate the information the first time they hear it.

Quantitative student feedback may take the form of a questionnaire, with answers given using a five-point Likert scale, with 1 being *Disagree strongly* and 5 being *Agree strongly* *(87)*. This type of questionnaire often appears in published articles about student reactions to curriculum change. However, it has one major drawback: the teacher wrote the questions and may not be thinking about the course the same way that the students are. When such questionnaires are used, the students should also be given the opportunity to write in their own free-response answers or comments.

One evaluation method that scientists are less likely to use because it yields "soft" data is qualitative assessment, such as interviews or focus groups with students *(88,89)*. It was the experience of one of us (DUS) that this form of assessment can yield information that is not obtainable by other means. In one physiology class students were given Likert scale pre- and posttests that asked about their ability to self-regulate their learning, their preferred type of instruction, and their confidence in their ability to think about and solve physiological problems. According to the quantitative test, there were no significant changes over the course of the semester. However, during end-of-semester interviews the same students expressed views that indicated that they had changed significantly. It appears that from this discrepancy that the students either did not reflect very deeply on their changes when they took the quantitative test or that perhaps the wording on the test did not match their vocabulary for describing the same events.

7. REPORTING CURRICULUM RESEARCH AND DEVELOPMENT

An innovative, effective, well-evaluated curriculum is as much a product of scholarship as are the findings of bench research *(90–92)*. Standards for the quality of the scholarship of teaching have been proposed by Glassick et al. *(93)*. Designing a curriculum that emphasizes integration requires both breadth and depth of knowledge about the subjects to be integrated. Curriculum development requires you to be well informed not only about the content of your subject but also about appropriate and effective teaching and learning strategies. Teachers must have the intellectual discipline to ensure that objectives, methods, and assessment are aligned and that curricular innovations are rigorously evaluated. All of these intellectual activities require creativity.

As with bench research, dissemination and peer review of the results and products of scholarship are expected in contemporary academia. In graduate school, physiologists learn how to report research findings. We develop expertise in presenting our work as posters and oral presentations, and we learn what is expected for written papers and how to select appropriate journals for publication of our findings. We should be doing the same for our scholarly work in physiology education.

The general structure of a paper reporting curriculum innovations parallels that of a research report *(94)*. A curriculum publication should include an introduction; a description of the curriculum, its development, and evaluation of its educational results; discussion, and conclusions. Specific guidance about the peer review process for such manuscripts is provided by Bordage *(95)*. Table 2 itemizes the elements editors and peer reviewers expect in a paper reporting curriculum development.

7.1. Venues for Reporting Results of Educational Innovations

As scientist-educators we need to inform ourselves about appropriate venues for reporting the results of our work so that our colleagues may benefit from our successes and avoid repeating our failures. Many professional associations now include posters and talks on education at their meetings. There are also numerous peer-reviewed journals that publish papers on life science education, including several that are exclusively online. One journal that publishes peer-reviewed articles specifically about physiology education is *Advances in Physiology Education*, published by The American Physiology Society. This and other journals that publish articles about life science education are listed in Table 3.

7.2. Human Subjects Approval for Educational Research

Most teachers never think of what they do in the classroom as research, but if we are changing the way we teach, collecting information on the effectiveness of those changes, reflecting on the results, then making further changes, we have gone through the same intellectual process as a scientist in the laboratory. In this instance, the classroom is the laboratory and our students are the subjects of the research. Thus, we should be aware that if we wish to publish findings from our educational experiments, we must follow the ethical guidelines for human subjects research *(96)* and obtain approval from the Institutional Review Board (IRB) of our university.

Table 2
Elements of a Curriculum Development Manuscript for Publication

Section of manuscript	Subsections
Introduction	• Justification of need for a curriculum innovation; educational problem it was designed to solve • Review of relevant literature • Importance or significance of the innovation • Educational goal of the curriculum
Development	• Description of overall process used for curriculum development • Decision process used for learning objectives and content • Decision process used for teaching, learning, and assessment methods
Description of the Curriculum	• Overview detailing major curriculum components and their relationship • Topics and examples • Classroom activities, assignments, and other teaching and learning methods • Formative and summative assessments
Results	• Description of curriculum evaluation strategy and methods used • Student performance data • Evaluation of curriculum by students • Evaluation of curriculum by creators and teachers
Discussion and conclusions	• Interpretation of results • Strengths of the curriculum • Deficiencies or limitations of the curriculum • Implications for future educational research and development

Fortunately, the US government considers research that is conducted in commonly accepted educational settings and that is being used to improve or assess curriculum or instructional techniques to be exempt from the usual regulatory oversight. The exact policy may vary from institution to institution, so the description below is intended as an example only.

At the University of Texas-Austin, if a teacher is collecting data on student performance so that he or she may improve his or her teaching and if the data collected will never leave the university, no human subjects review is required. If the instructor collects data for his or her own use, then later decides to publish it so that other instructors might benefit from the findings, he or she must submit a request to the IRB for a Category 4 Exemption. This exemption request states that the data were not collected for research purposes and that the proposed publication is a secondary use of existing information. On the other hand, if the teacher knows in advance that he or she is planning to publish the data, the university's guidelines for informed consent must be followed to provide assurance that the research will not have any effect on the students' grades in the class and that student confidentiality will be maintained *(97)*. Most refereed journals will require that the author submitting a paper on

Table 3
Journals for Publication of Articles on Education

Journal	URL	Notes
Advances in Physiology Education	http://advan.physiology.org	Published free online by the American Physiological Society. Refereed.[a]
American Biology Teacher	www.nabt.org	Published by the National Association of Biology Teachers. Refereed.
Biochemistry and Molecular Biology Education	www.bambed.org	Published by the International Union of Biochemistry and Molecular Biology. Free access 1 yr after publication date.
BioScene: Journal of College Biology Teaching	http://papa.indstate.edu/amcbt/bioscene.html	Published by the Association of College & University Biology Educators. Scanned images of the last 22 yr of the journal are available on the journal's web site. Refereed.
Bioscience	www.aibs.or	Published by American Institute for Biological Sciences. Includes a section on education. Refereed.
Cell Biology Education	www.cellbiologyeducation.org	Published free online by the American Society for Cell Biology. Refereed.[a]
Electronic Journal of Science Education	unr.edu/homepage/jcannon/ejse/ejse.html	A free, full-text electronic journal. Refereed.
International Journal of Science Education	www.tandf.co.uk/journals/	Web site provides access to tables of contents of past issues. Refereed.
Journal of Biological Education	www.iob.org	Published by the Institute of Biology. Refereed.
Journal of Clinical Problem-Based Learning	www.jclinpbl.org	Quarterly web publication. Free online access with registration.
The Journal of College Science Teaching	www.nsta.org/college	Published by the Society of College Science Teachers. Refereed.
Journal of Research in Science Teaching	www.josseybass.com	Published for the National Association for Research in Science Teaching.
The Electronic Journal of Science Education	http://unr.edu/homepage/jcannon/ejse/ejse.html	Refereed. Free online access.
Science	www.aaas.org	Published by the American Association for the Advancement of Science. Although primarily technical in nature, on occasion undergraduate education is addressed.

[a]Provides free registration for email notification of newly issued table of contents

educational research include in the paper a statement that explains the process used for institutional approval or exemption.

8. CONCLUSION

This book represents the leading edge of an exciting development in biological science: the re-emergence of integrative physiology as its organizing principle. The continuing development of this central role for our discipline depends on attracting and educating students who will be the next generation of physiologists. By adapting and applying our research skills to our classrooms, we can improve our abilities to teach and excite students. When we select teaching and learning methods for our courses, we should take advantage of new findings from education research about how we can best help our students learn. By creating and evaluating new integrative curricula and disseminating the results to colleagues, we contribute to our discipline's development as a profession that places high value on teaching as a form of scholarship.

ACKNOWLEDGMENTS

We would like to thank Ms. Kristina Schlegel for creating Fig. 1. We would also like to express our appreciation to Marilla Svinicki, PhD, The University of Texas Center for Teaching Effectiveness, Patti Thorn, PhD, Arizona State University, and Kenneth B. Roberts, MD, DPhil, Memorial University, for their support and advice on our educational endeavors.

REFERENCES

1. Tiberius, R. G., Smith, R. A. and Waisman, Z. (1998) Implications of the nature of "expertise" for teaching and faculty development, in *To Improve the Academy* (Kaplan, M., ed.), New Forums Press, Stillwater OK., **17**:123–138.
2. Mestre, J. (2003) *Transfer of learning: issues and research agenda*. National Science Foundation publication 03-212. Available at www.nsf.gov/pubs.
3. National Research Council, Committee on Undergraduate Biology Education to Prepare Research Scientists for the 21st Century. (2002) *BIO 2010: Transforming Undergraduate Education for Future Research Biologists*. The National Academies Press, Washington, D.C.
4. Silverthorn, D. U. (2003) Restoring physiology to the undergraduate curriculum: a call for action. *Adv Physiol Educ* **27**, 91–96.
5. Engelberg, J. (1995) Integrative physiology: some texts and methods of integrative study. *Advan. Physiol. Edu.* **269**, 55S–60S.
6. Richardson , R. S. (2003) Oxygen transport and utilization: an integration of the muscle systems. *Advan. Physiol. Edu.* **27**, 183–191.
7. Thompson, S. R., Ackermann, U. and Horner, R. L. (2001) Sleep as a teaching tool for integrating respiratory physiology and motor control. *Advan. Physiol. Edu.* **25**, 29–44.
8. Hall, J. E. (1999) Integration and regulation of cardiovascular function. *Advan. Physiol. Edu.* **277**, 174S–186S.
9. Granger, J. P. (1998) Regulation of extracellular fluid volume by integrated control of sodium excretion. *Advan. Physiol. Edu.* **275**, 157S–168S.
10. Ebert-May, D., Batzli, D. J. and Lim, H. (2003) Disciplinary research strategies for assessment of learning. *Bioscience* **53**, 1221–1228.
11. Hutchings, P., Babb, M. and Bjork, C. (2002) The scholarship of teaching and learning in higher education: an annotated bibliography. www.carnegiefoundation.org/elibrary/docs/bibliography.htm.
12. Wiggins, G. P. and McTighe, J. (1998) *Understanding by Design*. Association for Supervision and Curriculum Development, Alexandria, VA.
13. Hansen, P. A. (2002) Physiology's recondite curriculum. *Advan. Physiol. Educ.* **26**, 139-145.
14. Stokstad, E. (2001) Information overload hampers biology reforms. *Science* **293**, 1609.

15. Silverthorn, D. U. (2002) Developing a concepts-based physiology curriculum for bioengineering: a VaNTH project. *Proceedings of the Second Joint Engineering in Medicine and Biology and Biomedical Engineering Society Meeting* **3**, 2646–2647.
16. Bloom, B. S. (1956) *Taxonomy of Educational Objectives, Handbook I: Cognitive Domain*. D. McKay, New York, NY.
17. Anderson, L. W. and Krathwohl, D. R. (eds) (2001) *A Taxonomy for Learning, Teaching, and Assessing: A Revision of Bloom's Taxonomy of Educational Objectives*. Addison Wesley Longman, Inc., New York NY.
18. Levine, E. (2001) Reading your way to scientific literacy. *J. Coll. Sci. Teach.* **31**:122–125.
19. Modell, H. I. (2000) How to help students understand physiology? Emphasize general models. *Advan. Physiol. Educ.* **23**:101–107.
20. Ausubel, D. P. (1968) *Educational Psychology, A Cognitive View*. Holt, Rinehart and Winston, Inc., New York, NY.
21. Bransford, J. D., Brown, A. L. and Cocking, R. R. (eds) (2000) *How People Learn: Brain, Mind, Experience, and School (Expanded Edition)*, National Academy Press, Washington D.C.
22. Jungck, J. R. (1991) Constructivism, computer exploratoriums, and collaborative learning: constructing scientific knowledge. *Teaching Education* **3**, 151–170.
23. Gabel, D. (2003) Enhancing the conceptual understanding of science. *Educational Horizons* **81**, 70–76.
24. Michael, J. A. (1998) Students' misconceptions about perceived physiological responses. *Advan. Physiol. Educ.* **274**, S90–S98.
25. Rovick, A. A., Michael, J. A., Modell, H. I.,et al. (1999) How accurate are our assessments about our students' background knowledge? *Advan. Physiol. Educ.* **21**, S93–S101.
26. Pellegrino, J. W. and Glaser, R. (eds) (2001) *Knowing What Students Know: The Science and Design of Educational Assessment*. National Academy Press, Washington DC.
27. Mazur, E. (1997) *Peer Instruction: A User's Manual*. Prentice Hall, Upper Saddle River, NJ.
28. Carey, R. M. and Siragy, H. M. (2003) Newly recognized components of the renin-angiotensin system: potential roles in cardiovascular and renal regulation. *Endo. Rev.* **24**, 261–271.
29. Lavoie, J. L. and Sigmund, C. D. (2003) Minireview: overview of the renin-angiotensin system – an endocrine and paracrine system. Endocrinology **144**, 2179–2183.
30. Hollenberg, N. K., Fisher, N. D. and Price, D. A. (1998) Pathways for angiotensin II generation in intact human tissue: evidence from comparative pharmacological interruption of the renin system. *Hypertension* **32**, 387–392.
31. Bicket, D. P. (2002) Using ACE inhibitors appropriately. *Am. Fam. Physician* **66**, 461–468.
32. Liao, Y. and Husain, A. (1995) The chymase-angiotensin system in humans: biochemistry, molecular biology and potential role in cardiovascular diseases. *Can. J. Cardiol.* **11**, 13F–19F.
33. Carroll, R. G. (2001) Design and evaluation of a national set of learning objectives: the medical physiology learning objectives project. *Advan. Physiol. Educ.* **25**, 2–7.
34. Lemons, D. E. and Griswold, J. G. (1998) Defining the boundaries of physiological understanding: the benchmarks curriculum model. *Advan. Physiol. Educ.* **275**, S35–S45.
35. Bangert-Drowns, R., Kulik, J. and Kulik, C. (1991) Effects of frequent classroom testing. *J. Educ. Res.* **85**:89–99.
36. Black, P. and Wiliam, D. (1998) Assessment and classroom learning. *Assessment in Educ.* **5**, 7–73.
37. Angelo, T. and Cross, K. P. (1993) *Classroom Assessment Techniques*. Jossey-Bass Publishers, San Francisco.
38. Pelaez, N. J. (2002) Problem-based writing with peer review improves academic performance in physiology. *Advan. Physiol. Edu.* **26**, 174–184.
39. Guilford, W. H. (2001) Teaching peer review and the process of scientific writing. *Advan. Physiol. Edu.* **25**, 167–175.
40. Rao, S. P. and DiCarlo, S. E. (2000) Peer instruction improves performance on quizzes. *Advan. Physiol. Edu.* **24**, 51–55.

41. Seals, D. R. and Tanaka, H. (2000) Manuscript peer review: a helpful checklist for students and novice referees. *Advan. Physiol. Edu.* **23,** 52–58.

42. Lake, D. A. (1999) Peer tutoring improves student performance in an advanced physiology course. *Advan. Physiol. Edu.* **276,** 86S–92S.

43. Lightfoot, J. T. (1998) A different method of teaching peer review systems. *Advan. Physiol. Edu.* **274,** 57S–61S.

44. Etkina, E. and Harper, K. A. (2002) Weekly reports: student reflections on learning. An assessment tool based on student and teacher feedback. *J. Coll. Sci. Teach.* **31,** 476-480.

45. Parry, G. (2002) Privacy rights in the classroom: peer grading Supreme Court judgment 2002. *Education & the Law* **14,** 173–181.

46. Orsmond, P., Merry, S., and Reiling, K. (1996) The importance of marking criteria in the use of peer assessment. *Assessment and Evaluation in Higher Education* **21**:239–250.

47. Healey, M. (1999) Using peer and self-assessment for assessing the contribution of individuals to a group project. http://www.chelt.ac.uk/gdn/abstracts/a69.htm

48. Conway, R., Kember, D., Sivah, A. and Wu, M. (1993) Peer assessment of an individual's contribution to a group project. *Assessment and Evaluation in Higher Education* **18**:45–56.

49. Robinson, R. (2001) An application to increase student reading and writing skills. *Am. Biol. Teacher* **63**:474–476, 478–480.

50. Pelaez, N. J. (2001) Calibrated peer review in general education undergraduate human physiology, In *Proceedings of the Annual International Conference of the Association for the Education of Teachers in Science* (Rubba, P. A., ed.) Association for the Education of Teachers in Science, Costa Mesa, CA.

51. McIntosh, W. J. (1996) Assessment in higher education. *J. Coll. Sci. Teach.* **26**:52–53.

52. Webb, N. L. (1997) Determining alignment of expectations and assessments in mathematics and science education. *NISE Brief* **1**:1–8.

53. McNeal, A. P. and D'Avanzo, C. (1997) *Student-Active Science: Models of Innovation in College Science Teaching.* Saunders College Publishing, Fort Worth.

54. Bonwell, C. C. and Eison, J. A. (1991) *Active Learning: Creating Excitement in the Classroom.* ASHE-ERIC Higher Education Report No. 1. The George Washington University, School of Higher Education and Human Development, Washington D.C.

55. Siebert, E. D. and McIntosh, W. J. (2001) *College Pathways to the Science Education Standards.* National Science Teachers Association, Arlington, VA.

56. Michael, J. A. and Modell, H. I. (2003) *Active Learning in Secondary and College Science Classrooms: A Working Model for Helping the Learner to Learn.* Lawrence Erlbaum Associates, Inc., Mahwah, NJ.

57. Modell, H. I. (1996) Preparing students to participate in an active learning environment. *Advan. Physiol. Edu.***270**:69S–77S.

58. Huang, A. H. and Carroll, R. G. (1997) Incorporating active learning into a traditional curriculum. *Advan. Physiol. Edu.* **273**:14S–23S.

59. Lyman, F. (1981) The responsive classroom discussion, in *Mainstreaming Digest* (Anderson, A. S., ed.), University of Maryland College of Education, College Park, MD.

60. Kellum, K. K., Carr, J. E. and Dozier, C. L. (2001) Response-card instruction and student learning in a college classroom. *Teaching of Pyschol.* **28**:101–104.

61. The Foundation Coalition. (2003) Electronic response systems. www.foundationcoalition.org/publications/brochures/ers.doc

62. Judson, D. and Sawada, E. (2002) Learning from past and present: electronic response systems in college lecture halls. *J. Computers in Math. Sci. Teaching* **21**:167–181. www.aace.org/dl/files/JCMST/JCMST212167.pdf

63. Paschal, C. B. (2002) Formative assessment in physiology teaching using a wireless classroom communication system. *Advan. Physiol. Educ.* **26,** 299308.

64. Lopez-Herrejon, R. E. and Schulman, M. (2004) Using interactive technology in a short Java course: an experience report. *Proc. Innovation Technol. in Computer Sci. Educ.* www.ph.utexas.edu/~ctalk/bulletin/ITicse17.pdf

65. Johnson, D. W., Johnson, R. T. and Smith, K. A. (1991) *Cooperative Learning: Increasing College Faculty Instructional Productivity* (ASHE-ERIC Higher Education Report No. 4) The George Washington University, School of Education and Human Development, Washington, DC.

66. Svinicki, M. D. (ed) (1994) *Collaborative Learning: Beyond the Basics, New Directions for Teaching and Learning*. Jossey-Bass Publishers, San Francisco.

67. Slavin, R. E. (1995) *Cooperative Learning: Theory, Research, and Practice, 2nd edition*. Allyn & Bacon, Boston.

68. Smith, K. A. (1996) Cooperative learning: making "groupwork" work, In *Using Active Learning in College Classes: A Range of Options for Faculty* (Sutherland, T. E. and Bonwell, C. C., eds.), Jossey-Bass Publishers, San Francisco.

69. Michaelson, L. K., Fink, L. D. and Knight, A. (1997) Designing effective group activities: lessons for classroom teaching and faculty development, in *To Improve the Academy* (DeZure, D., ed.), New Forums Press, Stillwater OK , **16**:373–398.

70. Johnson, D. W., Johnson, R. T. and Smith, K. A. (1998*) Active Learning: Cooperation in the College Classroom.* Interaction Book Company, Edina MN.

71. Cooper, J. and Robinson, P. (1998) Small-group instruction in science, mathematics, engineering, and technology (SMET) disciplines: a status report and an agenda for the future. *J. Coll. Sci. Teaching* **27**:383–388. www.wcer.wisc.edu/nise/CL1/CL/resource/smallgrp.pdf

72. Adams, J. P., Brissenden, G., Lindell, R. S., Slater, T. F. and Wallace, J. (2002) Observations of student behavior in collaborative learning groups. *Astronomy Educ. Rev.* **1**. http://aer.noao.edu

73. Jensen, M. S. (1996) Cooperative quizzes in the anatomy and physiology laboratory: a description and evaluation. *Advan. Physiol. Edu.* **271**:48S–54S.

74. Wallace, J. D., Mintzes, J. J., and Markham, K. M. (1992) Concept mapping in college science teaching-What the research says. *J. Coll. Sci. Teach.* **22**:84–86.

75. Silverthorn, D. U. (1993) Teaching concept mapping. *Annals New York Acad. Sci.* **701**:139–141.

76. Chisman, J. K. (1998) Introducing college students to the scientific literature and the library. *J. Coll. Sci. Teach.* **28**:39–42.

77. Texas Information Literacy Tutorials (TILT) Interactive tutorial available at http://tilt.lib.utsystem.edu/

78. Palmer, D. H. (2000) Using dramatizations to present science concepts. *J. Coll. Sci. Teach.* **29**, 187–190.

79. Berg, P. (1971) *A Protein Primer* (video). Harper & Row, Publishers. http://videoserver.stanford.edu:8080/ramgen/realcontent3/medsol/imed/beckman/beckman.rm

80. Leonard, W. H. (1989) Research and teaching: ten years of research on investigative laboratory instruction strategies. *J. Coll. Sci. Teach.* **18**, 304–306.

81. Russell, C. P. and French, D. P. (2002) Factors affecting participation in traditional and inquiry-based laboratories. *J. Coll. Sci. Teach.* **31**, 225–229.

82. Modell, H. I., Michael, J. A., Adamson, T.,et al. (2000) Helping undergraduates repair faulty mental models in the student laboratory. *Advan. Physiol. Edu.* **23**, 82–90.

83. Woodhull-McNeal, A. P. (1992) Project labs in physiology. *Advan. Physiol. Educ.* **263**, 29S–32S.

84. McNeal, A. P., Silverthorn, D. U. and Stratton, D. B. (1998) Involving students in experimental design: three approaches. *Advan. Physiol. Educ.* **20**, S28–S34.

85. Krilowicz, B. L. and Downs, T. (1999) Use of course-embedded projects for program assessment. *Advan. Physiol. Edu.* **276**, 39S–54S.

86. Lechner, S. K. (2001) Evaluation of teaching and learning strategies. *Med. Educ Online* **6**, 4, www.med-ed-online.org

87. Learning Technology Dissemination Initiative. (2000) So you want to use a Likert scale? http://www.icbl.hw.ac.uk/ltdi/cookbook/info_likert_scale/

88. Glesne, C. and Peshkin, A. (1992) *Becoming Qualitative Researchers: An Introduction*. Longman, White Plains NY.

89. Merriam, S. B. (1998) *Qualitative Research and Case Study Applications in Education.* Jossey-Bass, San Francisco.

90. Society of College Science Teaching. (1998) The scholarship of college science teaching. http://a-s.clayton.edu/scst/TFSCHO5.PDF 1

91. Boyer E. L. (1990) *Scholarship Reconsidered: Priorities of the Professorate.* Carnegie Foundation for the Advancement of Teaching, Princeton, NJ.

92. Fincher, R. E., Simpson, D. E., Mennin, S. P.,et al. (2000) Scholarship and teaching: an imperative for the 21st century. *Acad. Med.* **75**, 887–894.

93. Reznich, C. B. and Anderson, W. A. (2001) A suggested outline for writing curriculum development journal articles: the IDCRD format. *Teach. Learn. Med.* **13**, 4–8.

94. Glassick, C. E., Huber, M. R. and Maeroff G. I. (1997) *Scholarship Assessed: Evaluation of the Professorate.* Jossey-Bass, San Francisco, CA.

95. Bordage, G. (2001) Reasons reviewers reject and accept manuscripts: the strengths and weaknesses in medical education reports. *Acad. Med.* **76**, 889–896.

96. U.S. Department of Education (2004) Protection of human subjects in research. http://www.ed.gov/about/offices/list/ocfo/humansub.html

97. University of Texas at Austin, Office of Research Support and Compliance (2003) http://www.utexas.edu/research/rsc/humanresearch/manual/appendixa/a6.html

98. Myers, M. J. and Burgess, A. B. (2003) Inquiry-based laboratory course improves students' ability to design experiments and interpret data. *Advan. Physiol. Edu.* **27**, 26–33.

99. Rivers, D. B. (2002) Using a course-long theme for inquiry-based laboratories in a comparative physiology course. *Advan. Physiol. Edu.* **26**, 317–326.

100. Bertram, J. E. A. (2002) Hypothesis testing as a laboratory exercise: a simple analysis of human walking, with a physiological surprise. *Advan. Physiol. Edu.* **26**, 110–119.

101. Kolkhorst, F. W., Mason, C. L., DiPasquale, D. M., Patterson, P. and Buono, M. J. (2001) An inquiry-based learning model for an exercise physiology laboratory course. *Advan. Physiol. Edu.* **25**, 45–50.

102. DiPasquale, D. M., Mason, C. L., Kolkhorst, F. W. (2003) Exercise in inquiry: critical thinking in an inquiry-based exercise physiology laboratory Course. *J. Coll. Sci. Teach.* **32,** 388–393.

103. National Center for Case Study Teaching in Science. http://ublib.buffalo.edu/libraries/projects/cases/ubcase.htm

104. Wheatley, J. (1986) The use of case studies in the science classroom. *J. Coll. Sci. Teach.* **15,** 428–431.

105. Engelberg, J. (1992) Complex medical case histories as portals to medical practice and integrative, scientific thought. *Advan. Physiol. Edu.* **263**, 45S–54S.

106. Herreid, C. F. (1994) Case studies in science—a novel method of science education. *J. Coll. Sci. Teach.* **23,** 221–229.

107. Herreid, C. F. (1996) Structured controversy: a case study strategy. *J. Coll. Sci. Teach.* **26**, 95–101.

108. Herreid, C. F. (1996) Case study teaching in science: a dilemma case on "animal rights." *J. Coll. Sci. Teach.* **25,** 413–418.

109. Herreid, C. F. (2003) The death of problem-based learning? *J. Coll. Sci. Teach.* **32**, 364–366.

110. Cliff, W. (2004) *Case Studies in Human Anatomy and Physiology.* Benjamin Cummings, San Francisco.

111. Berne, R. M. and Levy, M. N. (eds) (1994) *Case Studies in Physiology 3rd ed.* Mosby Yearbook, St. Louis MO.

112. Costanzo, L. S. (2001) *Physiology: Cases and Problems: Board Review Series.* Lippincott Williams & Wilkins Publishers, Philadelphia.